Kinetic Data Analysis
Design and Analysis of Enzyme and Pharmacokinetic Experiments

Kinetic Data Analysis
Design and Analysis of Enzyme
and Pharmacokinetic Experiments

Edited by
Laszlo Endrenyi

University of Toronto
Toronto, Canada

PLENUM PRESS · NEW YORK AND LONDON

Library of Congress Cataloging in Publication Data

Main entry under title:

Kinetic data analysis.

"Proceedings of a satellite symposium... organized in conjunction with
the XIth International Congress of Biochemistry, and held July 14, 1979,
in Toronto, Canada."—Verso of t.p.
Includes indexes.
1. Chemical reaction, Rate of—Congresses. 2. Enzymes—Congresses. 3.
Pharmacokinetics—Congresses. I. Endrenyi, Laszlo. II. International
Congress of Biochemistry (11th: 1979: Toronto, Ont.) [DNLM: 1. Enzymes—Congresses. 2. Kinetics—Congresses. 3. Models, Chemical—Congresses. 4. Pharmacology—Congresses. QU 135 K51 1979]
QP517.R4K56   541.3 94   81-120
ISBN 0-306-40724-8   AACR2

Proceedings of a satellite symposium on Design and Analysis
of Enzyme and Pharmacokinetic Experiments, organized in
conjunction with the XIth International Congress of
Biochemistry, and held July 14, 1979, in Toronto, Canada

© 1981 Plenum Press, New York
A Division of Plenum Publishing Corporation
233 Spring Street, New York, N.Y. 10013

PREFACE

Kinetic models have often served as useful examples in developing the methodology for the design and analysis of experiments involving mechanistic models. Thus, it is not surprising that these approaches have been applied quite successfully to kinetic observations.

Nevertheless, many ideas and methods were developed independently in various fields of science. More often than not, investigators working in one area have not been aware of relevant advances in others.

In order to facilitate the desirable exchange of ideas, a one-day symposium was held in Toronto in conjunction with the XIth International Congress of Biochemistry. Biochemists, pharmacologists, and statisticians came together and discussed many of the topics presented in this volume.

Participants in the symposium believed that it would be useful to publish a collection of the presentations together with some additional material. The present volume is the result. It is an attempt to convey some of the interdisciplinary concerns involving mechanistic, and especially kinetic, model building.

The coverage is by no means exhaustive: many principles, methods, and problems are not included. Even the applications are limited to biochemistry and pharmacology. Still, the symposium highlighted areas of current interest. These included questions of weighting, robust parameter estimation, pooled data analysis, model identification, and the design of experiments. These topics, which are of interest in many fields of science, are discussed also in the present volume.

The symposium was made possible by generous financial support from the International Union of Biochemistry as well as the following pharmaceutical companies: Ayerst, McKenna and Harrison, Inc., Boehringer Ingelheim (Canada) Ltd., Ciba-Geigy Canada Ltd.,

v

Fisons Corporation Ltd., McNeil Laboratories (Canada) Ltd.,
Parke-Davis and Company Ltd., Pfizer Company Ltd., and Poulenc Ltd.
I express my sincere gratitude to them also on behalf of the
speakers and participants in the symposium.

I wish to thank Professors B. Hess and K.G. Scrimgeour for
helpful advice and useful discussions concerning the organization
of the symposium. I am also grateful to Miss P. Arnold, Dr. F.Y.
Chan, and Mrs. A. Vassilev for their willing and effective help
in arranging the meeting. The typing of this volume is the
careful and excellent work of Mrs. L. Espeut and Mrs. L. Freund.
Mrs. M.J. Clements has efficiently assisted in organizing the
subject index. I sincerely thank all of them for their help.

Laszlo Endrenyi

CONTENTS

DESIGN OF EXPERIMENTS FOR KINETIC PARAMETER ESTIMATION

KINETIC MODEL IDENTIFICATION

COMBINATION OF EXPERIMENTS

KINETIC DATA ANALYSIS: METHODS AND APPLICATIONS

AN INTRODUCTION TO NONLINEAR LEAST SQUARES

Donald G. Watts

Queen's University
Department of Mathematics and Statistics
Kingston, Ontario, Canada   K7L 3N6

ABSTRACT

A brief introduction to linear and nonlinear least squares is presented. Emphasis is placed on the assumptions which must be satisfied in order for least-squares analysis to be valid. Interpretations of least squares in terms of sum of squares surfaces and geometry are used to provide insight into the procedure.

1. INTRODUCTION: THE PROBLEM AND A SOLUTION

A common problem in engineering and scientific work is that in which an investigator has a theoretical formulation for a phenomenon or response, which can be expressed as a function of k manipulated or control variables $\underline{u} = (u_1, u_2, \ldots, u_k)^T$ and p unknown constant coefficients or parameters $\underline{\theta} = (\theta_1, \theta_2, \ldots, \theta_p)^T$. That is, the response on the i-th setting is postulated to be

$$f(\underline{\theta}, \underline{u}_i) \qquad\qquad\qquad [1.1]$$

where $\underline{u}_i = (u_{i1}, u_{i2}, \ldots, u_{ik})^T$. Measurements can be made of the response at various settings of the variables, and the researcher desires to estimate the values of the parameters in the model [1.1] using data $\underline{y} = (y_1, y_2, \ldots, y_n)^T$.

For example, the Michaelis-Menten model of simple enzyme-catalyzed reactions predicts that the rate of formation of product, the "velocity" f, depends on the substrate concentration $\underline{u}$ according to

1

$$f = \theta_1 u / (\theta_2 + u) \quad . \hspace{4cm} [1.2]$$

In this equation, θ_1 and θ_2 are parameters which have the interpretation of the maximum velocity and the half-velocity concentration, respectively, such that when $u = \theta_2$, the velocity will be half the maximum. In order to determine values for the parameters, experiments can be carried out by setting the substrate concentration at a value, say u, and measuring the velocity y. Table 1 lists some data for such an experiment.

In such an experimental situation it is useful to assume that an observed value can be described as the sum of two components, a model for the phenomenon, and a model for the error or variation associated with that observation. We have then

$$Y_i = f(\underline{\theta}, \underline{u}_i) + Z_i \hspace{4cm} [1.3]$$

where Y_i is a random variable associated with an observation to be obtained at the setting $\underline{u}_i$ and Z_i is a random variable which describes the error or uncertainty associated with the experiment at $\underline{u}_i$.

If we can assume that the Z_i are distributed according to a Normal distribution, with each Z_i having an expected value of zero, a common variance σ^2, and being independent of all other Z's and of the control variables and response function, then statistical arguments using the likelihood or Bayesian approach dictate that the appropriate estimates for $\underline{\theta}$, given a set of observations $\underline{y}$ taken at experimental conditions $\underline{u}_i$, $i=1,2,\ldots,n$, are the values $\underline{\theta}$ which minimize

$$S(\underline{\theta}) = \sum_{i=1}^{n} [y_i - f(\underline{\theta}, \underline{u}_i)]^2 \quad . \hspace{3cm} [1.4]$$

These estimates, usually denoted by $\hat{\underline{\theta}}$, are called the least-squares estimates.

TABLE 1: Enzyme velocity and substrate concentration

Substrate concentration	u	2.0	2.0	.667	.667	.400	.400
		.286	.286	.222	.222	.200	.200
Measured velocity	y	.0615	.0527	.0334	.0258	.0138	.0258
		.0129	.0183	.0083	.0169	.0129	.0087

2.   THE ASSUMPTIONS OF LEAST SQUARES: THEIR IMPLICATIONS AND CONSEQUENCES

In the development of the least-squares solution to regression analysis, several important assumptions are made about the model function and the error structure. Because of the insistent influence of the error structure on our inferences, it is worthwhile to consider carefully all the assumptions made in the formulation of a problem, and their implications and consequences, as follows:

2.1. The Response Function: $f(\underline{\theta},\underline{u})$

The most important assumption about the model $f(\underline{\theta},\underline{u})$ is that it is the correct function! This may be absolutely true in some rare circumstances, but generally a more realistic view is that $f(\underline{\theta},\underline{u})$ is a good model to "entertain tentatively" (1,2).

The implications of the assumption are that all important variables are measured and taken account of in the model, or conversely, no important variables are omitted; also, that the effects of all the variables included in the model are correctly specified, or conversely, that no important variables are incorrectly incorporated in the model.

The consequences of this assumption are that the maximum number of parameters and the variables are specified and hence the dimensionality of the parameter space is limited. If more parameters are required or more or different variables are needed, then the minimum of the criterion for this limited parameter and variable space may not be the correct minimum for the data set.

2.2. Additive Error: $Y_i = f(\underline{\theta},\underline{u}_i) + Z_i$

The important assumption here is that the uncertainty Z is additive. The implication is effectively that the ignorance or errors enter simply and independently of the response. The consequence is that the density for Y_i is easily derivable and of the same form as the density for Z_i but translated by the amount $f(\underline{\theta},\underline{u}_i)$. Hence the likelihood function for $\underline{\theta}$ depends on the choice of density for Z.

2.3. Independence of $f(\underline{\theta},\underline{u}_i)$ and Z_i

The assumption that the uncertainty Z is independent of the model $f(\underline{\theta},\underline{u})$ is related to the assumptions 2.1 and 2.2. That is, independence of $f(\underline{\theta},\underline{u})$ and $\underline{Z}$ implies that if any important variables have been omitted, their effects in separate experiments will not be systematically related to $f(\underline{\theta},\underline{u})$ and hence they may be included in the random variable Z. This of course, is the main reason for using randomized experiments as suggested by Fisher (3) and for using designed experiments in which the controlled variables u are truly manipulated and not just observed, as pointed out by Box (4).

A further important implication of this assumption is that the u's are measured perfectly, or at least that errors in the u's are not systematically related to Z. The consequence is that the errors only occur in the y direction and so the estimation criterion need only involved deviations of y from the model function.

2.4. Independence of Z_i and Z_j

The assumption of mutual independence of the Z_i is one of the most crucial assumptions made since it implies, effectively, assumptions 2.1, 2.2 and 2.3 above. If the assumption of independence is not made, then further assumptions concerning the nature and form of the interdependence must be made. This can so seriously complicate the resulting analysis that no useable solution can be obtained. In addition, unless relatively pointed information as to the nature of the interdependence is available, it seems that the better approach is tentatively to assume independence and then test the validity of the assumption after the analysis is performed.

The main consequence of this assumption is that the joint density for the n random variables $Y_1, Y_2, \ldots, Y_n$ is simply the product of the individual densities and so the likelihood function for θ is very easily derived.

2.5. The Assumption of Normal Distribution

Since the form of density function for Z completely determines the form of the likelihood function for θ, this in turn affects the fit criterion which may influence the maximum likelihood estimates considerably. Thus choice of the density for Z is pivotal in any analysis. Choice of the Normal density can often be justified on the basis of the Central Limit Theorem [see for example Cramer (5)] which states that a random variable Z which is composed of the sum of several random variables, say W_j, each with possibly different densities, tends to have a Normal density. The practical importance of this theorem, then, is that if the uncertainty or errors on any one trial may be assumed to be the sum of many random effects of comparable magnitude, then the resulting error will be well-described by a Normal random variable. Thus, in many practical situations we may appeal to the Central Limit Theorem and make the assumption that Z has a Normal density.

The main implication of the Normal assumption is that the uncertainty arises from many contributing sources, all of comparable effect. The most important consequence of this assumption is that the likelihood function will consist of an exponential function with a sum of squares in the exponent. Hence the estimation criterion under the Normal assumption will be least squares. In addition, we may draw inferences in the form of confidence intervals for parameters or for the response at a particular design value.

2.6. Zero Mean of Each Z_i

The assumption that the uncertainty Z_i is distributed about zero implies that there is no systematic deviation or bias in the measurements, and consequently implies assumption 2.1. The converse does not follow, however; that is, assumption that the model is correct does not imply zero biases. The most important consequence of this assumption is that the number of unknown parameters is reduced by n since, in general, it is not necessary that each Z have the same mean.

2.7. Constant Variance of the Z_i

This assumption of constant scale parameter σ, or constant variance σ^2, although relatively minor theoretically, is extremely important practically. The main implication is that the measuring or experimental ability is equally good on all experimental runs. The important consequence is that the likelihood function is again simplified so that the number of unknown parameters is only $p+1$ (p parameters, θ, and one variance). In addition, the parameters σ^2 and θ become "separated" in the sense that the maximum of the likelihood function occurs at the same value of θ for any value of σ^2, and if we wished, we could estimate θ without bothering to estimate σ^2.

If we did not assume constant variance, we would have to get independent estimates of the individual σ_i's from replicate runs or other sources, or else make some further assumptions as to the way in which σ_i varies as a function either of the theoretical response $f(\theta, u_i)$ or of the variables u_i.

Most important, however, the assumptions of normality, of constant variance, and of independence imply that the simple least-squares criterion [1.4] is appropriate.

2.8. Summary of Assumptions

1. Correct model $f(\theta, u)$

2. Additive error: $Y_i = f(\theta, u_i) + Z_i$

3. Independence of $f(\theta, u_i)$ and Z_i

4. Independence of Z_i and Z_j

5. Normal errors

6. Zero mean of each Z_i

7. Constant variance of the Z_i

3. LEAST-SQUARES ANALYSIS

Once we are presented with a data set $\underline{y}$, the corresponding model function $f(\underline{\theta},\underline{u})$ and the experimental settings $\underline{u}_i$, $i=1,2,\ldots,n$, and we are willing to accept tentatively all the seven assumptions as being appropriate, then we are faced with two well-defined problems:

(i)   find the least-squares estimates, $\hat{\underline{\theta}}$, and
(ii) quantify their variability using confidence or likelihood regions.

Both these problems have straightforward solutions when the model function is linear in the parameters; that is when $f(\underline{\theta},\underline{u}_i)$ is of the form

$$f(\underline{\theta},\underline{u}_i) = \theta_1 f_1(\underline{u}_i) + \theta_2 f_2(\underline{u}_i) + \ldots + \theta_p f_p(\underline{u}_i) \qquad [3.1]$$

or, equivalently, when the partial derivatives do not depend on any of the parameters $\underline{\theta}$. For example, the Michaelis-Menten model $f=\theta_1 u/(u+\theta_2)$ is nonlinear in the parameters since

$$\frac{\partial f}{\partial \theta_1} = \frac{u}{u+\theta_2}$$

depends on θ_2 and

$$\frac{\partial f}{\partial \theta_2} = \frac{-\theta_1 u}{(u+\theta_2)^2}$$

depends on both θ_1 and θ_2.

In the following two subsections, we present analytical results for the linear and nonlinear case, and in subsequent sections discuss their interpretations in terms of sum-of-squares surfaces and geometry.

3.1. Linear Least Squares

3.1.1. <u>Calculating the Least-Squares Estimates</u>. In the linear case, we may let $f_j(\underline{u}_i)=x_{ij}$ in [3.1] so that the whole experiment may be modelled as

$$Y_1 = \theta_1 x_{11} + \theta_2 x_{12} + \ldots + \theta_p x_{1p} + Z_1$$

$$Y_2 = \theta_1 x_{21} + \theta_2 x_{22} + \ldots + \theta_p x_{2p} + Z_2$$

$$\vdots \quad\quad \vdots \quad\quad \vdots$$

$$Y_n = \theta_1 x_{n1} + \theta_2 x_{n2} + \ldots + \theta_p x_{np} + Z_n$$

or, in matrix terms,

$$\underline{Y} = \underline{X}\underline{\theta} + \underline{Z} \quad . \tag{3.2}$$

Given a set of observations $\underline{y}$, the least squares estimates are the values of $\underline{\theta}$ which minimize

$$S(\underline{\theta}) = \sum_{i=1}^{n} (y_i - \theta_1 x_{i1} - \theta_2 x_{i2} - \ldots - \theta_p x_{ip})^2$$

or, in matrix terms, which minimize

$$S(\underline{\theta}) = (\underline{y} - \underline{X}\underline{\theta})^T (\underline{y} - \underline{X}\underline{\theta}) \quad .$$

This is a quadratic form, and hence can be written as a constant plus a squared term,

$$S(\underline{\theta}) = (\underline{y} - \underline{X}\hat{\underline{\theta}})^T (\underline{y} - \underline{X}\hat{\underline{\theta}}) + (\underline{\theta} - \hat{\underline{\theta}})^T \underline{X}^T \underline{X} (\underline{\theta} - \hat{\underline{\theta}})$$

$$= S(\hat{\underline{\theta}}) + (\underline{\theta} - \hat{\underline{\theta}})^T \underline{X}^T \underline{X} (\underline{\theta} - \hat{\underline{\theta}}) \tag{3.3}$$

where $\hat{\underline{\theta}}$ satisfies

$$(\underline{X}^T \underline{X})\hat{\underline{\theta}} = \underline{X}^T \underline{y} \tag{3.4}$$

and $\hat{\underline{\theta}}$ is the least-squares estimate. Note that expressions [3.3] and [3.4] imply that the cross-product terms $\underline{X}^T(\underline{y} - \underline{X}\hat{\underline{\theta}})$ are identically zero.

3.1.2. Determining Confidence Regions. Suppose the true values of the parameters were θ^*. Then because of the seven assumptions, the sum of squares of θ^* will be distributed as a sum of squared normal random variables and hence $S(\underline{\theta}^*)$ will be distributed as σ^2 times a χ^2 with n degrees of freedom. Using Cochran's theorem (6), which states that a χ^2 with n degrees of freedom can be decomposed into two orthogonal components with p and n-p degrees of freedom respectively, Eq. [3.3] can be decomposed into a χ^2 with n-p degrees of freedom,

$$S(\hat{\underline{\theta}})/\sigma^2 \quad ,$$

and a χ^2 with p degrees of freedom,

$$(\underline{\theta}^*-\hat{\underline{\theta}})^T\underline{X}^T\underline{X}(\underline{\theta}^*-\hat{\underline{\theta}})/\sigma^2 \quad .$$

Thus the ratio

$$\frac{(\underline{\theta}^*-\hat{\underline{\theta}})^T\underline{X}^T\underline{X}(\underline{\theta}^*-\hat{\underline{\theta}})/p}{S(\hat{\underline{\theta}})/(n-p)} \qquad [3.5]$$

will be distributed as Fisher's F distribution with p and n-p degrees of freedom.

Hence we may test a given hypothesis at level α by comparing the test statistic value to the table value $F(p,n-p;\alpha)$. Alternatively, we may develop a 1-α confidence region for $\underline{\theta}$ as all those values of $\underline{\theta}$ for which

$$(\underline{\theta}-\hat{\underline{\theta}})^T\underline{X}^T\underline{X}(\underline{\theta}-\hat{\underline{\theta}}) \le pS(\hat{\underline{\theta}})F(p,n-p;\alpha)/(n-p) \quad . \qquad [3.6]$$

This defines a p-dimensional ellipsoid with center $\hat{\underline{\theta}}$ in the parameter space.

The estimated variance of the observations is the denominator of Eq. [3.5]:

$$s^2 = S(\hat{\underline{\theta}})/(n-p)$$

with n-p degrees of freedom.

3.1.3. An Example. As a simple example of linear least squares, we consider the concentration of polychlorinated biphenyls (PCB's) in Lake Cayuga trout as a function of age (7). The data are listed in Table 2.

TABLE 2: Lake Cayuga trout PCB concentrations and age

Age (years)	PCB concentration (ppm)			
1	0.6	1.6	0.5	1.2
2	2.0	1.3	2.5	
3	2.2	2.4	1.2	
4	3.5	4.1	5.1	
5	5.7			
6	3.4	9.7	8.6	
7	4.0	5.5	10.5	
8	17.5	13.4	4.5	
9	30.4			
11	12.4			
12	13.4	26.2	7.4	

A plot of the PCB concentration versus age reveals a curved relationship with increasing variance. Taking logarithms of the PCB concentration nicely stabilizes the variance and from a plot of logPCB versus age, Figure 1, it appears that a linear model of the form

$$logPCB = \beta_0 + \beta_1 age$$

is appropriate. For the logPCB and age data we have the vectors, matrices and products as shown in Table 3. Using these calculations in Eq. [3.4] gives

$$\hat{\underline{\theta}} = (\underline{X}^T\underline{X})^{-1}\underline{Xy} =$$

$$\begin{pmatrix} 28 & 155 \\ 155 & 1197 \end{pmatrix}^{-1} \begin{pmatrix} 41.046 \\ 315.05 \end{pmatrix} = \begin{pmatrix} .03147 \\ .25913 \end{pmatrix} \quad .$$

The residual sum of squares, $S(\hat{\underline{\theta}})=8.359$, so $s^2=.3215$ based on (28-2)=26 degrees of freedom (ignoring replications). Confidence regions may be calculated from 3.6 as

$$28(\theta_1-.0315)^2+310(\theta_1-.0315)(\theta_2-.2591)+1197(\theta_2-.2591)^2$$

$$= .643F(2,26;\alpha)$$

which defines an ellipse centered at (.0315,.2591).

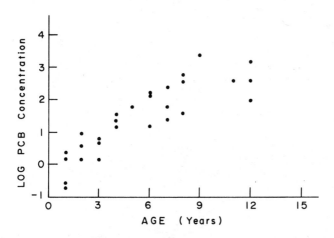

FIGURE 1: Log PCB concentration as a function of age

TABLE 3:   Calculations for logPCB data

$$
\underline{X} = \begin{pmatrix} 1 & 1 \\ 1 & 1 \\ 1 & 1 \\ 1 & 1 \\ 1 & 2 \\ 1 & 2 \\ 1 & 2 \\ 1 & 3 \\ 1 & 3 \\ 1 & 3 \\ 1 & 4 \\ 1 & 4 \\ 1 & 4 \\ 1 & 5 \\ 1 & 6 \\ 1 & 6 \\ 1 & 6 \\ 1 & 7 \\ 1 & 7 \\ 1 & 7 \\ 1 & 8 \\ 1 & 8 \\ 1 & 8 \\ 1 & 9 \\ 1 & 11 \\ 1 & 12 \\ 1 & 12 \\ 1 & 12 \end{pmatrix}
\qquad
\underline{y} = \begin{pmatrix} -.511 \\ .470 \\ -.693 \\ .182 \\ .693 \\ .262 \\ .916 \\ .788 \\ .875 \\ .182 \\ 1.253 \\ 1.411 \\ 1.629 \\ 1.740 \\ 1.224 \\ 2.272 \\ 2.152 \\ 1.386 \\ 1.705 \\ 2.351 \\ 2.862 \\ 2.595 \\ 1.504 \\ 3.414 \\ 2.518 \\ 2.595 \\ 3.266 \\ 2.001 \end{pmatrix}
$$

$$
\underline{X}^T\underline{X} = \begin{pmatrix} 1 & 1 & 1 & 1 & 1 & \ldots & 1 & 1 \\ 1 & 1 & 1 & 1 & 2 & \ldots & 12 & 12 \end{pmatrix}
\begin{pmatrix} 1 & 1 \\ 1 & 1 \\ 1 & 1 \\ 1 & 1 \\ 1 & 2 \\ \vdots & \vdots \\ \vdots & \vdots \\ 1 & 12 \\ 1 & 12 \end{pmatrix}
= \begin{pmatrix} 28 & 155 \\ 155 & 1197 \end{pmatrix}
$$

$$
(\underline{X}^T\underline{X})^{-1} = \begin{pmatrix} .12612 & -.01633 \\ -.01633 & .00295 \end{pmatrix}
$$

$$
\underline{X}^T\underline{y} = \begin{pmatrix} 41.046 \\ 315.05 \end{pmatrix}
$$

3.2. Nonlinear Least Squares

When the model is nonlinear in the parameters, there are no explicit analytical solutions for $\hat{\theta}$ or for the parameter confidence regions. In this case, we often use linear approximations, both to find the least-squares estimates by using an iterative linear approximation scheme, and to find an approximate confidence region.

3.2.1. Calculating the Least-Squares Estimates.

A method suggested by Gauss in 1809 (8) is still very useful. The approach is to form a linear Taylor's series approximation of the function near some initial guess values for the parameters. Under the assumption that near the starting value θ^0, say, the linear approximation is accurate, then we have for the response at the i-th setting

$$f(\underline{\theta},\underline{u}_i) \simeq f(\underline{\theta}^0,\underline{u}_i) + \frac{df_i}{d\theta_1}(\theta_1-\theta_1^0) + \frac{df_i}{d\theta_2}(\theta_2-\theta_2^0) + \ldots + Z_i$$

where f_i stands for $f(\underline{\theta},\underline{u}_i)$ and the derivatives are evaluated at θ^0. Substituting expression [3.7] into [1.3] and taking the first term over to the left-hand side gives

$$Y_i - f(\underline{\theta},\underline{u}_i) \simeq \frac{df_i}{d\theta_1}(\theta_1-\theta_1^0) + \frac{df_i}{d\theta_2}(\theta_2-\theta_2^0) + \ldots + Z_i \qquad [3.8]$$

Writing $Y_i-f(\underline{\theta},\underline{u}_i)=W_i$, $\dfrac{df_i}{d\theta_j}=v_{ij}$, and $\theta_j-\theta_j^0=\beta_j$ gives

$$W_i \simeq v_{i1}\beta_1 + v_{i2}\beta_2 + \ldots v_{ip}\beta_p + Z_i$$

and for the whole experiment we have

$$\underline{W} \simeq \underline{V}_0\underline{\beta} + \underline{Z} \qquad [3.9]$$

where the subscript on $\underline{V}$ indicates that the derivative matrix is evaluated at θ^0. But this is exactly of the same form as Eq. [3.2] and hence we may use the result [3.4] to get the apparent least-squares estimate based on a set of observations $\underline{y}$,

$$\underline{\beta}^1 = (\underline{V}_0^T\underline{V}_0)^{-1}\underline{V}_0^T\underline{w} \qquad [3.10]$$

where $\underline{w}=(w_1,w_2,\ldots,w_n)^T$ and $w_i=y_i-f(\underline{\theta}^0,\underline{u}_i)$. Adding the increment $\underline{\beta}^1$ to the current value θ^0 then gives the location of the apparent least-squares value of $\underline{\theta}$, using local information at $\underline{\theta}^0$, as

$$\underline{\theta}^1 = \underline{\theta}^0 + \underline{\beta}^1 \qquad . \qquad [3.11]$$

We may therefore try these new values in the model function and see if the sum of squares at $\underline{\theta}^1$ is the minimum, that is, if we have converged to the least-squares estimate. This requires a test for convergence, which can be done by determining whether the cross-product term on the k-th iteration,

$$\underline{v}_k^T[\underline{y}-f(\underline{\theta}^k,\underline{u})]$$

is zero or at least sufficiently small. For more details, see Bates and Watts (9).

If $S(\underline{\theta}^1)$ is less then $S(\underline{\theta}^0)$ but $\underline{\theta}^1$ is still not the least-squares estimate, then we repeat the iterative process starting from our improved estimate $\underline{\theta}^1$ to get an even better estimate $\underline{\theta}^2$, and so on. Sometimes the new sum of squares will be larger than the old value, in which case we have overshot the minimum. That is, the minimum is apparently in the direction of the increment called for, but it is not as distant as suggested. In such a case, a procedure suggested independently by Box (10) and by Hartley (11) can be used, namely to take only a fraction of the increment called for, say $\lambda=1/2$ or $1/4$ or so until the sum of squares at say $\underline{\theta}^k=\underline{\theta}^{k-1}+\lambda\underline{\beta}^k$ is less than that at $\underline{\theta}^{k-1}$. We may then continue the iterations from this new, better, point. Other methods such as Marquardt's compromise (12) can be helpful in obtaining convergence. For further details, see Chambers (13).

3 2 2. <u>Determining Parameter Confidence Regions</u>. Probably the most common method for obtaining confidence regions in the nonlinear case is to use the linear approximation. Thus, near $\underline{\theta}$ we have

$$y_i-f(\underline{\hat{\theta}},\underline{u}_i) \simeq v_{i1}(\theta_i-\hat{\theta}_1)+v_{i2}(\theta_2-\hat{\theta}_2)+\ldots+v_{ip}(\theta_p-\hat{\theta}_p)$$

and so

$$S(\underline{\theta}) = \sum_{i=1}^{n} [y_i-f(\underline{\theta},\underline{u}_i)]^2$$

$$\simeq \sum_{i=1}^{n} [y_i-f(\underline{\hat{\theta}},\underline{u}_i)-v_{i1}(\theta_1-\hat{\theta}_1)-\ldots-v_{ip}(\theta_p-\hat{\theta}_p)]^2$$

$$= \sum_{i=1}^{n} [y_i-f(\underline{\hat{\theta}},\underline{u}_i)]^2 + (\underline{\theta}-\underline{\hat{\theta}})^T\underline{V}^T\underline{V}(\underline{\theta}-\underline{\hat{\theta}})$$

$$= S(\underline{\hat{\theta}}) + (\underline{\theta}-\underline{\hat{\theta}})^T\underline{V}^T\underline{V}(\underline{\theta}-\underline{\hat{\theta}}) \quad .$$

But this is exactly the same form as Eq. [3.3] in the linear case and so we may use the same argument to arrive at the linear approximation confidence region

$$(\underline{\theta}-\underline{\hat{\theta}})^T\underline{V}^T\underline{V}(\underline{\theta}-\underline{\hat{\theta}}) \leq pS(\underline{\hat{\theta}})F(p,n-p;\alpha)/(n-p)$$

where, now, the derivative matrix evaluated at $\hat{\underline{\theta}}$ is simply denoted by $\underline{V}$. As before, this defines an ellipsoidal confidence region with center $\hat{\underline{\theta}}$ in the parameter space.

Exact likelihood and confidence regions can be obtained (13-15) but these require much computing and graphical displays, and are not very practical when the number of parameters is three or more. For a further discussion, see Bates and Watts (16,17).

3.2.3. An Example. As a simple example of nonlinear least-squares we consider the Michaelis-Menten enzyme kinetic model and the data given in Table 1. To determine the least-squares estimates, we must first obtain starting values for the parameters, which may be done as follows: since f tends to a limit of θ_1 when u gets very large, we use the largest average value (.057) as a starting value for θ_1. Similarly, θ_2 is the u value at which $f=.5\times\theta_1$, and since y is about .028 (=.5×.057) around u=.6, we use .6 as a starting value for θ_2.

Next, we calculate $f(.057,.6;u_i)$, $v_{i1}=\dfrac{df_i}{d\theta_i}$, and $v_{i2}=\dfrac{df_i}{d\theta_2}$ and the differences $w_i=y_i-f_i$ (Table 4).

TABLE 4: First iteration in estimating Michaelis-Menten parameters

i	x_i	f(.057,.6)	y_i	w_i	v_{i1}	v_{i2}
1	2.000	.0438	.0615	.01765	.769	-.0169
2	2.000	.0438	.0527	.00885	.769	-.0169
3	.667	.0300	.0334	.00339	.526	-.0237
4	.667	.0300	.0268	-.00327	.526	-.0237
5	.400	.0228	.0138	-.00900	.400	-.0228
6	.400	.0228	.0258	.00300	.400	-.0228
7	.286	.0184	.0129	-.00550	.323	-.0208
8	.286	.0184	.0183	-.00010	.323	-.0208
9	.222	.0154	.0083	-.00709	.270	-.0187
10	.222	.0154	.0169	.00151	.270	-.0187
11	.200	.0143	.0129	-.00135	.250	-.0178
12	.200	.0143	.0087	-.00555	.250	-.0178

Then,

$$\underline{V}_0^T \underline{V}_0 = \begin{pmatrix} 2.5370 & -.10155 \\ -.10155 & .00493 \end{pmatrix} \quad , \quad \underline{V}_0^T \underline{w} = \begin{pmatrix} .01305 \\ .00003 \end{pmatrix}$$

and so

$$\underline{\beta}^1 = (\underline{V}_0^T \underline{V}_0)^{-1} \underline{V}_0^T \underline{w} = \begin{pmatrix} .0307 \\ .6382 \end{pmatrix} \quad .$$

Then

$$\underline{\theta}^1 = \underline{\theta}^0 + \underline{\beta}^1 = \begin{pmatrix} .0877 \\ 1.238 \end{pmatrix} \quad .$$

At θ^0 the sum of squares is 6.1403×10^{-4}, while at θ^1 it is only $2.\overline{267} \times 10^{-4}$, which is smaller and so θ^1 is a better value than θ^0. We therefore move to θ^1 and repeat the calculations. Convergence is finally obtained at $\hat{\theta} = (.10579, 1.7077)^T$ with a sum of squared residuals of 1.980×10^{-4} based on 10 degrees of freedom (ignoring replications), and with

$$\underline{V}^T \underline{V} = \begin{pmatrix} .8992 & -.0320 \\ -.0320 & .00123 \end{pmatrix} \quad .$$

The $(1-\alpha)$ linear approximation confidence region is then the ellipse defined by

$$.899(\theta_1 - .106)^2 - .064(\theta_1 - .106)(\theta_2 - 1.708) + .00123(\theta_2 - 1.708)^2 =$$

$$= 1.624 \times 10^{-4} \quad .$$

4. OTHER INTERPRETATIONS OF LEAST SQUARES

4.1. Least Squares _via_ Sum-of-Squares Surfaces

The analytical procedures presented in Section 3 have informative interpretations in terms of sums-of-squares surfaces. To develop these, we consider the linear and nonlinear examples of the previous section.

4.1.1. The Linear Model. For a given experimental design and model combination, at any point $\underline{\theta}$ the sum of squares is given by Eq. [1.4]. In the linear case however, [1.4] can be written as expression [3.3], which defines an ellipsoidal paraboloid. For the linear model and data of Section 3.1.3 the paraboloid is shown in Figure 2 with some contours of constant sum of squares corresponding to horizontal slices through the paraboloid. Also shown are the images of those

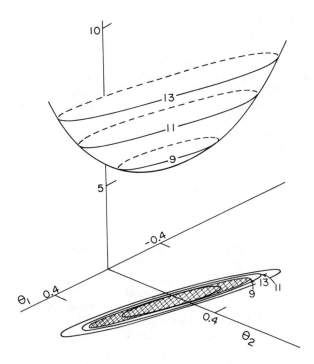

FIGURE 2:   The sum-of-squares surface for logPCB data.

contours projected onto the (θ_1, θ_2) plane.   It is seen from these
figures that the least-squares estimates are the values of θ_1 and
θ_2 at which the sum of squares is least, hence their name.

Furthermore, because the second derivative of $S(\underline{\theta})$ is constant,

$$\frac{\partial^2 S}{\partial \underline{\theta}^2} = \begin{pmatrix} \dfrac{\partial^2 S}{\partial \theta_1^2} & \dfrac{\partial^2 S}{\partial \theta_1 \partial \theta_2} \\[3mm] \dfrac{\partial^2 S}{\partial \theta_1 \partial \theta_2} & \dfrac{\partial^2 S}{\partial \theta_2^2} \end{pmatrix} = \underline{X}^T \underline{X}$$

this means that the sum of squares paraboloid is completely defined
by the coordinates of its center, $(\hat{\underline{\theta}})$, and its value there, $S(\underline{\theta})$,
and by its second derivative array $\underline{X}^T\underline{X}$.   Thus, if one were to use
a Taylor's series approximation to the model, the approximation

would be the same no matter what starting value were used to obtain
the local information.

Finally, because the confidence region is defined by the ratio
of the extra sum of squares due to not being at $\hat{\theta}$ relative to the
sum of squares at $\underline{\hat{\theta}}$, and the extra sum of squares is a constant
along any of the contour lines, we see that the confidence region
must be elliptical because the sum of squares contours are elliptical. The contour value for say, the $(1-\alpha)$ confidence region can
be obtained from Eq. [3.6] as

$$S(\underline{\theta}) = S(\underline{\hat{\theta}})[1+pF(p,n-p;\alpha)/(n-p)] \qquad [4.1]$$

For example, the 95% confidence region is shown as the shaded region
in the (θ_1, θ_2) plane in Figure 2.

4.1.2. The Nonlinear Model. No simply-described sum of squares
surface or contours occur in the nonlinear case. For example,
Figure 3 shows images of constant sums of squares contours calculated
for the nonlinear model and experimental design combination discussed in Section 3.2.3, from which it is seen that there is no
obvious simple family of curves describing the contours. Using the
expression [4.1] in the nonlinear case, it is possible to obtain a
good confidence region by finding the appropriate contour. Thus,
the shaded region enclosed by the 3.64 contour in Figure 3 gives a
95% likelihood region, which is also probably quite close to being
a 95% confidence region. The linear approximation confidence region,
given by Eq. [3.3], can be superimposed as shown in Figure 4, from
which it is seen that in this case, at least, the linear approximation confidence region is very different from that defined by the
sum of squares contour and hence is not a satisfactory approximation.

From this development, we see that the linear approximation
of the model at any point, say θ_o, corresponds to approximating the
sum of squares surface at θ_o by means of a paraboloid. The approximating paraboloid is such that at θ_o the approximate $\tilde{S}(\underline{\theta}_o)$ and
exact $S(\underline{\theta}_o)$ sum of squares are equal, and at θ_o the slopes and
curvatures of the approximate and true sum of squares surfaces are
also equal. Thus, our iteration procedure in which we approximate
the function by a linear Taylor's series corresponds, in the sum
of squares framework, to approximating the sum of squares surface
by means of a paraboloid. This is illustrated in Figure 5 which
shows two approximating ellipses on a path of convergence.

4.2. Least Squares *via* Geometry

The analytical procedures presented in Section 2 also have
informative interpretations in terms of geometry, as developed for
example by Fisher (18), Box and Coutie (19), Box, Hunter and Hunter
(20), Beale (21), and Bates and Watts (7,16,17). To illustrate the

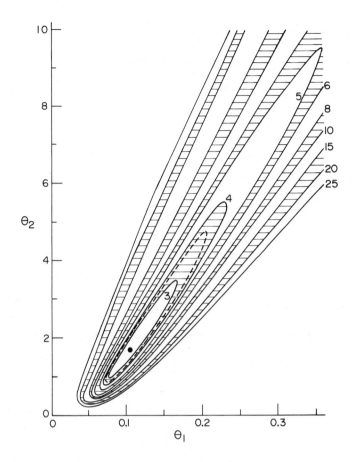

FIGURE 3:   Sum of squares contours for enzyme data.

geometrical aspects, we again use the linear and nonlinear examples
of the previous section, but in this instance we restrict ourselves
to only n=3 observations so we may plot things easily.

4.2.1. The Linear Model.   Consider the three observation pairs from
the example of Section 3.1.3.   In matrix terms we have

$$\underline{y} = \underline{X}\underline{\theta} + \underline{z} = \begin{pmatrix} -.511 \\ 1.386 \\ 2.595 \end{pmatrix} = \begin{pmatrix} 1 & 1 \\ 1 & 7 \\ 1 & 12 \end{pmatrix} (\theta_1, \theta_2) + \begin{pmatrix} z_1 \\ z_2 \\ z_3 \end{pmatrix} \quad .$$

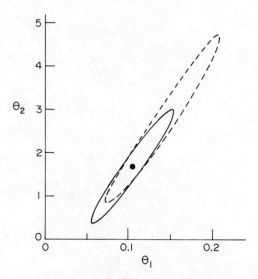

FIGURE 4:   95% linear-approximation confidence region and likeli-
            hood region for enzyme data .

The column vectors $\underline{y}=(-.511,1.386,2.594)^T$, $\underline{x}_1=(1,1,1)^T$, and
$\underline{x}_2=(1,7,12)^T$ each may be thought of as defining a point in an n=3
dimensional "response space", as shown in Figure 6. Furthermore,
the vectors $\underline{x}_1$ and $\underline{x}_2$ define a plane, and any point on that plane
can be reached by multiplying by $\underline{x}_1$ by θ_1 and $\underline{x}_2$ by θ_2. That is,
the expression $X\underline{\theta}$ also represents a mapping of the parameter plane
($\underline{\theta}$) onto a planar solution locus or solution plane $X\underline{\theta}$.

    The least-squares estimates correspond to the point $X\hat{\underline{\theta}}$ on the
solution plane which is closest to the observation point $\underline{y}$. But
since the vector $\underline{y}$ can be decomposed into a component in the solu-
tion plane, we see that the residual vector $\hat{\underline{z}}=\underline{y}-X\hat{\underline{\theta}}$ is orthogonal
or normal to X, that is $X(\underline{y}-X\underline{\theta})=0$, which is precisely the equation
[3.3] given earlier. Thus, the least-squares estimates $\hat{\underline{\theta}}$ are given
by the point which map to the point on the solution locus which is
closest to $\underline{y}$.

    If we consider another point on the solution plane, say $X\underline{\theta}$,
then the vector $\underline{y}-X\underline{\theta}$ may also be decomposed into two orthogonal
parts, $(\underline{y}-X\hat{\underline{\theta}})$ normal to the plane, and $X(\underline{\theta}-\hat{\underline{\theta}})$ in the plane. Thus
the sum of squares can be written

$$S(\underline{\theta}) = \|\underline{y}-X\underline{\theta}\|^2 = \|\underline{y}-X\hat{\underline{\theta}}\|^2 + \|X(\underline{\theta}-\hat{\underline{\theta}})\|^2$$
$$= S(\hat{\underline{\theta}}) + (\underline{\theta}-\hat{\underline{\theta}})^T X^T X (\underline{\theta}-\hat{\underline{\theta}})$$

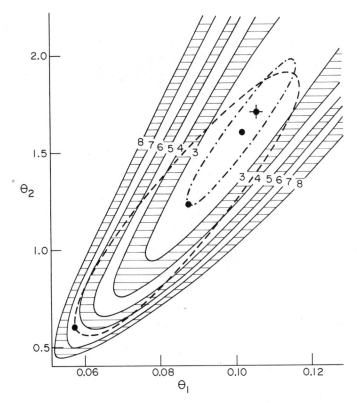

FIGURE 5:   Sum of squares contours and approximating ellipsoidal
contours for enzyme data: starting value (.057,.6) ------
                              first iteration (.088,1.24) —·—·—·—
                              least-squares estimate $\oplus$

which is equation [3.3]. The hypothesis test used to develop confidence regions corresponds, geometrically, to comparing the squared
length of the tangential component, $\|X(\theta-\hat{\theta})\|^2$, divided by the
dimensionality of the solution locus space, p, to the squared length
of the residual component $\|y-X\hat{\theta}\|^2$ divided by the dimensionality
of the orthogonal complement to the solution locus space, namely
n-p. That is, a point $X\theta$ is acceptable if the planar component is
not too large, where the scale of measurement is calibrated in
terms of the length of the residual vector. Thus, a (1-α) confidence region will consist of all values on the solution plane which
are within a radius $\rho=\{ps^2F(p,n-p;\alpha)\}^{1/2}$ of the point $X\hat{\theta}$. Finally,
because there is a simple one-to-one relationship between any point
$X\theta$ on the solution plane and its corresponding image point θ on the
parameter plane, there is a simple relationship between the points
on the circle of radius ρ centered at $X\hat{\theta}$ and the image points in
the parameter plane, which is

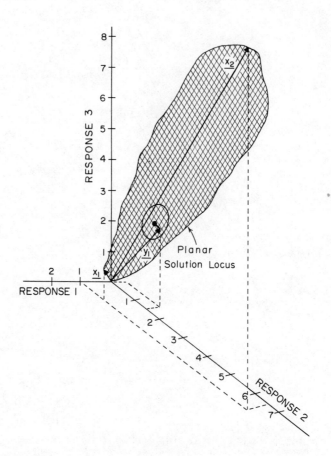

FIGURE 6:   The response space representation of least squares.

$$(\underline{\theta}-\hat{\underline{\theta}})^{T}\underline{X}^{T}\underline{X}(\underline{\theta}-\hat{\underline{\theta}}) = \rho^{2} \qquad .$$

The inverse mapping causes the solution plane circle to be mapped
into a parameter-plane ellipse.

The assumptions of normality and independence of the Z are
especially important here. Together they ensure that ratios of
simple sums of squares, corresponding to squared distances, are
appropriate. The assumption of a linear model causes the solution
locus to be a plane, and to have a simple mapping of straight,
parallel, equi-spaced parameter lines to straight, parallel equi-
spaced lines on the solution plane, and *vice versa* . Hence a sphere
in one space will map to an ellipsoid in the other.

4.2.2. The Nonlinear Model. There are dramatic differences between
the linear and nonlinear situations, as revealed geometrically,
because the solution locus is a curved surface rather than a plane,
and perhaps more importantly, because lines in the parameter space
map to curves on that curved surface. For example, Figure 7 shows
the solution locus for the Michaelis-Menten model $f(\underline{\theta},\underline{u})=\theta_1 u/(\theta_2+u)$
for the design $\underline{u}=(2,2/3,2/5)^T$. The surface was determined by set-
ting $\underline{u}$ equal to the design value and then calculating and plotting,
for specific values of θ_1 and θ_2, the point

$$\underline{n}(\theta) = \left(\frac{2\theta_1}{\theta_2+2}, \frac{.667\theta_1}{\theta_2+.667}, \frac{.4\theta_1}{\theta_2+.4}\right)^T .$$

Similarly, lines of constant θ_1 or θ_2 can be drawn. For example,
the line $\theta_2=0$ maps to a vector through the origin with direction
$(1,1,1)^T$, the line $\theta_2=1$ maps to a vector with the direction
$(2/3,2/5,2/7)^T$, and so on. A line θ_1=constant, however, maps to a
curve on the solution locus; for example, the line $\theta_1=1$ maps to a
curve with coordinates

$$\underline{n}(\theta_1=1,\theta_2) = \left(\frac{2}{\theta_2+2}, \frac{.667}{\theta_2+.667}, \frac{.4}{\theta_2+.4}\right) .$$

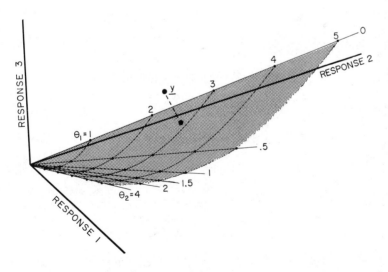

FIGURE 7: The solution locus for the Michaelis-Menten model.

It is clear from the figure that the solution locus is curved, and further, that the parameter curves are in general not straight, not parallel, and not equi-spaced. Hence, although there is a direct mapping and relationship which dictates a value for $\underline{\eta}(\theta)$ on the solution locus given a point $\underline{\theta}$ in the parameter space, there are not generally any simple analytic procedures for going in the reverse manner from a point $\underline{\eta}$ on the solution locus to the corresponding point in the parameter space. This makes the problems of finding the least-squares estimates and of finding confidence regions enormously difficult. If we imagine an observation vector $\underline{y}$ superimposed in Figure 7, we can see that the least-squares estimate will still be that point on the solution locus which is closest to $\underline{y}$, but finding that point, especially in terms of $\underline{\theta}$, is not easy.

The linear approximation method proposed by Gauss has an interesting and informative geometric interpretation, since it corresponds to two distinct approximations. First, the solution locus near the starting value is approximated by the tangent plane, so the orthogonal decomposition of the observation vector can be done. If this planar approximation is good, because the solution locus is very flat (or as discussed by Beale (21) and Bates and Watts (16,17), the solution locus has small intrinsic curvature), then the point on the tangent plane nearest to $\underline{y}$ will be close to the point on the solution locus which is nearest to $\underline{y}$, and hence we may obtain rapid convergence. However, we need some procedure for expressing points on the solution locus in terms of the parameters $\underline{\theta}$. The second approximation provides this, since it imposes a uniform coordinate system on the tangent plane so that the true parameter curves are approximated by a set of parallel, equi-spaced straight lines whose directions and spacing are specified by the columns of the matrix $\underline{V}$. Using the increment formula [3.10], therefore, enables us to determine roughly what direction and distance to go in the sample space in order to get to a point on the solution locus which is (apparently) closer to $\underline{y}$.

Unfortunately, however, the solution locus will curve away from the local tangent plane, and so the indicated direction will only be accurate locally. Second, because of the parameter-effects nonlinearity (16,17), which causes the loci of constant parameter values to curve when an increment is called for, the change in the coordinates on the solution locus will not be simple linear translation given by $\underline{V}\underline{\beta}_1$, but rather, the new point will have coordinates $\underline{\eta}(\underline{\theta}_0+\underline{\beta}_1)$: for $\underline{\beta}_1$ of good size, the vectors $\underline{\eta}(\underline{\theta}_0+\underline{\beta}_1)-\underline{\eta}(\underline{\theta}_0)$ and $\underline{V}\underline{\beta}_1$ may be very different! These two considerations therefore give a geometric explanation of why overshoot sometimes occurs, and why iterations must be used in nonlinear estimation.

Finally, we may also see why the accuracy of linear approximation confidence regions can vary widely from situation to situation,

since the confidence region contains effectively those values in
the parameter space whose image on the solution locus is not too
far from the observation point $\underline{y}$. In the linear case, the solution
locus points correspond to the intersection of the solution plane
with a sphere centered at $\underline{y}$ which results in a confidence disc on
the solution plane. Taking this confidence disc through the linear
mapping to the parameter plane produces an ellipse, as discussed
previously. In the nonlinear case, we have (roughly), the intersection of a curved solution locus with a sphere, which results in
an irregularly shaped curve on the solution locus, the coordinates
of which must be mapped back through the inverse of the nonlinear
mapping to produce an accurate confidence region. If the solution
locus is strongly curved (has large intrinsic curvature) and/or
if the parameter curves are strongly curved (have large parameter-
effects curvature), then the linear approximation confidence region
will be poor. If both the intrinsic and parameter-effects curvatures are small, however, so that, over the region of interest,
the solution locus is essentially flat and the parameter curves are
nearly straight, nearly parallel, and nearly equi-spaced, then the
linear-approximation confidence region will be accurate. For further discussion of the effects of nonlinearity on confidence regions,
see Bates and Watts (16,17).

5. CONCLUSIONS

Least-squares regression analysis is a powerful statistical
technique for the analysis of data. Like any other statistical
technique, the validity of the analysis depends on the appropriateness of the assumptions inherent in the procedure: that is, *IF
THE ASSUMPTIONS ARE NOT APPROPRIATE, THE ANALYSIS WILL NOT BE VALID*.
Special care should be exercised by all scientists who use regression
techniques to ensure that, for their data sets, the analysis is
valid because the assumptions are appropriate. They should, therefore, attach as much importance to analysis of the residuals and
the appropriateness of assumptions as they do to parameter estimates
and confidence regions for them.

REFERENCES

1. Box, G.E.P. and Jenkins, G.M. (1970) "Time Series Analysis
   - Forecasting and Control", Holden-Day, San Francisco-
   Cambridge-London-Amsterdam.
2. Box, G.E.P. and Tiao, G.C. (1973) "Bayesian Inference in Statistical Analysis", Addison-Wesley, Reading, Mass.
3. Fisher, R.A. (1935) "The Design of Experiments", Oliver and
   Boyd, Edinburgh-London, Hafner, New York.
4. Box, G.E.P. (1966) Technometrics, $\underline{8}$, 625-629.
5. Cramer, H. (1946) "Mathematical Methods of Statistics",
   Princeton. Univ. Press, Princeton, N.J.

6. Scheffe, H. (1959) "The Analysis of Variance", Wiley, New
    York-London-Sydney.
7. Bache, C.A., Serum, J.W., Youngs, W.D. and Lisk, D.J. (1972)
    Science, 117, 1192-3.
8. Gauss, C.F. "Theory of Least Squares", English Translation
    by Hale F. Trotter, Princeton University Statistical
    Techniques Research Group, Technical Report 5, 1957.
9. Bates, D.M. and Watts, D.G. (1980) "A Relative Offset Conver-
    gence Criterion for Nonlinear Least Squares", Queen's
    University, Department of Mathematics and Statistics,
    preprint 1979-14, Kingston, Ontario.
10. Box, G.E.P. (1958) Bull. Int. Stat. Inst. 36, 215-225.
11. Hartley, H.O. (1961) Technometrics, 3, 269-280.
12. Marquardt, D.W. (1963) SIAM J. Appl. Math. 11, 431-441.
13. Halperin, M. (1963) J. Roy. Stat. Soc. B25, 330-333.
14. Hartley, H.O. (1964) Biometrika, 51, 347-353.
15. Williams, E.J. (1962) J. Roy. Stat. Soc. B24, 125-139.
16. Bates, D.M. and Watts, D.G. (1980) J. Roy. Stat. Soc. B40,
    1-36.
17. Bates, D.M. and Watts, D.G. (1980) "Conservative Approximate
    Confidence Regions in Nonlinear Least Squares", Queen's
    University, Department of Mathematics and Statistics,
    preprint 1979-8, Kingston, Ontario.
18. Fisher, R.A. (1939) Ann. Eugenet. 9, 238-249.
19. Box, G.E.P. and Coutie, G.A. (1956) Proc. Inst. Electr. Eng.
    Suppl. 1, B103, 100-107.
20. Box, G.E.P., Hunter, W.G. and Hunter, J.S. (1978) "Statistics
    for Experimenters", Wiley, New York-Chichester-Brisbane-
    Toronto.
21. Beale, E.M.L. (1960) J. Roy. Stat. Soc. B22, 41-88.

STATISTICAL PROPERTIES OF KINETIC ESTIMATES

Carl M. Metzler

The Upjohn Company
Kalamazoo, Michigan   49001
U.S.A.

ABSTRACT

Most kinetic parameters of interest are related to the observed
data by nonlinear mathematical expressions. Thus the statistical
properties of the estimates are influenced by many factors: The
model and its degree of nonlinearity, the number of observations,
the values of the independent variable, the error structure, the
method of estimation, transformations of the parameter space, and
the initial estimates if the estimation procedure is iterative. For
most models, an asymptotic estimate of the covariance matrix can be
written which includes some of these factors. The validity of the
asymptotic estimate is questionable in the usual pharmacokinetic
study with few observations. Some small-sample properties of the
estimates can be inferred from Monte Carlo simulations, but the vast
variety of possible situations limits the usefulness of this approach.
A review of some Monte Carlo studies indicates that for many pharmacokinetic situations the bias of the estimates is small, but the
estimates of variability are poor. In general, estimates of the
parameters of linear systems have small-sample properties that are
well approximated by the asymptotic theory. Parameter estimates
from nonlinear systems are more difficult to compute and the statistical properties for small samples are not well approximated by
asymptotic theory.

1.   INTRODUCTION

It is widely believed, and has been stated in publications,
that computer methods, particularly least-squares estimation, are
better for estimating the parameters of kinetic models than are
graphical methods such as the "peeling" of exponential terms, often

used in pharmacokinetics, or the linearization of models, as is often done in enzyme kinetics. The reasons given are that computer methods are more objective and also provide statistical information about the estimates that graphical methods cannot give. Yet, in practice, this statistical information is often ignored. A review by Westlake (1) pointed out the need for giving more attention to the goodness of pharmacokinetic estimates.

There are three statistical properties of kinetic estimates that should be considered: bias, variability and correlation. It can be shown that bias is usually relatively small, but that variability can be very large and must be considered in applications of kinetic models. Correlation between estimates must be taken into account when making statements about differences in parameter estimates.

Most computer programs for nonlinear estimation provide at least the asymptotic estimates of the variance-covariance matrix. But it is usually not known how well these asymptotic estimates approximate the properties of estimates obtained with relatively small samples (experiments).

2. NONLINEAR PARAMETER ESTIMATION

An important part of much work in enzyme kinetics and pharmacokinetics is the estimation of the parameters of a kinetic model. Estimation is usually performed by "fitting" observed data to the proposed model. Often the model represents the data as a nonlinear (in the parameters) mathematical expression plus an error term. That is, the N observations are represented as

$$C_i = C(\underline{X}_i) = f(\underline{X}_i, \underline{\theta}) + e_i \quad , \qquad i=1,\ldots,N \qquad [1]$$

where $\underline{X}_i$ may be a vector of observed or controlled variables, $\underline{\theta}$ is an unknown vector of M parameters, $f(\cdot)$ is a nonlinear function and e_i is the error term.

In the early history of enzyme kinetics lack of computing power made nonlinear estimation prohibitive, and thus enzyme kinetic models were linearized (2,3). Pharmacokinetics, with its more recent development, has not resorted so heavily to linearizing its models, but since many pharmacokinetic models are sums of exponentials, much estimation was based on plotting on logarithmic paper. With the present availability of computing power, ease of computation is no longer a valid reason for linearizing models. This paper is concerned with the statistical properties of estimates obtained by using a nonlinear estimation procedure with a nonlinear model.

In many cases, there are data from more than one experiment and a set of estimates of $\underline{\theta}$ is obtained. The statistical analysis of these estimates is a separate problem which has been discussed elsewhere (4-6). Here I am concerned only with what can be said about the statistical properties of the estimate $\hat{\underline{\theta}}$ of $\underline{\theta}$ obtained from one set of data.

There are several criteria for fitting Eq. [1] to data, that is, for estimating $\underline{\theta}$. The most commonly used are least squares and maximum likelihood. For each of these criteria there have been many procedures suggested. I will not discuss the estimation procedures here; Draper and Smith (7) have discussed least-squares estimation. Bard (8) discusses many of the estimation procedures. Regardless of the method of estimation, the properties of $\hat{\underline{\theta}}$ are almost always obtained by linearizing $f(\cdot)$; that is, by approximating $f(\cdot)$ by a Taylor-series expansion which only uses the first-order terms. The statistical properties depend on at least the following eight factors.

3. FACTORS AFFECTING STATISTICAL PROPERTIES

3.1. The Nonlinearity of $f(\cdot)$

If $f(\cdot)$ is only a little nonlinear, then the approximation is good and the well-developed linear theory of regression can be applied to $\hat{\underline{\theta}}$. Beale (9) first discussed measures of nonlinearity for models of this type. Bates and Watts (10) have recently suggested other measures for evaluating the degree of nonlinearity of $f(\cdot)$.

3.2. The Number of Observations

As in most of statistics, the larger the sample size N, the better are the estimates in the sense that the variances will be smaller. In addition, Jennrich (11) has shown the properties of $\hat{\underline{\theta}}$ based on linearization to be only asymptotic, that is, they hold as N approaches infinity (becomes very large).

3.3. The Values of X

In addition to the number of observations, the values of the independent variable also determine the properties of $\hat{\underline{\theta}}$, in that poorly chosen $\underline{X}_i$ may not contain very much information about the parameters of interest.

3.4. The Distribution of the Errors

The distribution of the error term e_i also influences $\hat{\underline{\theta}}$. Most of the analytical results assume that the errors are independent with identical Gaussian distributions (in the case that the variance is estimated from the data). If the variance-covariance matrix is known (which it seldom is), then the error variances need not be identical.

3.5. The Value of θ

The true value of θ which generates the data also determines the properties of $\hat{\theta}$. As is always the case when estimating θ, the true value is not known, although something may be known about the probability of its lying in a given region of the parameter space.

3.6. The Method of Estimation

The algorithms and implementation of estimation influence the properties of $\hat{\theta}$, although all methods which obtain the least-squares solution should provide estimation with identical properties. However, methods may differ in their ability to reach the absolute minimum and thus may obtain different estimates.

3.7. Transformations of θ

Parameter estimation is often improved by transformation of the parameter space (7), that is, by reparameterization of the model. Some estimation programs effect limits on the parameter search by transformations (12). Any transformation of the parameter space will alter the statistical properties of $\hat{\theta}$.

3.8. Initial Estimates

All nonlinear estimation procedures are iterative, and some are sensitive to changes in the initial estimates of θ. Such procedures will provide estimates with properties different from those procedures that are robust in regard to selection of initial estimates.

4. ASYMPTOTIC STATISTICAL PROPERTIES

When the model is linearized, the N×M matrix of partial derivatives is formed, where

$$a_{ij} = \frac{\partial f(\underline{X}_i, \theta)}{\partial \theta_j} \quad . \tag{2}$$

Then $\underline{V} = \underline{A}^T \underline{A}$ is the information matrix, and asymptotically (as N gets very large) variance $(\hat{\theta}) = \sigma^2 \underline{V}^{-1}$, so that $\underline{C} = \underline{V}^{-1}$ is called the variance-covariance matrix of $\hat{\theta}$. Using s^2, the error sample variance as an estimate of σ^2, the following statements are true asymptotically:

$$\text{variance}(\hat{\theta}) = \underline{C}s^2 \quad , \tag{3}$$

$100(1-\alpha)\%$ confidence limits for each $\hat{\theta}_i$ are given by

$$\hat{\theta}_i \pm t(\alpha/2, N-M)\sqrt{c_{ii}}\; s \quad , \tag{4}$$

$100(1-\alpha)\%$ confidence regions for $\hat{\underline{\theta}}$ are defined by

$$(\underline{\theta}-\hat{\underline{\theta}})'\underline{C}(\underline{\theta}-\hat{\underline{\theta}}) = Ms^2 F(M, N-M, \alpha) \quad , \tag{5}$$

where $t(\cdot)$ and $F(\cdot)$ are the Student's t and Snedecor's F statistics with the appropriate degrees of freedom.

It can be seen that these asymptotic estimates of variances and confidence regions are functions of the second to fifth factors listed above, while the goodness of the approximation depends on the first factor. For a linear model, the confidence regions defined by Eq. [5] are hyper-ellipsoids; for nonlinear models they will not be. One way to assess the nonlinearity of a model is to compute level contours of the sums-of-squares surface and project them onto planes defined by pairs of parameters in $\underline{\theta}$. The amount by which these projections differ from ellipses indicates the nonlinearity of the model.

5.   MONTE CARLO INVESTIGATIONS

Most pharmacokinetic experiments have such a small number of observations that the validity of the asymptotic properties is in severe question. Small-sample properties of the estimates can be inferred from Monte Carlo studies. Unfortunately, there is a vast variety of possible situations, and generalization from a Monte Carlo study of nonlinear estimation must be quite limited. There have been a few such studies reported in the literature and they will be reviewed here along with a little additional work.

Bard (8) carried out a Monte Carlo study with the function $f(\underline{x},\underline{\theta})=\exp[-\theta_1 x_1 \exp(-\theta_2/x_2)]$ for 5 values of x_1 at each of 3 values of x_2 (N=15). He used parameter values $\underline{\theta}=(1000,1000)$ and simulated 100 experiments with normal error distribution and 100 experiments with uniform error. (Each at three different values of σ.) For these conditions and this model, the estimates had little or no bias (the sample bias was smaller than the standard deviation of the mean), while the estimates of variance were consistently underestimated, up to as much as 16%.

Duncan (13) examined two models that are much used in pharmacokinetics. They are

   (A) $f(t,\underline{\theta}) = \theta_1[\exp(-\theta_2 t) - \exp(-\theta_1 t)]/[\theta_1-\theta_2]$

and

(B) $f(t,\underline{\theta}) = 1 - [\theta_1 \exp(-\theta_2 t) - \theta_2 \exp(-\theta_1 t)]/[\theta_1 - \theta_2]$.

Duncan was interested in obtaining better confidence intervals for θ by applying jackknife methods, an idea that was not successful. But he did report the results of Monte Carlo studies with these models. His experimental points were t=0.5,1,2,4,8 and 16, each replicated 4 times (N=24). For Model A, σ^2=.003 and $\underline{\theta}$=(0.2,0.5); for Model B, σ^2=.001 and $\underline{\theta}$=(0.2,0.5),(1.4,0.4). Each experiment was simulated 500 times with random error, and Duncan reports the number of times that the approximated 95% confidence interval as defined by equations [4] or [5] contained the true values of θ. His conditions include errors that were either additive or multiplicative from normal, contaminated normal and uniform distributions. For additive errors the confidence intervals contained the true parameter values 89.4% to 98.8% of the time, with most of the situations being very close to 95%. For multiplicative errors the results were from 80.4% to 98.2%. Thus for additive errors, the structure assumed by the estimation procedure, the asymptotic results for these two models are quite good in sample size as small as N=24.

Duncan also reports simulation results for the model $F(X,\theta)=$ $1000 \cdot \theta_1^g$, where $g=\theta_2^X$. Again, for additive errors the asymptotic 95% confidence intervals are quite good, for multiplicative errors they do not do as well.

Metzler and Elfring (14) reported Monte Carlo studies of nonlinear standard curves, or calibration curves. Their main interest was in the statistical properties of the estimated concentrations. But they also obtained information on the statistical properties of the estimates of the parameters of two models. The first model they considered was the hyperbola,

(C) $f(x,\underline{\theta}) = x/(\theta_1 + \theta_2 x)$,

where x=.01,.15,.25,.40 and .55, with 4 replicates (N=20). The parameter values were $\underline{\theta}$=(.002,.000544), (.008,.00218), (.0005,.00717), (.005,.000136) and (.008,.000136). Each situation was simulated 1000 times with additive normal errors. The estimates were entirely unbiased and the average of the estimated standard deviations were within one percent of the standard deviations of the sets of 1000.

The other model investigated was

(D) $f(x,\underline{\theta}) = \theta_3/[1+\theta_1 \exp(-\theta_2 x)]$,

where x=.05,.1,.3,.5 and .6 with 4 replicates (N=20). The parameter values were at nine points in the region $1.75 < \theta_1 < 7.5$, $3 < \theta_2 < 12$, $\theta_3 = 1$. Errors were additive normal with σ=.01. This model was simulated

500 times for each combination of parameters. The results are
shown in Table 1. It can be seen that, with one exception, the
bias, if any, is quite small, and the average estimates of the
standard deviations of the estimates are very close to the standard
deviations of the 500 estimates. The one exception is $\underline{\theta}=(7.5,3.0,$
$1.0)$. With these parameter values, the value of x do not include a

TABLE 1: Estimated parameters of Model (D)
and their standard deviations

Parameter	Avg. of estimate	S.D. of estimate	Avg. estimate of S.D.
θ_1 = 3.5	3.596	.444	.402
θ_2 = 3.0	2.997	.255	.232
θ_3 = 1.0	1.022	.114	.102
σ^3 = .01	.00977	.00151	
θ_1 = 3.5	3.506	.0645	.069
θ_2 = 6.0	6.010	.158	.163
θ_3 = 1.0	1.000	.0112	.0114
σ^3 = .01	.00981	.00177	
θ_1 = 3.5	3.499	.101	.0986
θ_2=12.0	11.998	.317	.318
θ_3 = 1.0	1.00	.0038	.00392
σ^3 = .01	.00998	.00172	
θ_1 = 7.5	8.296	3.050	2.863
θ_2 = 3.0	3.001	.284	.293
θ_3 = 1.0	1.097	.455	.367
σ^3 = .01	.00979	.00164	
θ_1 = 7.5	7.525	.207	.191
θ_2 = 6.0	6.006	.173	.164
θ_3 = 1.0	1.001	.0211	.0195
σ^3 = .01	.00977	.00172	
θ_1 = 7.5	7.528	.218	.212
θ_2=12.0	12.023	.213	.207
θ_3 = 1.0	1.000	.00412	.00404
σ^3 = .01	.00974	.00177	

Model (D): $f(x,\underline{\theta}) = \theta_3/[1+\theta_1 \exp(-\theta_2 x)]$

x=.05,.1,.3,.5,.6;   4 replicates
σ=0.01;   N=20

wide enough range to provide much information about θ_1. The estimated standard deviations indicate that the estimate is poor (high
variance). This illustrates the influence of the third factor on
the properties of the parameter estimates.

Some additional simulation, not previously reported, has been
done with the "one-compartment open" model:

(E) $f(t,\underline{\theta}) = \theta_3\theta_1[\exp(-\theta_1 t) - \exp(-\theta_2 t)]/[\theta_2-\theta_1]$.

The values of t were .25,.5,1,2,3,4,5,6,8,10,12 and 14 (N=12). The
error was normal additive, but proportional to $f(\cdot)$, (σ=.05f). The
results of 250 simulations are shown in Table 2. In the table,
E(1/2) is the "elimination half-life", E(1/2)=(ℓn 2)/θ_2; A(1/2) is
the "absorption half-life", A(1/2)=(ℓn 2)/θ_1.

Again, for this situation, which is representative of many
pharmacokinetic models, the estimates have essentially no bias and
the average of the approximate standard deviations is close to the
standard deviation of the set of 250 estimates. The ranges of the
estimates are close to $\bar{X}\pm3$ standard deviations. Thus, there is no
evidence that the estimates are not normal with expected value equal
to the true parameter values and σ^2 well estimated by Eq. [3].

TABLE 2:   Estimated parameters of Model (E)
and their standard deviations

Parameter	Average estimate	S.D. of estimates	Avg. estimate of S.D.	Range of estimates
θ_1 =2.0	1.944	.157	0.136	1.58-2.53
θ_2 =.15	.1502	.00643	.00744	.132-.171
θ_3 =200.	200.54	7.179	5.677	178.7-218.7
E(1/2)=4.62	4.622	0.197	-	4.06-5.27
A(1/2)=.347	.350	.027	-	.274-.439
Area =1333.3	1335.5	26.78	-	1243-1432
Peak =162.12	162.24	3.94	-	150.3-173.0

Model E: $f(\underline{\theta},t) = \theta_3\theta_1[\exp(-\theta_1 t) - \exp(-\theta_2 t)]/[\theta_2-\theta_1]$
t=.25,.5,1,2,3,4,5,6,8,10,12,14
σ=.05f;   N=12

The pharmacokinetic models examined thus far are all models of linear systems, that is, the system can be described by a system of linear differential equations. In spite of the almost certainty that biological systems are not linear, their successful use with many drugs indicates that for those drugs they are a good approximation. For other drugs they are not a good approximation, and attempts have been made to find nonlinear compartment models that are better approximations. One class of nonlinear models assumes that elimination of drug from the system, rather than being a first-order process, is a "Michaelis-Menten" process. The well known Michaelis-Menten (M-M) equation has been widely and successfully used in enzyme kinetic studies in spite of some difficulties in the theoretical derivation (2,3). The M-M equation

$$v = V_{max} \, c_s/(K_m + c_s) \tag{6}$$

is used to estimate V_{max} and K_m from observations of the initial rate of reaction v at various substrate concentrations c_s. It has been adapted to pharmacokinetic models by assuming that the elimination rate from a compartment has a form similar to the right side of Eq. [6]. Thus, if $c(t)$ is the concentration of drug in a compartment, then the rate of change (derivative) of $c(t)$ is

$$\frac{dc(t)}{dt} = f_0(t,\underline{\theta}) - Vc(t)/[K+c(t)] \quad , \tag{7}$$

where $f_0(t,\underline{\theta})$ is a function determined by the input into the compartment. In the usual pharmacokinetic application of Eq. [7], attempts are made to estimate V and K along with the other model parameters, $\underline{\theta}$, by fitting the model to observed concentrations of drug at n sampling times after dosing.

The simplest such model is the one-compartment model with bolus i.v. input. For that model $f_0(t,\underline{\theta})$ of Eq. [7] is zero, and the model is

$$\frac{dc(t)}{dt} = -Vc(t)/[K+c(t)] \quad .$$

For this model, there is no solution which expresses $c(t)$ as an explicit function of t, and the solution is an implicit function.

For more complex models, say a one-compartment model with first-order input, a solution in closed form does not exist. For this model $f_0(t,\underline{\theta})$ is given by

$$f_0(t,K_A,D) = DK_A \, \exp(-K_A t) \quad ,$$

where D is a dilution factor, roughly equal to dose divided by a volume of distribution, and the absorption rate constant is K_A.

It is possible, however, to fit these models to observed data
by numerically integrating Eq. [7] to obtain points [t,c(t)] predicted by the model. In recent years many reports have been published of pharmacokinetic studies which obtained parameter estimates
by combining a nonlinear regression algorithm with numerical integration. Numerical integration of these models can be very expensive in terms of computer time, however, and when combined with
nonlinear regression algorithms which require many evaluations of
the solution, the expense of fitting models in this way can be
prohibitive. Also, experiences of many researchers indicate that
these methods may be unsatisfactory in terms of convergence and
final fit. These problems motivated an examination of the matema-
tical properties of these models (15) as well as a Monte Carlo study
of the properties of the estimates (16). For the simulation studies
some representative cases were chosen with the hope that they would
yield insight to the general problem. Table 3 lists the cases simulated. All models involved one compartment with either bolus input
or first-order absorption and M-M type elimination. Two types of
error structure were considered: additive error in which each simulated "observation" was the model-predicted value plus a normal
random number from the distribution N(0,.09), or proportional error
in which each "observation" was the model-predicted value plus the
product of the model value and a N(0,.0025) random variable. The
proportional error structure is equivalent to "5% error". This
error structure violates one of the assumptions of least-squares
estimation, and either a transformation should be done, or variable
weights used. The usual least-squares estimation was carried out,
however, in the belief that this reflects common practice where no
error analysis is done and the error structure is not known.

Each data set had N observations. The values of t were 0,.5,
1,1.5,2,3,4,5,6,7 and 10 for the bolus i.v. model, and .25,.5,1,
1.5,2,3,4,6,9,12 and 18 for the first-order input models. Each
model was simulated n times; that is, n sets of "observations" were
generated and for each set of N "observations" the parameters were
estimated by a nonlinear regression program. This program also
estimated the standard deviation for each estimate and the correlation matrix of the estimates.

The results of the Monte Carlo studies of Cases 1-6 can be
summarized in the following points: (i) Estimates of V and K are
highly skewed to the right, with many of the estimates being many
times larger than the true parameter values; (ii) estimates of V
and K are very highly correlated so that the estimates of one almost
determines the estimate of the other; and (iii) estimates of V and
K have very high variances, although the estimated variances as
computed by the computer program from individual data sets underestimate the variability. The correlations of estimates of V and
K with their respective standard deviation estimates are also high.

TABLE 3: Specifications of Michaelis-Menten models and simulations

Case	Model (input)	Parameter	Parameter values	Error structure	N	n
1	Bolus i.v.	C_0,V,K	2.0,0.22,0.11	Additive	11	200
2	Bolus i.v.	C_0,V,K	2.0,0.22,0.11	Proportional	11	200
3	First-order	K_A,V,K	1.5,2.0,5.0	Additivie	11	200
4	First-order	K_A,V,K	1.5,2.0,5.0	Proportional	11	200
5	First-order	K_A,V,K,D	1.5,2.0,5.0,10.0	Additive	11	200
6	First-order	K_A,V,K,D	1.5,2.0,5.0,10.0	Proportional	11	200
7	First-order (3 functions)	K_A,V,K,D_1,D_2,D_3	1.5,2.0,5.0,10.0,20.0,40.0	Proportional	33	100
8	First-order	K_A,V,K,D	1.5,2.0,5.0,10.0	Proportional	33	100
9	First-order	K_A,V,K,D	1.5,2.0,5.0,40.0	Proportional	33	100

The estimated variances of the estimates of V and K, as computed
by a nonlinear regression program, do not give a true picture of
the uncertainty in the estimates.

As a related point, the linear parameters, K_A and D, are well
estimated in the sense that the distributions of their estimates
are symmetric with small bias and small standard deviations. The
correlation of their estimates, while smaller than that of estimates
of V and K, is still large.

Estimation was best for the simple model, Cases 1 and 2 of
Table 3. But even for these cases the distributions were skewed to
the right with some very extreme values, and the correlations were
high.

For all these simulations, the initial estimates for the iterative nonlinear regression program were the true values of the parameters used to generate the data. A limited amount of simulation
with other starting values indicated that, for a particular data
set, the initial values affected the final estimates, but, averaged
over 200 simulations, initial estimates within 30% of the true
values did not make much difference. Initial estimates which differ
by more than 30% often converge to values far removed from the true
parameters.

It has been recognized that a M-M type model may be well approximated by a linear model for a single drug exposure, but observations
at more than one dose level will expose the nonlinearity of the
system. Thus, Case 7, Table 3, was an attempt to evaluate the
effect of simultaneous fitting of observations at three dose levels.
For Case 7 it was assumed that V and K remained the same when D=10,
20 and 40. Data were simulated for three experiments (D=10,20 and
40) and these three observed curves, with a total of N=33 observations were fitted simultaneously to one set of (K_A,V,K) and three
D's. This gave much better parameter estimation. The distributions
of the estimates of V and K were symmetric about the true values
with no extreme estimates. The correlation of the estimates of V
and K was reduced.

The improvement in estimation of V and K is not due only to
the larger sample sizes. To confirm this, Cases 8 and 9, Table 3,
also have 33 observations per data set, but in these cases the observations consisted of 3 replications at each of the sampling
times used in Cases 3 through 6. Although the variability was
markedly reduced and there were no extreme estimates, the distributions were still skewed to the right and the correlations were still
very high.

6.   CONCLUSIONS

These examples would suggest that for commonly used kinetic models of linear systems, with observations appropriately timed, the asymptotic theory is a good approximation, even with simple sizes as small as 12. More complex systems need larger experiments (more observations) to be as well approximated by the asymptotic results. But the dangers of generalization should not be forgotten. Since well-planned simulation studies can be simple and relatively inexpensive, it is often worthwhile to do one for the model, experiment and data being analyzed, thus avoiding the need to rely on the asymptotic theory, or on the results of another simulation study of a different, though perhaps similar, situation.

ACKNOWLEDGEMENT

Some of this material was presented at the International Titisee Conference, "Pharmacokinetics during drug development", Titisee, West Germany, October, 1978.

REFERENCES

1.   Westlake, W.J. (1973) J. Pharm. Sci. 62, 1579-1589.
2.   Marmasse, C. (1977) "Enzyme Kinetics", Gordon and Breach, New York.
3.   Laidler, K.J. and Bunting, P.S. (1973) "The Chemical Kinetics of Enzyme Action", 2nd ed., Clarendon Press, Oxford.
4.   Westlake, W.J. (1979) Biometrics, 35, 273-280.
5.   Boxenbaum, H.G., Riegelman, S. and Elashoff, R.M. (1974) J. Pharmacokin. Biopharm. 2, 123-148.
6.   Metzler, C.M. (1974) Biometrics, 30, 309-317.
7.   Draper, N.R. and Smith, H. (1966) "Applied Regression Analysis", Wiley, New York.
8.   Bard, Y. (1974) "Nonlinear Parameter Estimation", Academic Press, New York.
9.   Beale, E.M.L. (1960) J. Roy. Stat. Soc. B22, 41-76.
10. Bates, D.M. and Watts, D.G. (1980) J. Roy. Stat. Soc., B40, 1-36.
11. Jennrich, R.I. (1969) Ann. Math. Stat. 40, 633-643.
12. Metzler, C.M. (1979) Proc. 4th SAS Users Group Intl. Conf., pp. 75-83, SAS Institute, Inc., Raleigh, N.C.
13. Duncan, G.T. (1978) Technometrics, 20, 123-129.
14. Metzler, C.M. and Elfring, G.L. (1972) J. Stat. Comput. Simul. 1, 261-271.
15. Tong, D.D.M. and Metzler, C.M. (1980) Math. Biosci. 48, 239-306.
16. Metzler, C.M. and Tong, D.M. J. Pharm. Sci., submitted for publication.

ON PARAMETER REDUNDANCY IN CURVE FITTING OF KINETIC DATA

Jens G. Reich

Zentralinstitut für Molekularbiologie
Akademie der Wissenschaften der DDR
DDR-1115 Berlin-Buch
German Democratic Republic

ABSTRACT

Parameter redundancy arises when the proposed kinetic model is too detailed for the actual information content of the measurable data. Then an enormous set of totally different, but interdependent parameter values is able to explain the data, and parameter estimation becomes impossible. The paper shows that the defect may be studied before experiments are available. Numerical criteria are outlined which can be used to estimate redundancy and redundant parameter combinations, to predict standard deviations of estimated parameters, to optimize the experimental design for reducing redundancy.

1. INTRODUCTION

In the old days of biochemistry and pharmacology it was customary in the publications to present raw kinetic data directly. This was of high ethical standard, since it exposed the author to the most candid criticism of his results, but it made papers long and difficult to comprehend. In more recent times, then, the method of curve fitting came into use. Here the data are presented in the context of a mechanistic or descriptive model. They should fit to the correct (or, at least, to an acceptable) model, should determine the numerical range of its parameter values, and should show only small residual deviations which can be explained as random errors of measurement.

If applied with common sense and interpreted with caution, this method is elegant and facilitates understanding of an author's

concepts. There are, however, mathematical and statistical difficulties which require specific knowledge and sometimes even research efforts. I wish to treat, in this article, one particular
aspect of parametric curve fitting which is (at least in my experience) of great practical importance, but has not received due
attention of biochemical and pharmacological kineticists so far.
I have in mind the problem of *parameter redundancy*. I am going
to explain this concept in detail, so suffice it to state now that
this means a situation where the parameters have too much 'freedom'.
One and the same curve can, to reasonable accuracy, be produced by
totally different parameter sets. We have published examples
(1-3) of such cases from the field of enzyme kinetics (in particular allosteric models). Still more frequent is the phenomenon in
other fields, like compartmental kinetics, superposition of Gaussian
and Lorentzian peaks in spectral analysis and chromatography, and
in other disciplines where sums of exponential expressions are
analyzed. Everybody with practical experience knows that sums of
three or more exponentials can often be approximated by two or
even one component, and that the parameters (amplitudes and decay
constants, height and width parameter of a peak, *etc*.) are difficult to estimate and may easily vary, in accordance with certain
mutual restrictions, without any deterioration of the fit.

Two points must be emphasized from the start. The first is
that parameter redundancy is a defect which is entirely caused by
the mathematical structure of the model. It has nothing to do
with experimental accuracy. To a certain extent one may cope with
redundancy by very accurate measurements in the most sensitive
range of experimental design (measurements at early or late times,
in the range of small or high concentrations of effectors, *etc*.,
as the case may be), but this has obvious limits and helps only
in cases of moderate redundancy. Thus we have a defect of the
model, not of data. The second aspect is that the defect has
nothing to do with the question of whether or not the model is the
correct one. It may be entirely correct and consistent with the
data, and its descriptive or physico-chemical assumptions may be
fully justified, and nevertheless it may produce redundancy to
such an extent that even the decadic range of the parameter values
cannot be estimated from the data without further information. In
a sense, one may even say that the better the model, the more
refined its parameter structure, the more redundancy comes into
play. The model is a box, more or less 'black', and it is difficult, for informatic reasons, to elucidate its internal structure
from a limited output.

I intend to explain the concept in simple words, without
high-browed mathematics. I introduce, therefore, several simpli-
fications and restrictions which make the basic idea more cons-
picuous. In particular, I shall consider the case that the model
is a function, where a measurable response variable, called y,

depends in a unique manner on the value of an independent variable
x (concentration, time, *etc*.) and of a set of invariant unknown
parameters, p:

$$y = f(x, p_1, p_2, p_3, \ldots p_n).$$ [1]

We shall not treat, only occasionally mention, the case when
several independent x's may be varied by the experimenter, and
also when several simultaneous responses are recorded. We shall
assume that the function is smooth and reasonably bounded. The
same holds for the independent variable and parameters.

Our general strategy of kinetic analysis has been published
in detail (1-6) and critically evaluated by authors of reviews
and monographs [*e.g.*(7-9)]. The restrictions of the present outline intend to facilitate understanding. Readers with special
interest are referred to those papers.

2. SENSITIVITY

Consider, at first, two different models which are to be
compared:

$$y_1 = f(x,p) : \quad \text{model } f$$
$$y_2 = g(x,p) : \quad \text{model } g$$ [2]

and assume that they have the same function value for a certain
$p=p_0$ and $x=x_0$ (Figure 1).

Let us make an attempt to identify p_0 by a measurement of y,
subject to a certain error Δy. It is evident that model f is
better suited, since the influence of a change in p is more considerable than in the case of model g. Therefore a measured value
$y \pm \Delta y$ is compatible with a much larger interval of possible p-
values in the case g than in case f. The derivative $\partial f/\partial p$, we
note, is larger in absolute value than $\partial g/\partial p$. Since the sign of
the derivative is irrelevant for this sensitivity, we migh propose

$$b_f = \left(\frac{\partial f}{\partial p}\right)^2_{\substack{p=p_0 \\ x=x_0}} \quad \text{and} \quad b_g = \left(\frac{\partial g}{\partial p}\right)^2_{\substack{p=p_0 \\ x=x_0}}$$ [3]

as comparable measures of sensitivity of the models f and g with
respect to changes of the parameter p in the vicinity of p_0.

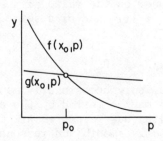

FIGURE 1:   Sensitivity of two models to a parameter, p.

Simple error analysis allows to estimate the error of p from the errors Δf and Δg of the measurements:

$$\Delta p_f = \left| \Delta f / \sqrt{b_f} \right| \qquad \text{and} \qquad \Delta p_q = \left| \Delta g / \sqrt{b_g} \right| \qquad . \qquad [4]$$

Relative errors are given by

$$\frac{\Delta p_f}{p} = \left| \frac{\Delta f}{f} \cdot \frac{f}{\sqrt{b_f} \cdot p} \right| \qquad [5]$$

with

$$\frac{\sqrt{b_f} \cdot p}{f} = \left| \frac{\partial \ln f}{\partial \ln p} \right| \qquad . \qquad [6]$$

An analogous expression can be written for model g. It is obvious that the uncertainty is higher in the case of g than f if the errors are equal. Later we shall consider generalizations of these formulae. Of course, more refined statistical arguments involving distributions could be introduced.

Now an experimental estimate is obtained not only for one value, x_0, of the independent variable. Rather, we measure, for a set of x_k-values, a complete curve and assume that the parameter value is practically constant during the series of measurements. It is the sensitivity in the whole region of x's of the model which gives rise to the proper information measure (Figure 2).

As a numerical criterion we introduce the average value of·[3] for many points of the curve

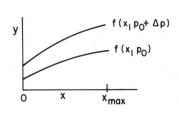

FIGURE 2:   Sensitivity of a response over an experimental
region.

$$\frac{1}{m} \cdot \sum_{k=1}^{m} \left[\frac{\partial f(x_k, p)}{\partial p} \right]^2_{p=p_0} \quad . \tag{7}$$

As a further refinement, for a very large number of exploration
points, $m \to \infty$, we get the integral

$$\frac{1}{X_{max}} \int_0^{X_{max}} \left[\frac{\partial f(x, p)}{\partial p} \right]^2_{p=p_0} dx \tag{8}$$

as a measure of sensitivity. We could introduce weight factors for
different x_k, but shall postpone this extension.

Let us pause here in order to deal with some potential arguments. One might call in question the utility of an information
measure whose calculation requires the value of the assessed parameter to be known in advance. Such a logical circle, however, is
inevitable in the case of a nonlinear model, and is no particular
defect of our treatment. If we wish, for instance, to estimate the
age of persons from their body height, it makes a decisive difference
in which region the parameter is expected: the body height of a child
tells something of his age, but not that of an adult. A preliminary
idea of the position of a parameter is required for a serious discussion of a nonlinear candidate model.

A second argument could be advanced that not only first derivatives determine the course of a function in its vicinity. However, exploration of first derivatives in a neightborhood gives the

same information as higher derivatives at a point. So we should have in mind regions rather than just one p_0.

A further objection could be that a vector of x's rather than one x is involved (several effectors, ligands, *etc.*). Such a case can be handled by summation or integration over a plane or in space rather than over an interval.

A final argument leads us further ahead on our main road. It says that not the sensitivity towards one parameter is problematic, but rather the redundancy of several parameters as an ensemble of variables in the model function.

3.   REDUNDANCY

Indeed, variability of several parameters poses problems which cannot be solved by the methods sketched so far. As an example of first inspection, consider the Michaelis-Menten equation

$$v = V_{max}S/(S+K_m) \quad .$$ [9]

It is easily seen that in the range of small S, where the function reduces to its approximation

$$v \simeq (V_{max}/K_m)S,$$

the two parameters cannot be estimated independently of each other. Only in the region of high S, where the approximation

$$v \simeq V_{max}$$

becomes valid, can we estimate V_{max} without having K_m. This, in turn, permits the estimation of K_m afterwards from the low-S points of the curve.

We speak of redundancy if parameters cannot be estimated in an independent manner. In the given case, the redundancy is removed by extending the measurements over the full region, *i.e.*, by design optimization. But there are models with redundant parameters even for an optimal design.

How can we extend our previous measures in order to cover this new case? Things become now a bit more difficult in notation, not in concept. We propose to study the case of two parameters at first. Again, we look at a linear expansion

$$f(x,p_1,p_2) - f(x,p_1^0,p_2^0) = \frac{\partial f}{\partial p_1} (p_1-p_1^0) + \frac{\partial f}{\partial p_2} (p_2-p_2^0)$$ [10]

and find an extended sensitivity measure

$$T = [f(x,p_1,p_2) - f(x,p_1^0,p_2^0)]^2$$

[11]

$$= \left(\frac{\partial f}{\partial p_1}\right)^2 (p_1-p_1^0)^2 + \left(\frac{\partial f}{\partial p_2}\right)^2 (p_2-p_2^0)$$

$$+ 2\left(\frac{\partial f}{\partial p_1} \cdot \frac{\partial f}{\partial p_2}\right) (p_1-p_1^0)(p_2-p_2^0) \qquad .$$

This expression may be extended to an interval of x as pre-viously. If the interval is densely covered, we get the integral

$$M_{ij} = M_{ji} = \frac{1}{X_{max}} \int_0^{X_{max}} \frac{\partial f(x,p_1,p_2)}{\partial p_i} \cdot \frac{\partial f(x,p_1,p_2)}{\partial p_j} \cdot dx \qquad [12]$$

$$\text{for } p_1 = p_1^0, \; p_2 = p_2^0$$

and may write the equation in T as quadratic form in matrix notation,

$$\underline{T} = \underline{\Delta p}' \cdot \underline{M} \cdot \underline{\Delta p}$$

[13]

with the vector

$$\underline{\Delta p} = \begin{pmatrix} p_1-p_1^0 \\ p_2-p_2^0 \end{pmatrix} \qquad .$$

[13a]

The matrix $\underline{M}$ contains two types of information:

(i) The *sensitivity* with respect to one parameter, collected as diagonal elements M_{ii}. These expressions are the same as in Eq. [8];

(ii) The "*cross-sensitivity*" with respect to two parameters varying together: the elements M_{ij}, $i \neq j$.

It is convenient to separate these components into two matrices by the transformation

$$\underline{S} = \underline{D}^{-1}\underline{M}\underline{D}^{-1} \qquad ,$$

[14]

where

$\underline{D}$ - a diagonal matrix, $D_{ii} = +\sqrt{M_{ii}}$, expressing sensitivity,
$\underline{S}$ - a scaled matrix with unity in the diagonal, expressing redundancy.

The symmetric matrix S has real positive eigenvalues. If one or more of these is very close to zero, then there exist parameter directions along which T will be practically zero. We call these directions redundant. A global measure of redundancy is the determinant of S, *i.e.*, the product of eigenvalues, which is largely determined by the smallest one of them. More specific information can be obtained by the orthogonal transformation of S (3).

4. APPLICATION OF INFORMATION MEASURES

I shall sketch only the basic ideas:

1.   Sensitivity measures of the principal form of expression [3] may be used to predict, in the spirit of Eq. [4], the standard error of a parameter, if it is the main or the only variable one:

$$s_i^2 = \overline{\sigma}^2/M_{ii} \qquad , \qquad\qquad [15]$$

where

$\overline{\sigma}^2$ - variance of the mean of measurements ($=\sigma^2/m$, variance divided by number of measurements);

M_{ii} - square of the mean deviation of f with changing p_i (see expression [12] for i=j).

This may be transformed to dimensionless relative errors as indicated above with Eq. [5].

2.   A generalization to the case of more than one variable (adjustable) parameter is possible (3):

$$s_i^2 = \overline{\sigma}^2 \cdot (M^{-1})_{ii} \qquad\qquad [16]$$

with M^{-1} being the inverse of matrix M. When parameters are skillfully scaled, the elements of D are not far from unity and the decadic range of the elements of M^{-1} is set by the reciprocal of the eigenvalues (in particular, of the smallest ones) of matrix S. Their square roots determine (roughly) the factor by which the parameter error is increased over the experimental scatter. For instance, if the smallest eigenvalue of S is 0.01, then we may expect the relative error of a parameter estimate to be 10-fold higher than the relative error of measurement of y. Again, weight factors may be included into this concept.

3.   A more global but more easily calculated estimate of the just mentioned factor is $(\det S)^{\frac{1}{2}}$. We shall have an example below.

4.   Orthogonal transformation of S may be used for a study of redundant parameter directions and for an indication about additional information which would remove redundancy (3).

5.   Redundancy, as we have seen, may be caused by design defects. If sums over design point x_k rather than integrals are evaluated for $\underline{M}$ (analogy to Eq. [7] instead of [8]), different designs may be tested for redundancy. There is a Russian monograph of Fedorov (10) on design optimization pursuing such ideas.

5.   EXAMPLE

The reader might welcome a concrete example. Let us study

$$y = ae^{-kt} + be^{-ht} \qquad\qquad [17]$$

where a,b, are amplitude parameters, and h,k are decay constants of an experimental decay of relaxation process. We assume to have access to y for arguments of $0 \leq t \leq 1$, where the time is scaled such that at t=1 the components are practically decayed. This simplifies expressions, since we can approximate integrals from zero to unity by those from zero to infinity. We find:

$$\partial y/\partial a = e^{-kt} \qquad\qquad \partial y/\partial b = e^{-ht}$$

$$\partial y/\partial k = -tae^{-kt} \qquad\qquad \partial y/\partial h = -tbe^{-ht} \qquad . \qquad\qquad [18]$$

The elements of $\underline{M}$ are:

$$M_{11} = \int \left(\frac{\partial y}{\partial a}\right)^2 dt = \frac{1}{2k}$$

$$M_{12} = M_{21} = \int \frac{\partial y}{\partial a} \cdot \frac{\partial y}{\partial b} dt = \frac{1}{h+k}$$

$$M_{13} = M_{31} = \int \frac{\partial y}{\partial a} \cdot \frac{\partial y}{\partial k} dt = -\frac{a}{4k^2}$$

$$M_{14} = M_{41} = \int \frac{\partial y}{\partial a} \cdot \frac{\partial y}{\partial h} dt = -\frac{b}{(h+k)^2}$$

$$M_{22} = \int \left(\frac{\partial y}{\partial b}\right)^2 dt = \frac{1}{2h}$$

$$M_{23} = M_{32} = \int \frac{\partial y}{\partial b} \cdot \frac{\partial y}{\partial k} dt = -\frac{a}{(h+k)^2}$$

$$M_{24} = M_{42} = \int \frac{\partial y}{\partial b} \cdot \frac{\partial y}{\partial h} dt = -\frac{b}{4h^2}$$

$$M_{34} = M_{43} = \int \frac{\partial y}{\partial h} \cdot \frac{\partial y}{\partial k} dt = \frac{2ab}{(h+k)^3}$$

$$M_{33} = \int \left(\frac{\partial y}{\partial k}\right)^2 dt = \frac{2a^2}{8k^3}$$

$$M_{44} = \int \left(\frac{\partial y}{\partial h}\right)^2 dt = \frac{2b^2}{8h^3} \qquad .$$

The elements of matrix $\underline{D}^{-1}$ come out as

$$(D^{-1})_{11} = \sqrt{2k}$$

$$(D^{-1})_{22} = \sqrt{2h}$$

$$(D^{-1})_{33} = 2k^{3/2}/a$$

$$(D^{-1})_{44} = 2h^{3/2}/b \qquad .$$

Matrix $\underline{S}$ results as follows:

$$\underline{S} = \begin{pmatrix} 1 & q & -1/\sqrt{2} & -q(1-\omega) \\ q & 1 & q\omega & -1/\sqrt{2} \\ -1/\sqrt{2} & q\omega & 1 & q^3 \\ -q(1-\omega) & -1/\sqrt{2} & q^3 & 1 \end{pmatrix}$$

with $\qquad q = \dfrac{2\sqrt{hk}}{h+k} \qquad$ and $\qquad \omega = \dfrac{k}{h+k} \qquad .$

It is noteworthy that this matrix no longer depends on amplitudes. We introduce the ratio of decay constants

$$r = h/k$$

which can be arranged, by definition, to be <1. Laborious manipulation of the matrix $\underline{S}$ yields the result

$$\det \underline{S} = \frac{1}{4} \left(\frac{1-r}{1+r}\right)^8 \qquad .$$

It is interesting to study $\det \underline{S}$ as function of r. Its value is between 0 and 0.25 (Table 1).

If we accept 0.01 as a limit (which means a factor of 10 in the transformation from measurement error to parameter error), then parameters are redundant if decay constants differ by less than a factor of 5.

TABLE 1: Variation of redundancy with the
ratio of exponents

r	det $\underline{S}$
1	0
0.9	1.2×10^{-9}
0.8	6.0×10^{-9}
0.7	2.4×10^{-7}
0.6	3.8×10^{-6}
0.5	3.8×10^{-5}
0.4	2.8×10^{-4}
0.3	1.8×10^{-3}
0.2	0.01
0.1	0.05
0.01	0.21
0.001	0.246
0	0.250

6. CONCLUSIONS

The ideas sketched in this paper permit investigation of
numerical criteria of parameter redundancy in a proposed model.
They can be extended to the optimization of experimental design.
It is possible to state which parameter combination is redundant,
and which independent information is required to remove reduncancy.
Kineticists should study their model before they invest money in
expensive experiments.

REFERENCES

1. Reich, J.G. (1970) FEBS Lett. 9, 245-251.
2. Reich, J.G. (1974) Studia Biophys. 42, 165-180.
3. Reich, J.G. and Zinke, I. (1974), Studia Biophys. 43, 91-107.
4. Reich, J.G., Wangermann, G., Falck, M. and Rohde, K. (1972)
      Eur. J. Biochem. 26, 368-379.
5. Reich, J.G., Winkler, J. and Zinke, I. (1974) Studia Biophys.
      42, 181-193.
6. Reich, J.B., Winkler, J. and Zinke, I. (1974) Studia Biophys.
      43, 77-90.
7. Wong, J.T.F. (1975) "Kinetics of Enzyme Mechanisms", Academic
      Press, London.

8. Garfinkel, L., Kohn, M.C. and Garfinkel, D. (1977) CRC Crit.
       Rev. Bioeng. 2, 329-361.
9. Henderson, J.F. (1978) In "Techniques in Protein and Enzyme
       Biochemistry", Vol. B1/II, pp. 1-43, Elsevier/North Holland,
       Amsterdam-New York-Shannon.
10. Fedorov, V.V. (1972) "Theory of Optimal Experiments", Academic
       Press, New York-London.

Remark: Copies of the series in Studia Biophysica are available
on written request to the author.

TANGENTIAL AND CURVILINEAR COORDINATES

IN NONLINEAR REGRESSION

A.T. James

The University of Adelaide
Department of Statistics
Adelaide, Australia   5001

ABSTRACT

The discrepancy between tangential and curvilinear coordinates
in nonlinear regression is illustrated by a calculation and plot
for a specific example. The implications for computing and inference
are discussed.

1.   INTRODUCTION

The problem of estimation of nonlinear regression parameters
and their computation has generated a very extensive literature,
which this paper will not attempt to review.

The older methods used expansion in a Taylor series, and methods   such as Marquardt's (1) [see also Fletcher (2) for a modifica-
tion] still retain the Taylor series as part of a two pronged
approach. Beale (3) distinguished two components of nonlinearity;
(i) essential nonlinearity or curvature of the solution locus, and
(ii) curvature of the coordinate system within the solution locus
(in excess of essential geodesic curvature). Since it can be removed by reparametrization of the solution locus, he called it removable nonlinearity. Ross (4) has suggested replacing the parameters by an equal number of fitted values to achieve this.

The removal of nonlinearity is important for both computation
and inference.

The present paper takes a very simple example and sets out to
illustrate the removable nonlinearity by explicit calculation of a
tangential and a curvilinear coordinate system. The two coordinate

51

systems can be compared from their superimposed plots in Figure 1.
Both are in a tangent plane to the solution locus at a point of
contact given by "initial values" for the parameters. Figure 1
shows a striking distinction between the two coordinate systems,
which many methods of computation and systems of inference take as
approximating each other. This distinction occurs even in a simple
very well-behaved example.

The example shows that fitted values calculated from the *linear*
formula, as shown in Table 1, may be remarkably accurate even in
the first stage of iteration.

If they are utilized to obtain updated values of the parameters
by some appropriate nonlinear process, *e.g.* by providing them as an
input of an initial value routine, values remarkably close to the
final values may be obtained in even one stage. Tangential coordinates are also important for inferential purposes.

In the paper of Bliss and James (5), it was pointed out on
p. 598 that the essential singularity, *i.e.*, the curvature of the
solution locus, may be judged by the magnitude of the difference,
$\sigma^{-2} \Sigma (y-Y) f_{\theta\theta}$, between realized and expected information as a fraction of the expected information $\sigma^{-2} \Sigma f_{\theta}^2$ where y has variance σ^2 and
expectation f, which is a nonlinear function of θ.

In the example of Wilkinson (6) quoted by them, the curvature
of the solution locus is negligible, but the curvature of the
coordinate system is appreciable.

The aim of the paper is to calculate the two coordinate systems numerically for this example and plot them for comparison. A
proper appreciation of the discrepancies between tangential and
curvilinear coordinate systems should help to simplify the problems
of computation and inference in nonlinear regression.

2. A SIMPLE MICHAELIS-MENTEN CURVE AS AN ILLUSTRATION

Let y be a vector of n independent observations whose i-th
component y_i has expectation $\eta(x_i;\beta,\delta)$ which is a function of a
determining variate x_i and a nonlinear function of parameters (β,δ).
The generalization to more than one determining variable and more
than two parameters is obvious.

Wilkinson's (6) example of the fit of a Michaelis-Menten curve
will be used as the illustration and results from the discussion
in the appendix to Bliss and James (5) will be quoted.

In this example

$$\eta(x;\beta,\delta) = \beta x/(x+\delta) \tag{1}$$

where $\beta=V_{max}$ is the maximum velocity and $\delta=K_m$ is the Michaelis-Menten constant.

The expectation vector $\underline{\eta}$ of $\underline{y}$ which has components $\eta(x_i;\beta,\delta)$ is a vector valued function $\underline{\eta}(\beta,\delta)$ which maps a region of the (β,δ) plane into a two-dimensional surface in the sample space R^n, called the solution locus (7).

Equation [1] may be linearized by taking reciprocals of η and x. Unweighted least-squares fitting of the double reciprocal plot yields initial estimates of

$$\beta_0 = 0.585323 \qquad \delta_0 = 0.440628 \quad .$$

A very rough approximation to η can be obtained from the linear term of the Taylor series. Since β is already linear we need only differentiate with respect to δ:

$$\eta(x;\beta,\delta) = \beta\left(\frac{x}{x+\delta_0}\right) + \beta\Delta\delta\left(-\frac{x}{(x+\delta_0)^2}\right) + \dots \tag{2}$$

$$= \beta_1 x^{(0)} + \beta_2 x_*^{(0)} + \dots$$

where

$$\beta_1 = \beta , \qquad \beta_2 = \beta\Delta\delta ,$$

$$x^{(0)} = x/(x+\delta_0), \qquad x_*^{(0)} = -x/(x+\delta_0)^2 \quad . \tag{3}$$

When calculated for the determining variable x_i corresponding to the i-th observation y_i, $x^{(0)}$ and $x_*^{(0)}$ are the respective components of two vectors $\underline{x}^{(0)}$, $\underline{x}_*^{(0)}$ which generate the tangent plane to the solution locus at (β_0,δ_0). (For other types of functions $\underline{\eta}(\beta,\delta)$, we would normally simply use the partial derivatives, $\underline{\eta}_\beta$ and $\underline{\eta}_\delta$, of $\underline{\eta}$ with respect to the parameters β and δ. In this case, $\underline{\eta}_\beta=\underline{X}$, but since $\underline{\eta}_\delta=\beta\underline{X}_*$ it is simpler to use $\underline{X}_*$ in place of $\underline{\eta}_\delta$ as one of the generators of the tangent plane to the solution locus.)

Hence the vector equation

$$\underline{\eta}_t = \beta_1\underline{x}^{(0)} + \beta_2\underline{x}_*^{(0)} \tag{4}$$

shows that (β_1,β_2) are the coordinates of a rectilinear coordinate system within the tangent plane. However, for comparison with a curvilinear system, we put

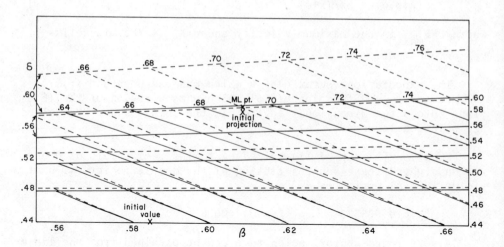

FIGURE 1: The tangential coordinates (broken lines) and curvilinear
coordinates (solid lines) of the tangent plane to the solution locus
at $\delta_0=0.44$. The solution locus is for a simple Michaelis-Menten
relation fitted to 6 data points with parameters $\beta=V_{max}$ and $\delta=K_m$.

The coordinate lines have been plotted by calculating rectangular
coordinates (u,v) relative to orthogonal vectors $(\underline{a},\underline{b})$ of unit
length. The first, $\underline{a}$, is a unit vector along the line $\delta=0.44$ and
the second, $\underline{b}$, is the unit vector in the tangent plane orthogonal
to $\underline{a}$. However, for simplicity, the vectors $\underline{a}$, $\underline{b}$ are not shown and
the (u,v) corrdinates have been erased. The initial projection of
the sample vector $\underline{y}$ upon the plane is shown (corresponding to the
first column of fitted values in Table 1), as also the projection
of the final fitted values corresponding to the maximum likelihood
estimates. The very slight discrepancy between the points is presumably due to curvature of the solution locus.

$$\beta = \beta_1, \qquad \delta = \delta_0 + \beta_2/\beta_1 \quad .$$

This will be called the tangential coordinate system.

On the other hand, the solution locus has (β,δ) as curvilinear
coordinates of the point $\underline{\eta}(\beta,\delta)$. Let $\underline{\eta}_c$ be the orthogonal projection of $\underline{\eta}(\beta,\delta)$ on the tangent plane. $\underline{\eta}_c(\beta,\delta)$ determines a curvilinear coordinate system within the tangent plane.

In order to bring the tangent plane into the plane of the
page for plotting, we let $\underline{a}$ and $\underline{b}$ be two orthogonal unit vectors
obtained by orthonormalizing $\underline{X}^{(0)}$ and $X_*^{(0)}$. For any point $\underline{\eta}=\beta\underline{X}(\delta)$,
with components $\beta x_i/(x_i+\delta)$, on the solution locus,

$$u = \beta\underline{a}'\underline{X}(\delta) \quad , \quad v = \beta\underline{b}'\underline{X}(\delta)$$

are the rectilinear coordinates of a point $\underline{\eta}_c$ in the tangent plane. In this way, the curvilinear coordinate system was plotted in Figure 1 as solid lines.

In the tangential coordinate system, β and δ determine a linear combination

$$\underline{Y} = \beta\underline{X}^{(0)} + \beta(\delta-\delta_0)\underline{X}_*^{(0)}$$

from which $u=\underline{a}'\underline{Y}$, $v=\underline{b}'\underline{Y}$ can be calculated for plotting. The tangential system was likewise plotted in Figure 1 as broken lines. The (u,v) coordinates have been erased to simplify the diagram.

Since β is a linear parameter, the coordinate lines for δ constant but β varying will be straight even for the curvilinear system. However for constant β but varying δ, the solid curvilinear coordinate lines curve away to the left of their tangents at $\delta=0.44$, which are the tangential coordinates represented by broken lines.

The two coordinate systems are in exact agreement at $\delta=0.44$. At $\delta=0.48$, the straight line of the tangential system is slightly above the solid straight line of the curvilinear system. For higher values of δ, the discrepancy becomes progressively more marked until the tangential coordinate line for $\delta=0.56$ almost coincides with the curvilinear coordinate line for $\delta=0.60$.

It follows that the true maximum likelihood estimate of $\delta=0.60$, approximately, is represented in tangential coordinates by $\delta=0.56$. On account of the fact that the curvilinear coordinate lines for constant β curve to the left of the tangential lines, the maximum likelihood estimate of $\beta=0.69$, approximately, is represented in tangential coordinates by $\beta=0.68$.

We see that, both tangential and curvilinear coordinate systems are valid and accurate "maps" of the tangent plane and either can be used for precise computational and inferential work, provided that tangential coordinates are not confused with curvilinear coordinates except within a very small neighbourhood of the initial value.

The maximum likelihood procedure yields a fitted value

$$\underline{Y}_t = b_1\underline{X}^{(0)} + b_2\underline{X}_*^{(0)} \quad .$$

If we solve the equations

$$b_1 = \hat{\beta} \quad , \quad b_2 = \hat{\beta}(\hat{\delta}-\delta_0)$$

for $(\hat{\beta},\hat{\delta})$, we obtain tangential coordinates.

One way to transform them into curvilinear coordinates, which are the required values, is to input the fitted values Y_t into an initial value routine instead of data, *e.g.* a double-reciprocal plot.

Results are shown in Table 1.

3.   CONCLUSIONS

1.   In a program using a Taylor series up to a linear term, an option be included to calculate fitted values from the linear term of the Taylor series.

TABLE 1:   Data and fitted values for a Michaelis-Menten hyperbola fitted to 6 data points

		1st iteration	2nd iteration	3rd iteration	4th iteration
(From double-reciprocal regression)	$\hat{\beta}$	0.58532	0.69506	0.69050	0.69040
	$\hat{\delta}$	0.44063	0.60283	0.59674	0.59655
From ML	$\hat{\beta}_1$	0.67730	0.69047	0.69040	0.69040
	$\hat{\beta}_1$	0.07821	0.00426	-0.00013	0.00000

x	y	Fitted values using linear equation [4]			
.138	.148	.1293	.1297	.1297	.1297
.220	.171	.1861	.1860	.1860	.1860
.291	.234	.2269	.2263	.2264	.2264
.560	.324	.3353	.3343	.3343	.3343
.766	.390	.3888	.3881	.3881	.3881
1.460	.493	.4887	.4902	.4901	.4901
Tangential	$\hat{\delta}_1$	0.55610	0.60900	0.59655	0.59655

In the first iteration, the double-reciprocal regression was applied to the data y. Subsequently it was applied to the fitted values, as shown, from the previous iteration. The $(\hat{\beta}_1,\hat{\delta}_1)$ are the tangential coordinates for $(\hat{\beta},\hat{\delta})$ for the iteration. The corresponding curvilinear coordinates are the $(\hat{\beta},\hat{\delta})$ at the top of the column for the next iteration (apart from a slight roundoff error).

2.   If the program has an initial value routine, a provision be
     made to feed such values into the initial value routine.
3.   On the assumption that the observations are homoscedastic
     multinormal, the linear Taylor series coefficients (b_1, b_2)
     will be very nearly multinormal.

REFERENCES

1.   Marquardt, D.W. (1963) J. Soc. Indust. Appl. Math. 11,
     431-441.
2.   Fletcher, R. (1971) "A Modified Marquardt Subroutine for
     Nonlinear Least Squares", U.K. Atomic Energy Authority
     Research Group Report, AERE-R6799, H.M. Stationery Office,
     London.
3.   Beale, E.M.L. (1960) J. Roy. Stat. Soc. B22, 41-76.
4.   Ross, G. (1975) Proc. 40th Congr. Int. Stat. Inst. 2, 585-593.
5.   Bliss, C.I. and James, A.T. (1966) Biometrics, 22, 573-602.
6.   Wilkinson, G.N. (1961) Biochem. J. 80, 324-332.
7.   Watts, D.G. (1980) In "Kinetic Data Analysis: Design and
     Analysis of Enzyme and Pharmacokinetic Experiments"
     (L. Endrenyi, ed.) pp. 1-24, Plenum, New York.

USE OF MULTIPLE LINEAR REGRESSION IN THE ANALYSIS

AND INTERPRETATION OF BIOKINETIC DATA

Allan J. Davison and A.H. Burr

Simon Fraser University
Departments of Kinesiology and Biosciences
Burnaby, B.C.
Canada V5A 1S6

1. INTRODUCTION

1.1. Description, Interpretation and Analysis of Kinetic Data

Those courageous enough to attempt to describe the kinetic
behaviour of biological systems face pitfalls, frustrations and
difficulties which can never be fully surmounted. Those who press
beyond these hazards to attempt to *explain* what they painfully and
incompletely describe, will find their intellectual resources
strained to and beyond their limits of capability, in terms not
only of the availability of existing theory but also its applicability and comprehensibility. This chapter offers to such des-
pairing souls some practical compromises in numerical techniques
which allow at least faltering progress toward the analysis and
mechanistic interpretation of kinetic phenomena.

Although ultimate rigour in estimating the coefficients of rate
equations still seems out of reach (1), standard computational
techniques for curve fitting have become widely available, inexpensive, and simple to use [*e.g.*, (2-5)]. Similarly, given a hypothetical mechanism for a process, standard methods are available
for predicting the kinetic consequences of the mechanism (6-8).
Less consideration has been given to the reverse question: What
conclusions can be drawn rigourously regarding the *mechanism* from
a kinetic description of a process? Even here, a beginning has
been made (9). An excellent update on standard methods for the
analysis of enzyme kinetic data has recently become available which
covers most of these topics at an introductory level (10).

Application of these methods should result in the development
of relatively complete kinetic descriptions of data by mechanistically interpretable rate equations in terms of all known variables.
However, this goal is rarely achieved. More often, data are analysed
with respect to only a single variable at a time (interactions
being ignored), and then described by logarithmic curves, exponential functions, polynomials, etc., which bear no relationship to
the underlying mechanisms. Simulations of these uninterpretable
relationships are often carried out using methods which give unreliable estimates of the coefficients, and fail to describe quantitatively either the validity of the models used or the reliability
of the coefficients derived. Such a waste of effort is no longer
excusable.

The major reasons for such inadequate analyses have been:

1. The large number of experimental variables in most biological
systems, which may deter an investigator who is accustomed to using
a ruler and a sheet of graph paper as his main analytical tools;

2. The confusing proliferation in the literature of papers which
describe more and more complex numerical methods in the pursuit of
greater and yet greater statistical rigour;

3. The failure of research workers to realize how important it is
to apply curve fitting to *mechanistically interpretable* equations
(a term we shall use repeatedly to mean equations of a form which
can be derived from, or compared with those derived from plausible
mechanisms for the process under investigation).

The present paper attempts to introduce the method of multiple
linear regression for the least-squares fitting of such equations
to experimental data. The assumptions limiting the applicability
of the methods will be stated, and shown to be readily testable.
More complex methods are available when the standard methods are
inapplicable or when the rate equations to be fitted are nonlinear
[(1,5,10,11), etc.], and the reader is referred to these articles,
to advanced textbooks, and to other chapters in this book, for
their description.

1.2. <u>Approaches and Problems in the Numerical Analysis of Kinetic
Data</u>

It is common when analysing the complex kinetics of biotrans-
formations, that the dependent variable being followed is found to
be a function of many independent variables. Usually, for example,
enzymic reaction velocity (y) is a function of the concentrations
of its substrates (x_1, x_2), and of a variety of other factors
$(x_3, x_4, \ldots, etc.)$ which may be functions of other substrates, acti-

vators, and inhibitors, pH, ionic strength, *etc*. The most common case, and the easiest to solve, is the one when y is a "linear" function of the other variables, *i.e.*, the relationship between the variables can be expressed as an equation of the form:

$$y = b_1 \ f_1(y,x_1,x_2,\dots) + b_2 \ f_2(y,x_1,x_2,\dots) + \dots$$

Note that while it is a common misconception that in order to be fitted by multiple linear regression, equations need to be soluble in terms of y. It is in fact no more difficult (albeit more hazardous) to find least-squares solutions when y is embedded in any (or all) of the functions on the right-hand side, as in the above example.

Such equations may readily be derived by analysis of proposed mechanisms using techniques such as that of King and Altman (6), Cleland (8), *etc.*, which yield the functions $f_1, f_2, \dots, etc.$ These functions relate to the concentrations of the Debye-Hückel theory, linear free energy relationships, absolute rate theory, *etc.*, yield further functions $f_3, f_4, \dots, etc.$ which relate to physico-chemical characteristics of the system such as ionic strength, temperature, *etc.* While the equations may become complex, the methods for their derivation soon become familiar to the practising kineticist. After the resulting hypothetical equations have been derived, their applicability can be tested and the coefficients can be estimated, by a variety of now standard techniques, such as linear or nonlinear, weighted or unweighted least-squares, parametric or non-parametric curve-fitting techniques.

In this paper we shall:

1.   Review briefly the use of unweighted and weighted multiple linear regression for obtaining the coefficients which correspond to the best fit of such equations to available experimental data;

2.   Suggest computational and graphical methods for coping with the multiplicity of variables usually involved.

2.   MULTIPLE LINEAR REGRESSION

2.1. Curve Fitting

In general, one has as data a set of measurements of the dependent variable y_i, and corresponding values of independent variables $x_{1i}, x_{2i}, \dots$, and wishes to find their functional relationship. One usually has in mind at least one type of hypothetical relation-

TABLE 1:   Types of models commonly used in fitting rate data

---

Linear model in one or several variables:[*]

I $y = b_0 + b_1 x_1 + b_2 x_2 + \ldots + b_h x_h$

Linear model in one or several powers of x:

II $y = b_0 + b_1 x + \ldots + b_p x^p$

Nonlinear models in one variable:

III $y = b_0 + b_1 \sin k_1 x + b_2 \cos k_2 x + \ldots$

IV $y = b_1 e^{kx}$

V $y = b_0 + b_1 e^{k_1 x} + b_2 e^{k_2 x} + \ldots$

VI $y = b_1 x / (k_1 + x)$

---

[*] $x_1, x_2, \ldots, x_h$ could represent independent variables of different functions of the variables as, for example, in Model II.

---

ship or model (Table 1). The problem of fitting the model to the data involves estimating the coefficients that best fit the data, then checking the results by various methods to ensure that a good fit has been obtained.

2.2. Types of Models and Transformations

Note in Table 1 that only linear coefficients are present in the linear models and that the nonlinear coefficients $k_1, k_2, \ldots,$ differ in being nonlinearly related to y. The nonlinear Model VI will be recognized as the Michaelis-Menten equation; IV and V with negative k's are single and multiple first-order kinetic equations.

Such linear models have the advantage that their coefficients can be estimated from the data in a straightforward way using multiple linear regression as described below. Nonlinear coefficients must be solved by an iterative algorithm. Whereas a linear model with only 2 coefficients can be fitted to the data using only a hand calculator, nonlinear models must be fitted by computer.

Fitting to nonlinear models falls outside the purview of this article and has been extensively treated elsewhere [*e.g.*, (1,5,10,11)].

Fortunately, many models with nonlinear coefficients can be
transformed into linear ones. The well-known logarithmic transformation converts Model IV into the form

$$y' = \log y = b_1 + b_2 x \quad .$$

Often it is not realized how easily complicated functions,
such as ratios of polynomials or hyperbolic functions, can be rearranged to a linear form. In general, equations of the form

$$y = \frac{b_1 \cdot f_1(y,x_1,x_2,\ldots) + b_2 \cdot f_2(y,x_1,x_2,\ldots) + \ldots}{c_1 \cdot g_1(y,x_1,x_2,\ldots) + c_2 \cdot g_2(y,x_1,x_2,\ldots) + \ldots}$$

can be rearranged to the form:

$$y = b_1 \cdot f_1(y,x_1,x_2,\ldots) + b_2 \cdot f_2(y,x_1,x_2,\ldots) + \ldots$$
$$- y \cdot c_1 \cdot g_1(y,x_1,x_2,\ldots) - y \cdot c_2 \cdot g_2(y,x_1,x_2,\ldots) - \ldots$$

Other methods are commonly used to transform the Michaelis-
Menten equation (Model IV), notably those applied by Lineweaver
and Burk, Eadie and Hofstee, and Hanes, and described, for example,
in Cornish-Bowden (10). However, a common error in carrying out
such transformations is to fail to weight the transformed data
appropriately, or to realise that data in which all points were
equally reliable before transformation will no longer be equally
reliable afterwards. Such omissions will certainly give imprecise
estimates.

2.3. Use of Multiple Linear Regression in Curve Fitting

The procedure used in the least-squares method fitting of a
model to data is relatively straightforward and involves the following steps:

(i)   choosing a model,
(ii) calculating the coefficients (b_i), and
(iii) evaluating the quality of fit.

Each step should be considered equally important, otherwise the
results may be misleading.

2.3.1. <u>Choosing a Model</u>. Usually one begins by choosing the simplest
model that could be expected, given the circumstances. For example,
in studying the denaturation rate of a protein, Model IV (transformed) might be examined first. If step (iii) indicates that
the model may not be appropriate, a second or multiple first-order

model (Model V) could be tried. In studying enzyme kinetics (initial rate as a function of substrate concentration), Model VI (transformed) would usually be chosen first, followed by a more complete model if this basic model proves unsatisfactory.

Obviously, if initial rates are collected over a limited range of substrate concentration ($x \ll k$), one could be misled by an excellent fit to the simple linear model $y=(b_1/k_1)x$. Thus, attention must also be paid to suitability of the data as well as the model. A rule of thumb often applied in determining Michaelis (binding, or dissociation) constants, is that data should be collected over a range of concentrations spanning the range one-fifth to five times the expected Michaelis constant.

2.3.2. Calculating the Coefficients. For linear models, the simplest and most direct method for fitting kinetic equations to data (*i.e.*, determining the coefficients) is by multiple linear regression. To accomplish this, the matrix approach described in references (4,5) is versatile, easy to use, and yields the desired coefficients directly.

2.3.3. Prerequisites for Satisfactory Curve Fitting. Some assumptions of the method of multiple linear regression, regardless of number of linear coefficients fitted, are [*e.g.*, (3-5,11)]:

1.   The chosen model is the correct one. If not, the analysis should reveal this, and the proposed model can be rejected with statistical confidence.

2.   The distribution of errors in the y_i is approximately normal or Gaussian with mean approximately equal to zero. The accuracy of this assumption will be revealed when the residuals are plotted as described hereunder. Errors in the independent variables are considered negligible compared to errors in y_i. This assumption will not be tested further, since measurement of the actual quantities determined by the investigator can usually be done to within ±0.5%.

3.   All y_i have approximately the same variance or standard deviation. This too will be revealed in the residual plot.

4.   The errors in y_i are independent of each independent variable and independent of each other.

To be confident that the calculated b_i are good estimates of the true parameters, all of these assumptions must be met to a degree. Two ways of ascertaining that this is so are: (i) examining the residuals, and (ii) testing for quality of fit.

2.3.4. <u>Examining the Residuals</u>. The residuals, RES=y_i-Y_j, are
plotted as a function of x or y. Several possible outcomes are
illustrated in Figure 1. Violations of assumptions (1) and (3)
show up graphically. Clearly a more suitable model is needed in
Fig. 1b. Fits in Figures 1c and 1d would be improved by suitable
weighting. Deviations from Gaussian error distribution (assumption
2) can be detected by plotting the linearized cumulative frequency
distribution of the residuals (3), but this is pointless for small
N (less than 50). An additional advantage to examining residuals
is the ease by which outliers can be identified. As a result,
residual plots have been usefully considered and applied for enzyme
kinetic experimentation [*e.g.*, (10,12)].

2.3.5. <u>Testing for the Quality of Fit</u>. Mannervik and his associates
have discussed repeatedly criteria for model identification and for
discriminating between alternative models (15-17). These are reviewed in this volume (18). They include the requirement for
physically or biologically meaningful estimated values of the parameters which should, in addition, have relatively small standard
errors.

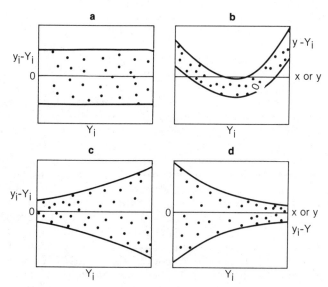

FIGURE 1:   Common plots of residuals: Displayed as $(y_i-Y_i)/[\mathrm{Var}(y_i)]^{\frac{1}{2}}$,
plotted as a function of Y_i. (a) Appropriately weighted,
(b) showing systematic deviation from the fitted curve,
(c) improperly weighted (low values of y_i too heavily
weighted), (d) high values of y_i too heavily weighted.

In the presence of replicate observations, the adequacy of a
model can be judged by comparing the variance estimated from the
residuals of the regression with the variance calculated from the
replicates. Similarly, two models having differing number of parameters can be distinguished on the basis of the extra sum of squares
principles which compares the residual variances of the two models
(3,4).

3. GRAPHICAL DISPLAY OF FUNCTIONS OF MANY INDEPENDENT VARIABLES

3.1. Graphical Display of Goodness of Fit

In the simplest case, one obtains a preliminary fit of the
dependent variable (y) to a single independent variable (x), and
then experimental and calculated values of y can be plotted as a
function of x, and the "goodness of fit" can be examined. Likewise,
the necessary plot of the residuals as a function of x is easily
drafted. When y is found to be the sum of a series of functions
of many independent variables, it is necessary to plot the residuals
against *each* of the functions, so that the particular inappropriate
function(s) which are the sources of remaining systematic deviations, can be identified and modified. This is time consuming,
but not particularly difficult, and it should always be done. A
somewhat intractable problem remains, however, that of how to de-
monstrate graphically the goodness of fit which has been achieved.
The problem lies in the compression of graphs in 3 or any number
of dimensions onto the two dimensions which paper will allow.

One way is to plot the experimental and predicted values of
y as a function of any of the primary independent variables. However, examination of the plot reveals that the "scatter" present
in the experimental values of y, due to interactions among the
various independent variables, is duplicated with some precision
in the calculated values, thus completely concealing the quality
of fit. If there are only two independent variables, triangular
graphs may be used. These are easy to draw, but hard for the un-
initiated to interpret. Alternatively, three-dimensional graphs
may be used, which are easy to interpret but difficult to plot.
Commonly, however, there are more than two independent variables,
and then no convenient method for graphic display is available.
The equations can be compressed onto one pair of axes at a time,
and families of curves can be plotted, but these give little visual
indication of the overal quality of fit. While the remaining variability can be guantitatively estimated from the residual mean
square deviation of the predicted points, the presence of any remaining *systematic* deviation cannot be detected by this means. A
possible approach is to plot the calculated values Y_i, as a function of the experimental values y_i (and as an assessment of remaining error to plot the residuals as a function of Y_j). What

follows is an alternative solution to this problem, which is facile, and general enough to be extended to virtually any number of independent variables. This method offers conceptual advantages over the method just described, in displaying the manner in which the independent variables interact and the magnitude of such interactions.

3.2. Correction of the Primary Independent Variable for the Contributions of the Other Variables

The proposed solution is to express the dependent variable as a function of a single composite value, to which the dependent variable is directly proportional, and into which the weighted contributions of all the independent variables are incorporated. Traditionally, when the dependence of y on x_1 and x_2 has been determined at each of (say) 4 different values of x_1 and 5 values of x_2, the technique used has been to compress the 3-dimensional graph of these values along one of the three dimensions so that 5 graphs of y as a function of x_1 are obtained on a single pair of axes (one graph at each value of x_2). The graph representing the 3-dimensional array of points has thus been compressed along the x_2 axis to yield 5 separate graphs. Use of the method proposed here, displays instead the same twenty points as a single curve representing y as a function of x_1 corrected for the contribution of x_2 or indeed any other independent variables. This method is thus analogous to a compression of the graph along the best surface through the 3-dimensional array of the data in space, so that they come to lie along a single curve in the xy plane. The computational technique which will achieve this is relatively simple, universal in its application, and easily incorporated into any multiple regression subroutine. The method sometimes yields mechanistic insights into the interaction between the factors determining y, and it allows graphic display of the precision with which the value of y is estimated by the equation being fitted to the data. Various hypothetical relationships can thus be readily compared.

In the following analysis, the most general case is considered, so that the equations may seem complex. However, it should be emphasized that in any actual curve-fitting situation the computation of the composite independent variable is far simpler than the mathematics of the regression analysis used, and no more complex than the calculation of predicted values of z by standard procedures. The simplicity of the method in practice is reflected in the brevity of the APL statements used to accomplish the necessary computations, as illustrated in a computer program which is available on request.

3.3. General Mathematical Treatment

The relationship between the variables can be expressed in a linear form:

$$y = b_1 \cdot f_1(y, x_1, x_2, \ldots) + b_2 \cdot f_2(y, x_1, x_2, \ldots) + \ldots$$

(where b_j are unknown coefficients, and f_j are known functions).

The equation is then solved as described above for the coefficients of the equation, b_j, yielding a string of values comprising the estimated coefficients, B_j, which provide the best possible fit of the equation to the experimental data.

The modification which allows graphical display of the composite independent variable by correcting $b_1 \cdot f_1$ (the major function determining y) for the extent to which each of the other coefficients $b_2, b_3, \ldots$, and function $f_2, f_3, \ldots$, contributes to the value of y, so that it becomes an estimate of effective f_1 (F_e).

Thus,

$$y = b_1 \cdot F_e(y, x_1, x_2, x_3, \ldots)$$

where

$$F_e = f_1(y, x_1, x_2, \ldots) + \frac{b_2}{b_1} \cdot f_2(y, x_1, x_2, \ldots)$$

$$+ \frac{b_3}{b_1} \cdot f_3(y, x_1, x_2, \ldots) \quad .$$

Now, by comparison of this equation with the original regression equation fitted, the best fit to a plot of y as a function of F_e is a straight line through the origin and having slope of b_1.

It simplifies interpretation if the function chosen for adjustment to F_e is the major contributing function f_1. Then the quantity F_e is a corrected value of the function to which y was hopefully (but not necessarily) most closely proportional. However, it is now corrected in a quantitative way for each of the other functions on which y depends. For example, in determining enzyme velocities in the presence of variable concentrations of both substrate and a competitive inhibitor, the function yielded describes an "effective" concentration of substrate, *i.e.*, the actual concentration of substrate diminished by the extent to which it is displaced from the enzyme by competition with inhibitor.

Plotting y and Y as a function of F_e yields a cluster of points about a straight line. The spread of the points represents quantitatively the imprecision of the equation used in the regression analysis, so that the RSS is identical to that of the multiple linear regression analysis. In such a graph, in contrast to a graph of y and Y against any of the independent variables, the known systematic errors have been corrected for, and the scatter of the points reveals directly any variance unaccounted for in the equation fitted. *Random* errors, due to statistical variability in the measurements, can thus be readily distinguished as can *systematic* errors resulting from inadequacies in the equation being fitted.

At the same time, a graphic comparison can be made of the goodness of fit of the current equation with other equations competing as candidates in fitting the data. Comparison of a graph of y and Y as a function of f_1 (as in Figure 2a), with a graph as a function of F_e (as in Figure 2b), reveals the extent to which the scatter of points in the former graph, is due to the other functions in the overall equation. If F_e is obtained in stepwise fashion, correcting for one additional function at a time, then the contribution of each separately (and all together) to the systematic deviations from linearity in the original plot is graphically displayed. The method allows the calculation of these contributions. A major advantage of the above method is that the calculated values of the dependent variable are displayed as a function of a single composite independent variable which has some mechanistic significance, in that each of the primary variables contributes to it to an extent quantitatively revealed in the intermediate computations. If the variable

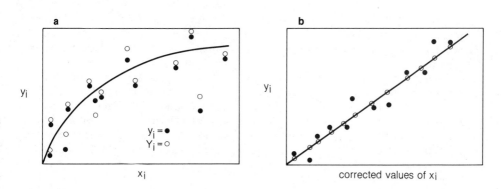

FIGURE 2: Comparison of y_i with Y_i, displayed as a function of f_1 (the major function determining y). (a) As calculated, (b) after correction for all the other functions contributing to y, *i.e.*, as a function of F_e.

to be corrected for the contribution of the others *is* in fact the
major term contributing to the calculated value, then tabulation
of the above-mentioned intermediate computed values may reflect the
extent to which the primary variable is influenced by contributions
from each of the other variables over the whole range of conditions
studied.

3.4. Correction of Dependent Variable for Physico-chemical Conditions

As a modification, to simplify analysis and presentation of
kinetic data, it is recommended that the function chosen as the
primary independent variable be the most dominant chemcial species
present and that it be corrected as described for the contributions
of other *chemical* variables, but that the functions corresponding
to the *physical* characteristics of the system $(g_1, g_2, \ldots)$ be applied

not as corrections to a dependent variable, but as corrections to
the independent variable, by subtracting $c_k g_k (y, x_1, x_2, \ldots)$ from the

values of y and Y (c_k being the coefficients of the functions g_k).

In this way, the rate corrected for the physical characteristics
of the reaction will be plotted as a function of the most dominant
chemical characteristic corrected for the other chemical characteristics of the system. The tabulated values of the contributions
of each of the variables is stored in the memory of the computer
when the program provided is used, and plotting these values on
the same axes as the experimental and calculated data, shows graphically the contribution of each of the subsiduary variables considered
over the range of values of the primary variables plotted.

4. CONCLUSIONS

The most commonly used techniques for curve fitting by multiple
linear regression have been presented for linear equations, and for
those made linear by algebraic transformations. The computational
techniques described represent practical and useful methods for
determining the accuracy with which the kinetic implications of
proposed mechanisms can be fitted to actual experimental data.
Thus they comprise the criteria by which mechanisms may be tested,
and excluded if they do not fit the data within experimental error
at the desired limits of confidence. They also yield the numerical
values of the coefficients in the rate equations describing the
data.

Analytical methods for displaying the goodness of fit have been
described which allow graphical presentation of all the results on
a single two-dimensional plot, no matter how great the number of
experimental variables, nor how complex the interactions between
them. The method involves correction of a single primary dependent
variable and an independent variable for the contributions of all

the other variables, and thus yields in tabular form the magnitudes of the contributions of each of the variables considered.

5. REFERENCES

1. Nimmo, I.A. and Atkins, G.L. (1979) Trends Biochem. Sci. 4, 236-239.
2. Cleland, W.W. (1967) Adv. Enzymol. 29, 1-32.
3. Daniel, C. and Wood, F.S. (1971) "Fitting Equations to Data", Wiley, New York.
4. Draper, N.R. and Smith, H. (1966) "Applied Regression Analysis", Wiley, New York.
5. Watts, D.G. (1980) In "Kinetic Data Analysis: Design and Analysis of Enzyme and Pharmacokinetic Experiments" (L. Endrenyi, ed.) pp. 1-24, Plenum, New York.
6. King, E.L. and Altman, C. (1956) J. Phys. Chem. 60, 1375-1378.
7. Gutfreund, H. (1965) "An Introduction to the Study of Enzymes", Blackwell, London.
8. Cleland, W.W. (1963) Biochim. Biophys. Acta, 67, 104-137.
9. Davison, A.J. (1966) Nature, 210, 1161-1162.
10. Cornish-Bowden, A. (1979) "Fundamentals of Enzyme Kinetics", Butterworths, London-Boston.
11. Jennrich, R.I. and Ralston, M.L. (1979) Ann. Rev. Biophys. Bioeng. 8, 195-238.
12. Mannervik, B., Gorna-Hall, B. and Bartfai, T. (1973) Eur. J. Biochem. 37, 270-281.
13. Mannervik, B. (1975) Bio Systems, 7, 101-119.
14. Endrenyi, L. (1974) In "Mathematical Models of Metabolic Regulation" (T. Keleti and S. Lakatos, eds.) Akademiai, Budapest, pp. 11-30.
15. Bartfai, T. and Mannervik, B. (1972) FEBS Lett. 26, 252-256.
16. Bartfai, T. and Mannervik, B. (1972) In "Analysis and Simulation of Biochemical Systems" (H.C. Hemker and B. Hess, eds.) pp. 197-209, North-Holland/American Elsevier, Amsterdam.
17. Bartfai, T., Ekwall, K. and Mannervik, B. (1973) Biochemistry, 12, 387-391.
18. Mannervik, B. (1980) In "Kinetic Data Analysis: Design and Analysis of Enzyme and Pharmacokinetic Experiments" (L. Endrenyi, ed.) pp. 235 - 270, Plenum, New York.

MULTIVARIATE STATISTICAL METHODS IN PHARMACOKINETICS

Roberto Gomeni and Carla Gomeni

Synthelabo, L.E.R.S.
Department of Clinical Research.
Statistics and Pharmacokinetics Unit
58, rue de la Glacière
75013 Paris, France

ABSTRACT

The purpose of the present paper is to provide a description
of some multivariate statistical methods for comparing the performance of a drug in different formulations or doses within a given
population and between different populations. In multivariate
statistics all the parameters evaluated for each subject are simultaneously analysed taking into consideration their possible interrelationships. Three typical trials taken from the literature are
analysed using multivariate analysis of variance completed by discriminant analysis. Graphical display of the parameters in discriminant space show that the interpretation of the results may be
quite simple and meaningful, even to a nonstatistical scientist who
must take a decision based on the analysis.

1. INTRODUCTION

Pharmacokinetics is a scientific discipline which deals with
the mathematical description of the biological processes affecting
drugs and affected by drugs. The most popular method to characterize
a pharmacokinetic drug profile is compartmental analysis (1,2).
Several numerical methods and computer programs are now available
(3-6) to estimate the parameters which best fit the observed data
to a defined model. When the parameters so found are statistically
consistent, they are assumed to characterize the drug at the dose

and formulation used, for a given human population (newborn, elderly, healthy volunteers, patients, *etc*.).

The purpose of the present paper is to provide a description of some multivariate statistical methods for comparing the performances of a drug in different formulations or doses within a given population and between different populations. It is generally accepted that speaking of a pharmacokinetic profile of a drug in an undefined population is nonsensical. In fact, for the concurrence of variables such as genetic differences (7), physiopathological conditions (8-11), drug interactions (12,13) and age (14,15), the pharmacokinetic profile of various drugs may be very different in each individual patient. For this reason, it is important to define correctly the pharmacokinetic drug profile for each homogeneous group of subjects, in order to administer the drug in the most effective form and in the correct dosage regimen, adjusted to the patients' ages and diseases. After a preliminary data analysis, we obtain a sequence of parameters: area under concentration-time curve, absorption, disposition and elimination half-life, volume of distribution, total body and renal clearances, lag time, availability, transfer rate constants, *etc*. Actually, this set of parameters is considered like a set of independent univariate characteristics equivalent to measures of variables with normally distributed zero mean error and are analysed by using the appropriate statistical models (16-19).

In this paper, we would like to characterize each subject in a given population by the vector of the variables, *i.e.*, by the combination of all the variables together, like a point in an n-dimensional space, where n is the number of variables considered. In other words, we would like to substitute the multivariate statistical technique to the univariate criterion for data description, comparison and classification.

The most important difference between univariate and multivariate methods is that in the multivariate methods all parameters are simultaneously analysed considering their possible interrelationships (covariance). In fact, the pharmacokinetic parameters are often assumed independent but this is not exactly true; for example, it is well known that elimination half-life, volume of distribution and total body clearance can be mutually correlated.

2. MULTIDIMENSIONAL METHODS

In general, multivariate procedures are concerned with trials in which several variables have been assessed for each subject in each population under study. The multivariate generalization of ANOVA (analysis of variance) studies the group differences and tests the null hypothesis of equality of population means (vectors) on the assumption of equality of dispersion (variance-covariance matrices).

This can be done using a global test for the difference of group means. If this test yields a significant result, the analysis has to be carried on following two different ways:

1. Using multivariate contrast technique to isolate those groups whose difference between various means were most important in contributing to the overall significance test;

2. Using univariate ANOVA to test the differences among groups as regards each variable.

The multivariate analysis of variance (MANOVA) is completed by the discriminant analysis. In fact, if the groups exhibit statistical differences, the problem is to see in which direction the population means differ. The discriminant analysis yields a relative spatial configuration of the groups in reduced subspaces which provides information of directionality of group differences. From a practical point of view, this method is a useful tool for visualizing the location of the groups in one- or two-dimensional space. On the basis of the previous analysis (MANOVA), the classification technique permits the estimation of the probability of an individual being a member of each of a number of groups on the basis of his variable values. This procedure allows an *a posteriori* probabilistic partition of all the subjects in the different populations, to check the predictive validity of the variable system used.

This technique can also be applied to compute the group membership of a sample of new subjects characterized by the same parameters set.

3. MULTIVARIATE ANALYSIS OF VARIANCE

The cross-over design is widely used in pharmacokinetic trials, because the biological variation among subjects is usually considerable and the cross-over design enables one to remove subject effect from the error variance, thus obtaining a more sensitive test. In a cross-over trial, each subject receives each formulation or dose in turn, and the order of administration of the treatments is very often arranged in a Latin square.

Therefore, we shall present the multivariate analysis of variance in this general model; it is obvious that this analysis can be easily simplified if the experimental design involves just one-way or two-way classification. The statistical model takes the form:

$$\underline{X}_{ijk} = \underline{M} + \underline{S}_i + \underline{F}_j + \underline{P}_k + \underline{e}_{ijk} \quad .$$

$\underline{X}_{ijk}$ is the vector of computed or observed variables for the i-th subject, the j-th treatment and the k-th period, and $\underline{X}$, $\underline{M}$, $\underline{S}$, $\underline{F}$, $\underline{P}$,

e, are n-dimensional vectors where n is the number of variables considered; $\underline{M}$ is the vector of population means; $\underline{S}_i$ is the subject effect, $\underline{F}_j$ is the treatment effect and $\underline{P}_k$ is the period effect; $\underline{e}_{ijk}$ is the error which is supposed to be normally distributed with zero mean and homogeneous dispersion. Generalizing the classical analysis of variance (20), the full partition of the total sum of squares and cross-products for the Latin square model is then:

$$\underline{T} = \underline{A}_S + \underline{A}_F + \underline{A}_P + \underline{W}$$

where:

- $\underline{T}$ is the total sample sum of squares and cross-products matrix

$$\underline{T} = \sum_{i=1}^{a} \sum_{j=1}^{a} \sum_{k=1}^{a} (\underline{X}_{ijk} - \underline{m})(\underline{X}_{ijk} - \underline{m})'$$

(a is the Latin square dimension, $\underline{m}$ is the vector of the sample means, symbol ' denotes a transposed matrix);

- $\underline{A}_S$, $\underline{A}_F$, $\underline{A}_P$ are, respectively, the subject, treatment, period sum of squares and cross-products matrices defined as:

$$\underline{A}_S = a \sum_{i=1}^{a} (\underline{m}_i - \underline{m})(\underline{m}_i - \underline{m})'$$

$$\underline{A}_F = a \sum_{j=1}^{a} (\underline{m}_j - \underline{m})(\underline{m}_j - \underline{m})'$$

$$\underline{A}_P = a \sum_{k=1}^{a} (\underline{m}_k - \underline{m})(\underline{m}_k - \underline{m})'$$

($\underline{m}_i$ is the mean vector of the i-th subject values, $\underline{m}_j$ the means vector of the j-th treatment values, $\underline{m}_k$ the mean vector of the k-th period values);

- $\underline{W}$ is the error matrix for the model computed as:

$$\underline{W} = \underline{T} - \underline{A}_S - \underline{A}_F - \underline{A}_P \quad .$$

Several statistics (21) are available to test the null hypothesis of common group means. This is done by assuming that the parameter vector has multivariate normal distribution with the same dispersion (variance-covariance matrix) for each group. In this paper we use the Wilk's test of maximum likelihood, which has the advantage of being approximated by the well-known Fisher's F.

The MANOVA Wilk's test criteria are:

- for the hypothesized period effect:

$$\text{Lambda}_P = \frac{\det(\underline{W})}{\det(\underline{A}_P + \underline{W})} \quad ;$$

- for the hypothesized treatment effect:

$$\text{Lambda}_F = \frac{\det(\underline{W})}{\det(\underline{A}_F + \underline{W})} \quad .$$

The Lambda statistic has the distribution of $U_{n,q,z}$ (21) where: n=number of variables, q=a-1 and z=degrees of freedom of the residual variance. In order for W to be nonsingular (with probability 1) we must require $n \leq z$.

4. DISCRIMINANT ANALYSIS

The methods of multiple discriminant analysis result in the reduction of the multiple variables to one or more weighted combinations (discriminant functions) having maximum potential for distinguishing among members of different treatment groups. The first discriminant function is the weighted composite which provides the maximum average separation among the treatment groups relative to variability within the groups. The second discriminant function is the weighted composite, uncorrelated with the first, which provides the maximum average separation among treatment groups, and so on. The maximum number of possible discriminant functions is either equal to the number of variables or one less than the number of groups, whichever is smaller.

The first discriminant function, which maximizes the ratio of the among treatment groups sum of squares to the error sum of squares, is given by

$$\underline{y}^1 = \underline{v}'\underline{X}$$

where $\underline{v}$ is the unknown vector to be determined. The among-treatment (A_F^1) and error sum of squares ($\underline{W}^1$) of the function assume the following form:

$$\underline{A}_F^1 = \underline{v}'\underline{A}_F\underline{v}$$
$$\underline{W}^1 = \underline{v}'\underline{W}\,\underline{v}$$

The unknown vector $\underline{v}$ can be found by maximizing the ratio of the two quadratic forms:

$$\alpha^1 = \frac{\underline{v}'\underline{A}_F\underline{v}}{\underline{v}'\underline{W}\,\underline{v}}$$

under the hypothesis $\underline{v}^{1'}\underline{v}^1 = 1$. The maximum values α^1 and the associated vector $\underline{v}^1$ are shown (22) to be the largest eigenvalue and the associated eigenvector of the matrix product

$$\underline{W}^{-1}\underline{A}_F \quad .$$

Successive discriminant functions are defined by the vectors associated with the remaining eigenvalues. The proportion of the j-th discriminant function to the total discriminating power can be computed as:

$$d_j = \frac{\alpha_j}{\sum\limits_{k=1}^{n} \alpha_k} \quad .$$

On the basis of the d values, it is possible to decide how many discriminant functions have to be retained and to appreciate their relative importance.

5. CLASSIFICATION PROCEDURES

Classification procedures are used to assign a subject to one of the populations considered, minimizing the errors of misclassification. A number of methods for the classification of individuals on the basis of multivariate measurement profiles are reported (23). We have used Geiser's method (24) because it provides probabilistic classification statistics for a MANOVA-oriented strategy and is based on a small-sample theory. The estimated probability that individual i is a member of the j-th group is given by

$$P_{ij} = q_j \left[\frac{N_j}{N_j+1}\right]^{\frac{n}{2}} * \left[\frac{\Gamma\frac{1}{2}(df+1)}{\Gamma\frac{1}{2}(df-n+1)}\right] * \left[1 + \frac{N_j(\underline{X}_i-\underline{m}_j)'\underline{D}^{-1}(\underline{X}_i-\underline{m}_j)}{(N_j+1)df}\right]^{-\frac{1}{2}(df+1)}$$

where:

n = dimensions of variables vector

N_j = number of subjects in j-th group

q_j = N_j/(total number of subjects)

df = degrees of freedom of error matrix $\underline{W}$

$\underline{D}$ = $\underline{W}$/df (variance-covariance error matrix)

$\underline{X}_i$ = observed vector of individual i

$\underline{m}_j$ = mean of the j-th treatment group

Γ = Gamma function   .

The classification rule assigns individual i to the group for which the probability P_{ij} is maximal. When these decisions have been made for all subjects in the sample, a table can be arranged in which the rows are actual groups, the columns are assigned groups and the cell entries in a row are the frequencies of assignments of each type for an actual member of that group. The percentage of correct classifications can also be evaluated, either for each group or for the total population.

6. APPLICATION TO EXPERIMENTAL DATA

Multivariate statistics have been used to re-examine three typical trials previously analysed by univariate techniques:

1. Pharmacokinetics and effects of a drug in two classes of uraemic patients and in normal subjects;
2. Pharmacokinetics of a drug during human development;
3. Bioavailability trials.

6.1. <u>Pharmacokinetics and Effects of Propranolol in Terminal Uraemic Patients and in Patients Undergoing Regular Dialysis Treatment (25)</u>

Propranolol blood and plasma levels were measured after a single oral dose of 40 mg in patients with chronic renal failure, in patients undergoing regular dialysis treatment and in healthy volunteers. The following parameters have been evaluated: absorption rate constant (K_a), area under concentration-time curve (AUC), availability (F), elimination half-life $(t_{1/2})$, volume of distribution (V_d), total body clearance (Cl). Means and standard errors are reported in Table 1. The univariate parameter analysis has given the following results:

1. AUC and F are significantly higher in group A than in groups B and C, F is significantly higher in group B than in group C;
2. Cl is significantly reduced in group A with respect to groups B and C;
3. K_a is significantly higher in groups A and B than in group C;
4. No significant differences are found for V_d and $t_{1/2}$.

The major conclusion is that the first-pass hepatic metabolism of orally administered propranolol is considerably reduced in patients with renal failure not in regular dialysis.

TABLE 1:   Pharmacokinetic parameters of propranolol (mean ± standard
           error) in patients and volunteers

Group	K_a (hr^{-1})	AUC (ng/ml.hr)	F (%)	$t_{1/2}$ (hr)	V_d (L/kg)	Cl (L/min/kg)
A	1.07 [4] ±0.05	809 [2] ±250	62 [1] ± 7	3.8 ±0.9	3.0 ±0.4	0.70 [3] ±0.13
B	1.27 [4] ±0.08	197 ± 40	32 [4] ± 4	2.6 ±0.2	3.5 ±0.4	1.16 ±0.1
C	0.81 ±0.06	118 ± 2	21 ± 2	3.1 ± 0.3	3.3 ±0.2	1.04 ±0.03

The univariate tests show the following differences:

(1) $p < 0.001$ in respect to groups B and C;
(2) $p < 0.01$ in respect to groups B and C;
(3) $p < 0.05$ in respect to groups B and C;
(4) $p < 0.05$ in respect to group C.

Group A: Patients not in regular dialysis;
Group B: Patients undergoing regular dialysis treatment;
Group C: Healthy volunteers.

    The multivariate treatment of data confirms the univariate
conclusions and shows a very important result: the dialysis treatment has the effect of normalizing patients with renal failure.

    The MANOVA shows in fact a global difference ($p < 0.05$) between
normal subjects and patients not in regular dialysis but, on the
contrary is not capable of identifying any significant global
difference between normal subjects and patients on regular dialysis.
This can be seen clearly from the graphical group representation
in the bi-dimensional discriminant space (Fig. 1).   This result of
the multivariate analysis is in agreement with a second series of
observations reported in the same paper.

    In this additional trial, propranolol kinetics was evaluated
during the day of dialysis and the day after, in 5 patients.   Under
the effect of dialysis, the kinetic parameters are significantly
modified, approaching the values of normal subjects.   The multivariate procedure allows one to evaluate also the correlation matrix

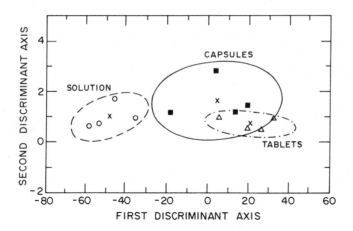

FIGURE 1: Discrimination between pharmacokinetic parameters of
propranolol among groups of patients with chronic renal failure.
Means of each group are represented by crosses. The ellipses represent 95% confidence intervals. MANOVA shows a significant
difference ($p<0.05$) between patients not on dialysis and normal
subjects.

of the variables; it can be of interest to observe that AUC, F and
Cl are significantly correlated, as well as $t_{1/2}$ and V_d. Moreover,
the classification test proves that the chosen parameters allow a
good identification of uraemic patients and normal subjects, respectively; there is in fact 100% of correct classification for
groups A and C, and 67% for group B.

6.2. <u>Digoxin Pharmacokinetics During Human Development</u> (26)

Digoxin was administered intravenously to three different
groups of patients: 5 newborns (3 to 9 days), 10 infants (1 to
11 months) and 3 children (2 to 5 years). The group of newborns
(A) received a dose ranging from 7 to 21 µg/kg. The group of
infants was divided into two subgroups; the first (B) received
doses ranging from 17.2 to 30 µg/kg, and the second (C), doses of
about half that administered to the first subgroup; finally, the
group of children (D) received doses comparable to the one administered to the first infant subgroup.

The following pharmacokinetic parameters have been used to
characterize the different groups: elimination half-life ($t_{1/2}$),
volume of distribution (V_d), total body clearance in units per
kilogram (Cl), and the total body clearance in units per m^2 (Cl_m).
Means and standard errors are reported in Table 2.

TABLE 2:   Digoxin pharmacokinetic parameters in children of various
ages (mean ± standard error)

Group	Dose (g/kg)	$t_{1/2}$ (hr)	V_d (L/kg)	Cl (ml/min/kg)	Cl_m (ml/min/m^2)
A	11.1 ± 2.6	69 ±25	7.53[1] ± 0.91	1.83 [1] ± 0.56	24.7 [1] ± 8.6
B	22.8 ± 2.2	18 ± 3	16.29 ± 2.10	10.74 [2] ± 0.74	187.4 [3] ±15.3
C	9.9 ± 0.5	33 ± 4	13.18 ± 1.48	4.50 ± 0.47	75.4 ± 7.4
D	17.3 ± 1.8	37 ± 9	16.10 ± 0.73	5.83 ± 1.73	138.7 ±45.3

The univariate tests show the following differences:

(1) $p < 0.01$ with respect to groups B and D;
(2) $p < 0.01$ with respect to groups C and D;
(3) $p < 0.01$ with respect to group C.

Group A: Newborns;
Group B: Infants at high dose;
Group C: Infants at low dose;
Group D: Children.

___

The univariate parameter analysis has given the following
results:

1.   Clearance values rise within the first 2 months of age and
     remain elevated up to 11 months; a consistent drop in values
     of about a half is noted in the following 2 to 3 years when
     the comparable dosage of digoxin are employed;
2.   The volume of distribution appears to be increased by age
     and/or dose;
3.   On the basis of this analysis it is not possible to separate
     the effects of dose and age on the pharmacokinetic parameters.

The multivariate approach confirms the univariate conclusions
about clearance and volume, but allows a better comprehension of
dose and age effect on digoxin pharmacokinetics. The MANOVA shows
in fact a global difference between newborns and children ($p < 0.05$),

and between infants at high dose and all the other groups (p<0.001).
The dose effect on digoxin pharmacokinetics is clearly demonstrated
by the difference between the two groups of infants treated at
different doses. The dose effect can also be very well appreciated
from Figure 2, where the cluster corresponding to the high-dose
infant group is completely separated from the clusters corresponding
to the other three groups.

The age effect can be seen from the difference between the
infants at high dose and children groups, which received not significantly different doses. It can be noted that in this trial only
the clearance/kg and the clearance/m^2 are significantly correlated.
The classification procedure assigned correctly 80% of the subjects
in the newborn group and 100% in the other groups.

6.3. Bioavailability of Digoxin Pharmaceutical Formulations (27)

In a cross-over design study, four dogs received 250 µg of
digoxin in three formulations: capsules, tablets and elixir. The
capsules and tablets were compared with the elixir of digoxin, which
is supposed to possess complete bioavailability. To test the bioavailability of the drug,the following parameters have been considered: peak plasma concentration (C_{max}), peak time (t_{max}) and area under
the concentration-time curve from 0 to 240 min. (AUC). Means and
standard errors are reported in Table 3. The univariate parameter
analysis has given the following results:

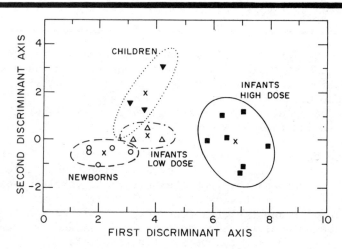

FIGURE 2: Discrimination between pharmacokinetic parameters of
digoxin among children of various ages. Means of each group are
represented by crosses. The ellipses represent 95% confidence
intervals. MANOVA shows a significant difference between newborns
and children (p<0.05) and between infants at high dose and all the
other groups (p<0.001).

TABLE 3:   Pharmacokinetic parameters of various digoxin formulations
(mean ± standard error)

Group	C_{max} (ng/ml)	t_{max} (min)	AUC (ng/ml/min)
A	4.1 [1] ±0.4	22 [2] ± 4	475 [2] ± 53
B	3.2 [1] ±0.4	45 ± 6	268 ± 23
C	1.8 ±0.2	52 ± 4	207 ± 21

The univariate tests shows the following differences:

(1) $p < 0.05$ with respect to group C;
(2) $p < 0.05$ with respect to groups B and C.

Group A: Digoxin solution;
Group B: Digoxin capsules;
Group C: Digoxin tablets.

---

1.   Peak concentration: a significant difference is found between
     solution and tablets and between capsules and tablets;
2.   Peak time: the solution values are significantly lower than
     the capsule and tablet values;
3.   AUC: the solution values are significantly higher than the
     capsule and tablet values.

On the basis of these results, the authors could assess that
digoxin capsules have a higher bioavailability than tablets. The
multivariate approach confirms the univariate results about each
parameter, but does not show any globally significant difference
between capsules and tablets. The only difference pointed out by
MANOVA is between elixir and capsules ($p < 0.01$) and between elixir
and tablets ($p < 0.01$) as can be seen in Figure 3.

The results of multivariate analysis, on the basis of all the
information available, is not in agreement with authors' conclusions
based only upon the difference found in the peak concentration
values. Moreover the MANOVA results seem to be much more reliable
because they have been obtained by considering the significant

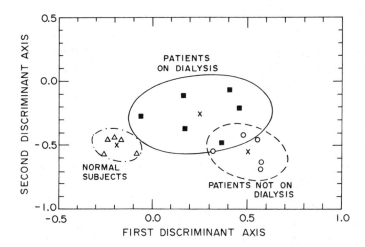

FIGURE 3:   Discrimination between pharmacokinetic parameters of
various digoxin formulations. Means of each group are represented
by crosses. The ellipses represent 95% confidence intervals.
MANOVA shows a significant difference between solution and capsules
(p<0.01) and between solution and tablets (p<0.01).

correlation (p<0.01) existing between peak concentration and area
under concentration-time curve.

7. CONCLUSIONS

1.   The problem of making a decision as to whether or not the differences between two or more populations are statistically significant, is a critical point in pharmacokinetic analysis. Several
parameters are always available, but very often the treatment groups
can differ only in respect to some of them; moreover, the differences
can be in opposite directions. In these cases, a decision, on the
basis of univariate analysis, can be very difficult and even arbitrary.

2.   Multivariate statistics provides a tool that allows one to
summarize a body of experimental parameters more effectively than
would be possible with repetitive application of univariate methods.

3.   Graphical representation of the parameters in discriminant
space stimulates hypothesis forming, and allows one to organize
experimental material that is otherwise difficult to classify by
univariate methods.

4.   The three typical trials previously analysed show that the
interpretation of the results in multivariate analysis may be quite
easy and meaningful even to nonstatistical scientists who must make

decisions based on the analysis.

5.   The analysis presented in this paper has been completed by
using the interactive graphic programs, MAOV, MAOV2, DISCRIM, CLASS.
These programs are written in FORTRAN IV and are currently being
used on a Digital Equipment PDP 11/70 computer. Copies of the
source code on nine-track tape and complete listings may be obtained
by writing to the authors.

REFERENCES

1. Gibaldi, M. and Perrier, P. (1975) "Pharmacokinetics", Dekker,
        New York.
2. Wagner, J.G. (1975) "Fundamentals of Clinical Pharmacokinetics",
        Drug Intelligence, Hamilton, Ill.
3. Metzler, C.M. (1969) "A User's Manual for NONLIN", Upjohn Co.
        Technical report 7292/69/7292/005, Kalamazoo, Mich.
4. Pedersen, P.V. (1977) J. Pharmacokin. Biopharm., 5, 513-531.
5. Berman, M. and Weiss, M.F. (1974) "User's Manual for SAAM",
        Public Health Service, N.I.H., Bethesda, Md.
6. Gomeni, C. and Gomeni, R. (1978) Comput. Biomed. Res. 11,
        345-361.
7. Wilding, G., Paigen, B. and Vesell, E.S. (1977) Clin. Pharmacol.
        Ther. 22, 831-842.
8. Benet, L.Z., ed. (1974) "The Effect of Disease States on Drug
        Pharmacokinetics", Amer. Pharmaceut. Assoc., Washington,
        D.C.
9. Dettli, L. (1976) Clin. Pharmacokin., 1, 126-134.
10. Nies, A.S., Shand, D.G. and Wilkinson, G.R. (1976) Clin.
        Pharmacokin. 1, 135-155.
11. Pang, K.S. and Rowland, M. (1977) J. Pharmacokin. Biopharm. 5,
        625-653.
12. Morselli, P.L., Garattini, S. and Cohen, S.N., ed. (1974)
        "Drug Interaction", Raven, New York.
13. Sellers, E.M. and Koch-Weser, J. (1970) Clin. Pharmacol. Ther.
        11, 524-529.
14. Morselli, P.L., ed. (1977) "Drug Disposition During Development"
        Spectrum, New York.
15. Triggs, E.J. and Nation, R.L. (1975) J. Pharmacokin. Biopharm.
        3, 387-418.
16. Metzler, C.M. (1974) Biometrics, 30, 309-317.
17. Westlake, W.J. (1974) Biometrics, 30, 319-327.
18. Westlake, W.J. (1973) J. Pharm. Sci. 62, 1579-1589.
19. Westlake, W.J. (1973) In "Current Concepts in Pharmaceutical
        Sciences; Dosage Form Design and Bioavailability" (J.
        Swarbrick, ed.) Lea & Febiger, Philadelphia, Pa.
20. Armitage, P. (1971) "Statistical Methods in Medical Research",
        Chap. 8, Wiley, New York.
21. Anderson, T.W. (1958) "An Introduction to Multivariate Statis-
        tical Analysis", Chap. 8, Wiley, New York.

22. Overall, J.E. and Klett, C.J. (1972) "Applied Multivariate Analysis", Chap. 9, 10, McGraw-Hill, New York.
23. Cooley, W.W. and Lohnes, P.R. (1971) "Multivariate Data Analysis", Chap. 10, Wiley, New York.
24. Geisser, S. (1964) J. Roy. Stat. Soc. 26, 807-817.
25. Bianchetti, G., Graziani, G., Brancaccio, D., Morganti, A., Leonetti, G., Manfrin, M., Sega, R., Gomeni, R., Ponticelli, C. and Morselli, P.L. (1976) Clin. Pharmacokin. 1, 373-384.
26. Morselli, P.L., Assael, B.M., Gomeni, R., Mandelli, M., Marini, A., Reali, E., Visconti, U. and Sereni, F. (1975) In "Basic and Therapeutic Aspects of Perinatal Pharmacology", (P.L. Morselli, S. Garattini and F. Sereni, eds.) Raven, New York.
27. Ghirardi, P., Catenazzo, G., Mantero, O., Merotti, G.C. and Marzo, A. (1977) J. Pharm. Sci. 66, 267-269.

TESTS FOR THE BEHAVIOUR OF EXPERIMENTAL ERRORS

Laszlo Endrenyi and Francis H.F. Kwong

University of Toronto
Department of Pharmacology and
Department of Preventive Medicine and Biostatistics
Toronto, Ontario M5S 1A8, Canada

ABSTRACT

The effectiveness (power) of three methods testing experimental error behaviour is evaluated for hyperbolic kinetic responses. The tests utilize residuals of unweighted nonlinear regression, and discriminate between two limiting error conditions: constant absolute error and constant relative error. An approach based on the transformation of the response is least suitable, a method assessing the slope of absolute residuals is better. Most powerful is the F_k-test which takes the ratio of first k and last k squared residuals in an ordered sequence of predicted responses. The F_k-ratio is contrasted with tabulated values of the F-statistic having (k,k) degrees of freedom. The F_k-test is useful also when only few observations are available.

1.   INTRODUCTION

Kinetic and many other kinds of experiments are often evaluated very effectively by applying the method of least squares. This technique, in common with various other procedures, makes assumptions about several features of the observational error. The outcome of the data analysis is particularly sensitive to correctly assuming the dependence of the errors on the magnitude of the response or responses.

Most frequently, their independence is assumed. In many experimental systems this may be a reasonable hypothesis, while in other systems the assumption is quite certainly not warranted. The error behaviour in some enzyme kinetic systems was assessed in recent years. In studies of aspartate aminotransferase (1),

bovine liver Type III hexokinase (2), and glutathione S-transferase A (3-5), acetylcholinesterase (6) and alcohol dehydrogenase
(6), the experimental error was found to be approximately proportional to the reaction velocity. In the study of Storer *et al*.
(7), in two investigated systems, enzyme kinetic measurement errors
either included two components, one of which was independent of
and the other proportional to the reaction velocity, or only the
latter component was detected. Storer *et al*. (7) also demonstrated
that results obtained by assuming a mixture of the two error conditions just mentioned were intermediate between those yielded by
calculations based on considering these error behaviours separately.
Errors of reaction velocities in other enzymic systems, such as
butyryl cholinesterase (4) and glutathione reductase (4), also
show an intermediate structure.

As a result, errors independent of or proportional to the magnitude of the responses (when either the absolute sizes of the
errors or their values relative to the responses, respectively,
are constant throughout the range of experimentation) may be considered as limiting cases (8,9) which occur frequently also by
themselves. Consequently, in order to fulfil the requirements of
data analysis, an investigator should recognize whether the experimental data are compatible with one or the other of these error
behaviours?

The error structure can be evaluated by various procedures
which require fairly careful, and not always simple considerations
utilizing, for instance, extensive and replicate observations.
Examples of such procedures are given in the quoted studies.
However, it is desirable to assess the behaviour of errors also
when there are not many observations available. For this purpose,
we shall compare the effectiveness of three methods evaluating
error structures in hyperbolic kinetic systems.

2. COMPUTATIONAL PROCEDURE

2.1. Simulation of the Michaelis-Menten Equation

The Michaelis-Menten equation

$$v = Vc/(K_m + c)$$

(where v is the velocity, c the substrate concentration, V the
asymptotic velocity, K_m the Michaelis constant) was selected as
an example of enzyme kinetic models which express the dependence
of the reaction velocity on substrate, inhibitor and modifier
concentrations.

Simulated observations following this model were arranged
according to various experimental designs. Either 5 or 10 points

were spaced in logarithmical (geometrically increasing) intervals
between true relative velocities (v/V) spanning, in 4 ranges,
between either 0.05 or 0.20 at the low end, and 0.80 or 0.95 at the
high end. In additional simulated experiments with 20 or 30 points,
the true relative velocities ranged from 0.05 or 0.20 to 0.80.

At each substrate concentration, normally distributed observations were simulated which had theoretical average values
equalling the true velocities at those concentrations, and theoretical standard deviations (σ) corresponding to either small or
large experimental errors. In the case of small errors, $\sigma = 0.01V$
and $\sigma_i = 0.02v_i$ were assumed for the simulation of constant absolute and relative errors, respectively. (The subscript i denotes
a quantity related to a given observation only.) The corresponding
large errors assumed $\sigma = 0.05V$ and $\sigma = 0.10v_i$, respectively.
Note that the two types of errors have identical magnitudes when
the relative velocity v/V = 0.5.

The computer simulated observations were evaluated by unweighted nonlinear regression. Following the optimization for
each simulated experiment, the residuals, *i.e.*, the differences
between observed reaction velocities and those predicted from the
least-squares parameters, were evaluated. One of the methods, to
be described, requires also least-squares estimation based on the
logarithms of the velocities.

2.2. Comparison of Methods Assessing Error Behaviours: Power and Reduced Power

The effectiveness of various methods was evaluated by simu-
lating experiments on the assumption of one of the error structures
and by noting the frequency at which the hypothesis of constant
absolute errors was rejected at a given probability level (α).

For instance, in the presence of observations having constant
errors, this error hypothesis would be rejected by all tests, in
principle and apart of random fluctuations, with a probability of
α (Fig. 1a). With observations having constant relative errors,
the constant-error hypothesis would be rejected at higher and also
different relative frequencies which would indicate the differing
powers of the various methods. Thus, the power of a test reflects
the ability and effectiveness of a method to discriminate between
the assumptions of the two error behaviours: The more powerful
of two tests rejecting the correct hypothesis with the same low
probability of α, will have a high probability for rejecting the
incorrect hypothesis.

In practice, however, the stated principle applies to linear
models. In nonlinear models, such as the Michaelis-Menten equation,
the constant-error hypothesis is not necessarily rejected with a

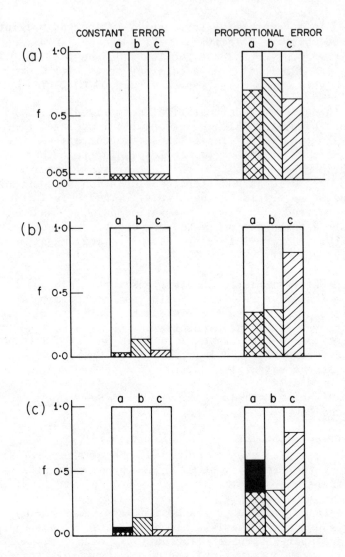

FIGURE 1: Methodology for comparing the powers of 3 methods
(a, b and c) which test for the constancy of the error variance.
f is the relative frequency (ranging from 0 to 1) of rejecting
the null hypothesis that the variance is constant. It is low when
the hypothesis is actually correct, and high when the true error
condition is different, such as if the relative errors are constant.
(a) The tests are performed at the 5% significance level ($\alpha = 0.05$).
Ideally, each of them rejects the null hypothesis when it is
actually correct, at a relative frequency of f = 0.05. The rejection frequencies for the alternative hypothesis (constant relative error) measure directly their power. Here, Method b is most,
and Method c is least powerful. (b) Actually, for nonlinear models,

probability of α even in the presence of observations having the
corresponding error structure (Fig. 1b). As a result, comparison of the powers of the various methods is more difficult. It
is, for example, not certain in Figure 1b whether Method a or
Method c is more powerful: The latter rejects more frequently,
for measurements with *either* error behaviour, the constant-error
hypothesis and, therefore, may not be judged as either the less
or the more powerful of the two procedures. In contrast, Method
c is certainly more powerful than Method b because it rejects
the hypothesis less frequently when this is correct (with constant
errors) and more frequently when this is incorrect (with constant
relative errors, Fig. 1b).

Such uncertainties in comparing powers of various methods
can be corrected by adjusting, for the execution of each test,
the probability level (α) in such a way that the hypothesis of
constant absolute errors is rejected, when it is actually correct,
with an identical *observed* relative frequency of α'. In Figure 1c,
for instance, the significance level (α) for Method a is raised
until the observed relative frequency of rejecting the constant-
error hypothesis equals that found for Method c. The rejection
frequency increases simultaneously also for experiments with constant relative errors but not sufficiently to reach the frequency
level of Method c (Fig. 1a). Thus, it is possible to claim now
that the adjusted or *reduced power* of Method c is higher than that
of Method a.

3.   METHODS ASSESSING ERROR BEHAVIOURS

3.1.   Plot of the Absolute Residuals Against the Predicted Velocities

The behaviour of experimental errors can be evaluated by
plotting the absolute values of the residuals against the predicted velocities (Fig. 2). If the errors tend to be, apart of
random fluctuations, constant over the whole range of experimentation then these residuals are expected to occupy a horizontal
band. If, on the other hand, the absolute errors are proportional
to the velocities (as when the relative or percentage errors tend

---

the rejection frequencies do not equal 0.05, and each other, even
if the tests are performed at the 5% significance level. As a
result, it is possible to state only that Method c is more powerful now than Method b. However, one can not decide whether Method
a has more or less power than either Method b or Method c?
(c) The significance levels of the tests are adjusted in order to
be able to compare their powers. In the figure, the significance
level is raised for Method a. Consequently, it becomes possible
to conclude that it is more powerful than Method b, but that it
has less power than Method c.

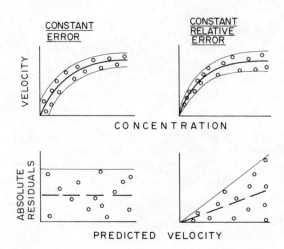

FIGURE 2: Hyperbolic response and absolute residuals at the two limiting error conditions. They are approximately constant with constant errors, and they tend to increase, linearly with the response, when the relative error is constant.

to remain constant) then the plot of the absolute residuals reflects this increasing trend.

Thus, the slope observed in this plot can be used for the quantitative assessment of the error structure (10). In the presence of constant absolute errors it should be approximately zero, whereas with constant relative errors it should be significantly larger than zero. Its statistical significance, at a given probability level, can be evaluated by comparing the magnitude of a calculated Student's t-value (the ratio of the observed slope and its standard error) with the corresponding tabulated, theoretical value.

3.2. Ratio of Sum of Squared Residuals (the F_k-test)

In the presence of approximately constant relative errors, the absolute residuals corresponding to the largest velocities tend to be larger than those belonging to the smallest velocities. Therefore, an F_k-value calculated as the ratio of the sums of the squares of the last and of the first k residuals, respectively, should exceed unity. In contrast, with constant absolute errors, this ratio should randomly fluctuate around the value of 1 and should, in fact, not deviate from it substantially (11).

The statistical significance of the calculated F_k can be evaluated, at a given probability level, by comparing it with tabulated values of the F-distribution with (k,k) degrees of freedom.

Naturally, the outcome and effectiveness of the test depends on the selection of k. For example, with 10 observations, k = 1 and k = 2 yield

$$F_1 = e_{10}^2 / e_1^2$$

and

$$F_2 = (e_9^2 + e_{10}^2) / (e_1^2 + e_2^2)$$

and rely on only 2 and 4 observations, respectively, and there-fore, may have limited reliability. Therefore, k = 3 yielding

$$F_3 = (e_8^2 + e_9^2 + e_{10}^2) / (e_1^2 + e_2^2 + e_3^2)$$

is expected to provide more useful results. Going further, however, k = 4 and 5 bring in additional points from the middle part of the observational range and, therefore, are likely to lower the sensitivity of the test.

3.3.   Transformation of the Dependent Variable

Box and Cox (12) suggested that the effectiveness of regres-sion calculations can be evaluated by performing them not only with the original response variable (the reaction velocity, v, in our case) but also with its power transformations, v^n . Conse-quently,

$$SS_n = \frac{1}{n^2 \cdot v_g^{2n-2}} \Sigma (v^n - \hat{v}^n)^2$$

is computed with various exponents, n, and the smallest value is selected. (Here v_g is the geometric mean of the velocities.)

In terms of error behaviours, constant absolute errors are represented by n = 1 and, therefore, by $SS_1 = SS$, whereas con-stant relative errors correspond to n = 0 which, as pointed out by Box and Cox (12), is evaluated from

$$SS_0 = v_g^2 \Sigma (\ln v - \widehat{\ln v})^2 \qquad .$$

(The approximate equivalence of constant proportional, relative errors and of regressions involving the logarithmic responses with constant additive errors can be seen from the Taylor expan-sion of the logarithmic expression involving an observations, the corresponding error and the value predicted by the model.)

After assuming the validity of one of the error behaviours, the possible plausibility (lack of statistical significance) of

its alternative was evaluated at a given probability level.
Regression computations were performed on the basis of both
error structures, and

$$A = (N.|\ln SS_1 - \ln SS_0|)^{\frac{1}{2}}$$

was compared with critical, tabulated values of the cumulative
standardized normal distribution (z). Thus, whenever the value
of A exceeds z, the plausibility of the error condition giving
rise to the larger of SS_0 and SS_1 is declared to be improbably
low.

4. RESULTS: COMPARISON OF THREE METHODS EVALUATING ERROR BEHAVIOUR

The effectiveness (power) of three methods assessing the two
behaviours is illustrated, for N = 5 and 10 observations, in
Tables 1 and 2, respectively. For each experimental condition
and each method, the frequency (in 50 simulated experiments) of
rejecting the hypothesis of constant absolute errors is noted.
This frequency should be low in experiments actually having constant errors, and high in the presence of observations for which
the relative of percentage error is constant.

The significance level of the method involving ratios of
squared residuals was held as close to 5% as possible for k = 2
with 5 observations and k = 3 with 10 observations. The theoretical significance levels of the other methods were varied
until suitable contrasts of the reduced powers could be obtained.
The comparable frequencies of rejecting the constant-error hypothesis are indicated for each method. Those judged to provide,
for a given experimental design, the best discrimination between
the two error structures are marked with an asterisk.

In general, ratios of sums of squared residuals were the best
discriminators between experiments with constant absolute and constant relative errors. The optimal number of residuals included
from the high and low ends of the concentration range in the numerator and denominator of the calculated F_k-statistics was k = 2
and 3 for investigations with N = 5 and 10 observations, respectively. However, the power decreased only slowly on the inclusion
of medium-ranged readings. The optimally discriminating F_k-tests
had the additional favourable property of rejecting the hypothesis
of constant absolute errors, in simulated experiments which were
actually based on this error structure, with a relative frequency
which was very close to the assumed significance level of $\alpha = 0.05$.

The method of Box and Cox (12) utilizing transformed observations showed by far the poorest discrimination between the two
error behaviours; This conclusion could be reached from assessing
the power either at the $\alpha = 0.05$ or at the 0.01 significance level.

TABLE 1: Number of experiments[a] with 5 observations rejecting the hypothesis of constant absolute errors

Range v_1 to v_5	Error σ	Slope of abs. residuals				F_k-test			Response transform.	
		t				F_2	F_1		A	
		30%[b]	15%	10%	5%	5%	10%	5%	5%	1%
0.05– 0.80	0.01	4*[c,d]		0	0	4	6		5	5
	0.02v_i	42*[e]		23	23	37	18		16	10
	0.05		1	0	0	1*	5		2	1
	0.10v_i		30	24	19	37*	19		14	10
0.05– 0.95	0.01			2		2*	4		6	6
	0.02v_i			21		41*	18		22	12
	0.05		3	1		2*	7		6	4
	0.10v_i		29	17		40*	18		20	11
0.20– 0.80	0.01	6*	2	0		6*	6		10	7
	0.02v_i	22*	22	12		22*	10		13	6
	0.05	6*	1	0		6*	13	5	11	7
	0.10v_i	23*	19	12		23*	16	8	13	5
0.20– 0.95	0.01		8	3		4*	6		12	8
	0.02v_i		20	11		24*	11		15	8
	0.05		7	1		4*	5		11	8
	0.10v_i		21	13		24*	11		15	8

(a) Out of 50 computer simulated experiments.
(b) Significance level (α) of the test which equals the probability of rejecting a null hypothesis when this is actually correct. In the present case, the hypothesis assumes the constancy of the underlying absolute observational errors. Therefore, the significance level is the probability (theoretical relative frequency) of rejecting the hypothesis of constant errors when it is in fact correct, *i.e.* in the first line of each pair of experimental conditions.
(c) Underlined figures are the rejection numbers at the significance level selected for the comparison of each test method.
(d) Rejections even though the assumption of constant errors is correct.
(e) Rejections because proportional-error assumption is correct.
*   Rejection number by the most powerful test method(s) at the given experimental condition.

TABLE 2: Number of experiments [a] with 10 observations rejecting the hypothesis of constant absolute errors.

Range v_1 to v_{10}	Error σ	Slope of abs. residuals t 10%[b]		7.5%	5%	F_5 5%	F_4 5%	2.5%	F_3 5%	Response transform. A 5%	1%
0.05- 0.80	0.01	$2^{(c,d)}$			1	2	3		2*	5	4
	0.02v_i	$41^{(e)}$			38	46	46		48*	35	30
	0.05	3			2	3	5		2*	5	2
	0.10v_i	42			37	46	46		48*	35	28
0.05- 0.95	0.10	4			1	2	3		2*	4	3
	0.02v_i	42			35	44	45		47*	37	31
	0.05	3			2	4	4		2*	5	5
	0.10v_i	42			37	45	46		48*	36	30
0.20- 0.80	0.01	4	2		1	2*	4		3	11	8
	0.02v_i	31	28		22	30*	29		27	22	15
	0.05	4			1	3*	4	3	3	11	8
	0.10v_i	30			25	30*	32	21	29	21	14
0.20- 0.95	0.01	4		2		4	4	3	2*	10	7
	0.02v_i	31		24		28	32	22	30*	23	18
	0.05	4		2		4	4	3	2*	11	8
	0.10v_i	32		23		30	33	25	30*	23	15

(a)   Out of 50 computer simulated experiments.
(b)   Significance level (α) of the test as explained in the text and in the footnote of Table 1.
(c)   Underlined figures are the rejection numbers at the significance level selected for the comparison of each test method.
(d)   Rejections even though the assumption of constant errors is correct.
(e)   Rejections because proportional-error assumption is correct.
*   Rejection number by the most powerful test method(s) at the given experimental condition.

Furthermore, the test rejected the constant-error hypothesis when this was actually correct, too frequently in comparison with the assumed significance level.

The effectiveness (power) of the test relying on the slope of the absolute residuals was intermediate between those of the other two methods. Generally speaking, the deviation from the optimal power was not very large. However, the hypothesis of constant absolute errors was rejected, in simulated experiments actually having such an error structure, much less frequently than expected on the basis of the significance level of the test.

The magnitude of the experimental errors had little effect on the power of any of the test methods.

The behaviour of the squared-residual-ratio test was further studied in simulated experiments of 20 and 30 observations each. For two experimental designs, 100 simulated experiments were performed. The significance levels were adjusted until the rejection rate of the observed constant-error hypothesis was 0.05. Inclusion of $k = 5$ and 8 terms in the numerator and denominator yielded the most powerful discrimination between the two error structures with 20 and 30 readings, respectively. However, moderate departures from these optimal numbers, especially in the upward direction, did not substantially lower the power of the test.

The results were very similar to the ones presented above when relative frequencies of rejecting the hypothesis of constant relative errors (and not of constant absolute errors) were recorded.

The favourable performance of the F_k-test has been demonstrated also under different conditions and with other competing tests (13).

5.   EVALUATION OF ERROR BEHAVIOUR

5.1. Use of Replicates for Assessing Error Behaviour

The relationship between experimental errors and response values can be assessed very effectively with the help of replicate observations. The procedures described for the residuals can be applied also to standard deviations and variances of velocities which are repeatedly measured at each concentration. Plots of standard deviations against predicted or observed responses are analogous to the residual plots. The formal tests described in the present study are similarly applicable. In fact, quantities based on replicate readings have the additional advantage of being evaluated before the least-squares calculations, instead of relying on them as is the case with the residuals.

Mannervik and his co-workers (3-5) have applied replicate measurements carefully and usefully for the evaluation of error behaviour. However, whether the replicates were actually repeated readings (3) or suitably pooled values (4), their approach required a substantial number of observations.

Otherwise, however, care should be exercised when repeated measurements are utilized. In particular, variances and weights should, generally, not be estimated from them separately at each concentration [e.g., (14)] since the calculated values can show large uncertainties (7,8).

Nevertheless, these replicates can be usefully applied in the tests for the behaviour of the experimental error which are described in this work. The reason is that the tests combine information from the various replicates and, thereby, have a sufficiently firm foundation.

In the same way, variances and weights relying on pooled information can be applied usefully. For instance, when the difference of duplicate readings is plotted against their average, the smoothed difference provides a reasonable estimate which is proportional to the standard deviation at any concentration. In addition, the slope of the plot (or of the corresponding double-logarithmic diagram) could provide a measure for the error behaviour, in similarity to the approach of Mannervik *et al.* (3-5).

5.2. Alternative Descriptions of Error Behaviour

While the two error behaviours of constant absolute error (constant variance) and constant relative error (constant coefficient of variation) prevail in many experimental systems, they may be considered as limiting conditions. Thus, if the variance (σ^2) of an observation is related to its magnitude (v) according to (3-5)

$$\sigma^2 = k_i v^a \quad ,$$

then $a = 0$ corresponds to constant absolute errors and $a = 2$ to constant relative errors. From this description of the error behaviour, a can be evaluated by means of a logarithmic linearization (3,4,15).

On the other hand, if the variance is assumed to have two components (3,9,16):

$$\sigma^2 = k_o + k_2 v^2 \quad ,$$

then the two limiting error conditions are given by $k_o = 0$ and $k_2 = 0$, respectively.

These two descriptions of error behaviours introduce flexibility and detail. However, their assessment requires several data points. The methods considered in the present work, including the most powerful among them, the ratio of sum of squared residuals, can be applied also when only few observations are available, as is usually the case.

5.3. Executing the Test for Error Behaviour

Among the tests being considered, the one calculating a ratio of sums of squared residuals, the F_k-test, has the highest power and is, therefore, preferred.

The residuals are based on unweighted nonlinear regression. In an ordered sequence of the predicted response values, the corresponding first k and terminal k residuals are selected and squared. Sums of both sequences are taken, and the one obtained from the last values is divided by the one gained from the early residuals. If the resulting F_k-ratios are larger than tabulated F-statistics at a given significance level (*e.g.* $\alpha = 0.05$ or 0.01) with (k,k) degrees of freedom (Table 3), then the hypothesis of

TABLE 3:   Critical values for the F_k-test

n	k	$F_{(k,k)}$ $\alpha=0.05$	0.01
5-8	2	19.00	99.00
9-12	3	9.28	29.46
13-16	4	6.39	15.98
17-20	5	5.05	10.97
21-24	6	4.28	8.47
25-28	7	3.78	6.99
29-32	8	3.44	6.03
33-36	9	3.18	5.35
37-40	10	2.98	4.85

n: Number of observations

k: Number of first and last residuals included in the F_k-test.

$F_{(k,k)}$: F-statistic with (k,k) degrees of freedom at a significance level of α.

of constant error variance is rejected and replaced, in the present case, by the assumption of constant relative errors.

k, the number of residuals taken from each end, can be obtained from

$$k = \text{Int}[(n+3)/4] \qquad .$$

The calculated number is in agreement with the results shown in Tables 1 and 2: k = 2 and 3 for N = 5 and 10 observations, respectively. The expression satisfies also other, less extensive simulations, yielding k = 5 and 8 for N = 20 and 30 readings, respectively.

The critical F-statistics used in the test are listed in Table 3 for various numbers of observations.

6.   CONCLUSIONS

The F_k-test is the most powerful among the three methods evaluated for discriminating between two limiting error behaviours (constant variance and constant relative error) in hyperbolic kinetic experiments. The test forms the ratio of the sums of the last k and first k squared residuals in a sequence of increasing predicted responses. The residuals are based on unweighted nonlinear regression.

The method can be applied also when only few observations are available.

The number of terms (k) and critical values of the F_k-statistics are shown in Table 3.

REFERENCES

1. Haarhoff, K.N. (1969) J. Theoret. Biol. 22, 117-150.
2. Siano, D.B., Zyskind, J.W. and Fromm, H.J. (1975) Arch. Biochem. Biophys. 170, 587-600.
3. Askelöf, P., Korsfeldt, M. and Mannervik, B. (1976) Eur. J. Biochem. 69, 61-67.
4. Mannervik, B., Jakobson, I. and Warholm, M. (1979) Biochim. Biophys. Acta, 567, 43-48.
5. Mannervik, B. (1980) In "Kinetic Data Analysis: Design and Analysis of Enzyme and Pharmacokinetic Experiments" (L. Endrenyi, ed.) pp. 235-270, Plenum, New York.
6. Nimmo, I.A. and Mabood, S.F. (1979) Anal. Biochem. 94, 265-269.
7. Storer, A.C., Darlison, M.G. and Cornish-Bowden, A. (1975) Biochem. J. 151, 361-367.

8. Endrenyi, L. (1974) In "Mathematical Models of Metabolic
    Regulation" (T. Keleti and S. Lakatos, eds.) pp. 11-30,
    Akadémiai, Budapest.
9. Cornish-Bowden, A. (1980) In "Kinetic Data Analysis: Design
    and Analysis of Enzyme and Pharmacokinetic Experiments"
    (L. Endrenyi, ed.) pp. 105-120, Plenum, New York.
10. Glejser, H. (1969) J. Am. Stat. Assoc. 64, 316-323.
11. Goldfeld, S.M. and Quandt, R.E. (1965) J. Am. Stat. Assoc.
    60, 539-547.
12. Box, G.E.P. and Cox, D.R. (1964) J. Roy. Stat. Soc. B26,
    211-252.
13. Hedayat, A., Raktoe, B.L. and Talwar, P.P. (1977) Commun.
    Stat. A6, 497-506.
14. Ottaway, J.H. (1973) Biochem. J. 134, 729-736.
15. Wagner, J.G. (1975) "Fundamentals of Clinical Pharmacokinetics",
    pp. 280-290, Drug Intelligence Publ., Hamilton, Ill.
16. Endrenyi, L. and Dingle, B.H. (1980) In "Pharmacokinetics
    During Drug Development - Data Analysis and Evaluation
    Techniques" (G. Bozler and J.M. van Rossum, eds.)
    Gustav Fischer, Stuttgart, in press.

ROBUST ESTIMATION IN ENZYME KINETICS

Athel Cornish-Bowden

University of Birmingham
Department of Biochemistry, P.O. Box 363
Birmingham   B15 2TT, England

ABSTRACT

Although the method of least squares is optimal for fitting
equations to data under idealized conditions, it gives poor results
if these conditions are violated. It is highly sensitive to the
presence of outliers in the data and it requires correct weighting.
The direct linear plot is a method of plotting the Michaelis-Menten
equation in parameter space that leads to robust estimates of the
parameters, $i.e.$, estimates based on few assumptions about the
statistical structure of the data. It performs well with the
Michaelis-Menten equation, and similar methods can be applied with
equal success to other two-parameter models, but it cannot easily
be applied to models with more than two parameters to be estimated.
For this reason there is interest in other robust methods, such as
biweight regression, which resembles least-squares regression in
which decreased or zero weight is assigned to outlying observations.
Such methods provide good protection against outliers but need
correct weights to work satisfactorily.

1.   INTRODUCTION

Parameter estimation in enzyme kinetics has commonly been discussed in the context of the Michaelis-Menten equation:

$$v = Vs/(K_m+s) \tag{1}$$

in which v is the initial rate observed at substrate concentration
s, and V and K_m are the parameters to be estimated. The most widely used method for estimating V and K_m depends on the fact that if

experimental error is ignored Eq. [1] can be transformed into the
following expression:

$$1/v = (1/V) + (K_m/V)(1/s) \tag{2}$$

which indicates that a double-reciprocal plot of $1/v$ against $1/s$
should be a straight line with slope K_m/V and intercept $1/V$ on the
$1/v$ axis. This plot has the rather dubious advantage of white-
washing the data, *i.e.* it tends to make bad observations less
noticeable than they are in most alternative plots of the same
data. This property probably accounts for its enduring popularity,
as Dowd and Riggs (1) pointed out.

Lineweaver and Burk (2) are commonly blamed for the introduction of the double-reciprocal plot, though they were aware of the
need for proper weighting in the fitting of transformed data. In
more detailed work published separately (3,4) they correctly pointed
out that if equal weighting of v values is appropriate then the
corresponding $1/v$ values need weights proportional to v^4. (Unlike
nearly everyone since, they obtained experimental evidence about
the proper weights to be used, and did not confuse an assumption
about weighting with knowledge of it.) Unfortunately, this part
of their work has been read by hardly any of those who "quote"
them. The typical practice is instead to draw a straight line by
eye or to fit Eq. [2] by unweighted linear regression, which is
probably worse than fitting by eye because it gives a spurious
suggestion of precision and numeracy to a bad method.

The worst aspects of the double-reciprocal plot have been
mitigated in recent years by increasing use of methods for fitting
Eq. [1] directly or by fitting Eq. [2] or another linear transformation with weights calculated from those thought to be appropriate
for v (5,6). The aim is to find values of V and K_m that minimize
the sum of squares, SS, defined by

$$SS = \sum_{i=1}^{n} w_i [v_i - Vs_i/(K_m + s_i)]^2 \tag{3}$$

where the subscript i denotes the i-th of n observations, and w_i
is the weight of v_i, which should be inversely proportional to the
variance of v_i. Under certain defined conditions, this criterion
for estimating V and K_m is optimal, in the sense that it minimizes
the variances of the estimates. This is less of an advantage than
it may seem because in a real experiment one does not know whether
the defining conditions actually apply to the data one is analysing.
The principal assumptions that must be satisfied for a least-squares
calculation to provide minimum-variance estimates are the following
[e.g., (7)]:

1. Only one variable (v_i) is subject to error, the other (s_i)
    being known exactly.
2. The errors in v_i are normally distributed.
3. The correct weights w_i are known: this requires that the
    functional form of the dependence on v_i of the variance of v_i
    be known.
4. The errors are random rather than systematic.

I shall be concerned in this article with the effects of violation
of the second and third of these: the first is probably a fair
approximation to reality in most enzyme kinetic experiments, and
the fourth requires major study on its own.

   One serious departure from a normal distribution is the occurrence of *outliers*, or anomalously erratic observations. These
should not be confused with the small proportion of highly erratic observations predicted by the normal distribution: for example, in a normal distribution with variance σ^2, about two-thirds
of the errors are numerically less than σ and about 1% are numerically greater than 3σ. These 1% of erratic observations are not
outliers and require no special treatment, because their occurrence
is fully allowed for by least-squares theory. Suppose instead,
however, that 80% of errors are numerically less than σ and 5% are
numerically greater than 3σ. Then it is reasonable to regard the
more erratic observations as outliers and to expect them to have a
severely harmful effect on least-squares estimation. When outliers
are present, one can usually get much better estimates than those
provided by the method of least squares by using a method that
depends more heavily than least squares on the observations with
small errors.

   Ideally, the weight w_i assigned to a rate v_i should be inversely proportional to $\sigma^2(v_i)$, the variance of v_i. Unfortunately,
however, not only are the variances usually unknown but the form
of the dependence of $\sigma^2(v_i)$ on v_i is usually unknown as well.
Some assumption is therefore required, and it is usual to choose
between the following two extremes: if $\sigma^2(v_i)$ is independent of
v_i, a circumstance I shall refer to as *simple errors*, we should
put $w_i=1$ for all i; if $\sigma^2(v_i)$ is proportional to v_i^2, which I shall
call *relative errors*, we should put $w_i=1/v_i^2$. With simple errors
the *standard deviation* of the v_i values is constant; with relative
errors the *coefficient of variation* (or standard deviation expressed
as a percentage) is constant. It is worth labouring this point a
little because there is evidence of widespread misunderstanding of
it: it is not uncommon to see (or hear) that values of a measured
variable were found to be scattered within a constant percentage
of their calculated values and that *therefore* equal weights were
used. This is wrong: the observation implies a constant coefficient of variation and hence requires unequal weighting. Equal

weighting of v_i values implies that their standard deviation, measured in mol 1^{-1} s^{-1}, mM min^{-1}, etc., but *not* in percent, is constant.

Simple and relative errors imply very different sets of weights: for example, if the v_i span a 5-fold range, the proper weights are all equal with simple errors but span a 25-fold range with relative errors. It follows that the results of a least-squares calculation are sensitive to the weighting assumption that is made, and if the assumption is wrong the optimal properties of the method are lost.

Because of the usual absence of genuine knowledge about the error structure of data, there has been considerable interest in recent years in *robust* methods of estimation. These are methods that are a little less accurate than least-squares estimation when the least-squares assumptions are true, but are more accurate — sometimes much more accurate — when they are false; *i.e.*, they are insensitive to violation of these assumptions. In enzyme kinetics, the direct linear plot (8-10) provides one approach to robust regression. Others are becoming available that are more generally applicable and are possibly better in other respects (11,12). In this article, I shall examine the properties of least-squares and robust estimators of the Michaelis-Menten parameters and will discuss how robust methods may be developed for a wider range of problems.

2. THE DIRECT LINEAR PLOT

Most of the commonly used ways of plotting data give rise to plots in *observation space*, *i.e.*, each axis refers to one of the observed variables or a transformation of it. It is also possible, however, to plot data in *parameter space*, whereby the axes refer to the parameters to be estimated and the observations appear not as points but as lines (or in complex cases as surfaces or hypersurfaces). The advantage of plotting in parameter space is that the relationships between the observations and the parameter values tend to be expressed more clearly, and the process of estimation is more direct and thus more comprehensible. A plot of the Michaelis-Menten equation in parameter space may be devised by recasting Eq. [1] as follows:

$$V = v_i + (v_i/s_i)K_m \quad . \tag{4}$$

This shows that if one draws K_m and V axes and plots an observation (s_i, v_i) as a straight line with intercepts $-s_i$ on the K_m axis and v_i on the V axis, the resulting line relates all possible K_m and V values that satisfy the observation exactly (8). A second observation (s_i, v_i) may be plotted as a second line, and the point of intersection of the two lines provides the only pair of values

$(K_{m(ij)}, V_{(ij)})$ that satisfy both observations exactly. In the absence of experimental error, the third and subsequent lines would all pass through the same point, but in reality one would obtain a family of intersection points as illustrated in Figure 1, each pair of observations (s_i, v_i) and (s_j, v_j) defining a pair of parameter estimates as follows:

$$K_{m(ij)} = (v_j - v_i)/[(v_i/s_i) - (v_j/s_j)] \qquad [5]$$

$$V_{(ij)} = (s_j - s_i)/[(s_j/v_j) - (s_i/v_i)] \quad . \qquad [6]$$

It may happen that these expressions define parameter values with enormous errors, especially if s_i and s_j are nearly equal. For this reason it is inappropriate to try to obtain best estimates by any sort of arithmetic averaging process. But the middle inter-sections ought to be acceptably precise and so the *medians* [or middle values of the ranked sets of $K_{m(ij)}$ and $V_{(ij)}$] should be acceptable estimates of K_m and V (9). They should be little af-fected by the presence of a small proportion of outliers among the

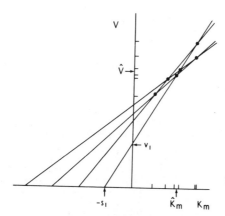

FIGURE 1: Direct linear plot for obtaining median estimates of K_m and V in the Michaelis-Menten equation. Each observation (s_i, v_i) is represented by a straight line with intercepts $-s_i$ and v_i on the K_m and V axes, respectively. Each pair of lines intersects to provide an estimate $K_{m(ij)}$ of K_m and an estimate $V_{(ij)}$ of V, as marked on the axes. The medians of the two series are taken as the best estimates. For clarity, only four observations are shown in this illustration, but for best results it is inadvisable to use the method with fewer than eight observations. For treatment of replicate observations, and intersections that do not occur in the first quadrant, see the text.

observations and so have little effect on the medians. Furthermore,
the determination of the medians involves no weighting and so lack
of knowledge of the variances of the observations is of no importance. One might hope, therefore, that this approach to the estimation of V and K_m would be robust against failure of the assumptions
required by the least-squares method. This hope is, in general,
satisfied (9), though more recent work has brought to light a number
of complications that need to be considered.

3. REFINEMENTS TO THE DIRECT LINEAR PLOT

It is clearly impossible to obtain meaningful values of V and
K_m from Eqs. [5] and [6] if s_i and s_j are equal. Nonetheless, the
lines on the direct linear plot corresponding to such a duplicate
pair of observations do intersect (unless v_i and v_j are also equal)
on the K_m axis, and thus provide a spurious pair of values
$K_{m(ij)}=-s_i=-s_j$, $V_{(ij)}=0$. Such an intersection is meaningless and
should be omitted from consideration when determining the medians.
The total number of intersections to be used is thus equal to the
maximum possible number, $\frac{1}{2}n(n-1)$, only if there are no replicate
observations.

A second problem that arises when there are replicate observations, is the need to decide whether it is advantageous to average
the replicate v values before analysis. In least-squares estimation
it makes no difference (provided that each mean is weighted in proportion to the number of observations that it contains); but in the
direct linear plot it does make a difference, and it turns out that
it is more important to maximize the number of intersections than
to maximize the precision of each. Consequently, it is better *not*
to average the v values from replicate determinations before analysis (13). This is especially important if there are fewer than
about eight different s values in the experiment.

Although the procedure outlined above works well if all of the
intersections occur in the first or second quadrants of the direct
linear plot, *i.e.*, all $V_{(ij)}$ values are positive, bias in both V
and K_m can arise if there are intersections in the third quadrant,
i.e., $V_{(ij)}$ and $K_{m(ij)}$ both negative. (Intersections in the fourth
quadrant do not need to be considered as they are impossible if
negative v_i values are not admitted.) The bias arises because the
dependent variables v_i and v_j occur in the denominators of the
expressions for $V_{(ij)}$ and $K_{m(ij)}$, as discussed more fully elsewhere
(10,14). In the graphical method, the bias can be avoided by
treating intersections in the third quadrant as having very large
positive values of both coordinates. The negative values are thus
treated as "beyond infinity" rather than as "below zero" (10). As

a computational method, it is more convenient and more rigorous to carry out the estimation in terms of $1/V$ and K_m/V, for which the corresponding expressions do not contain v_i and v_j in their denominators:

$$1/V_{(ij)} = [(s_j/v_j)-(s_i/v_i)]/(s_j-s_i) \qquad [7]$$

$$K_{m(ij)}/V_{(ij)} = [(1/v_i)-(1/v_j)]/[(1/s_i)-(1/s_j)] \qquad [8]$$

The median estimates of $1/V$ and K_m/V can readily be converted into the corresponding estimates of V, K_m and any other combinations required.

If there are more than a few observations, it is laborious to determine accurate medians from a plot, though it is not difficult to obtain approximate values. In any case, for processing large amounts of data it is clearly advantageous to express the method as a computer program. A program written in FORTRAN that takes account of all of the considerations discussed above and also determines confidence limits for the parameters (13,15) is available from me on request. In addition, Henderson (16) has published a slightly modified version of the same program in full.

4. PERFORMANCE OF MEDIAN ESTIMATES IN MONTE CARLO TRIALS

Extensive Monte Carlo trials have shown that median estimates of the Michaelis-Menten parameters have the anticipated robust properties. When the least-squares assumptions are true, the median estimates are on average less accurate than the least-squares estimates — as expected — but the disadvantage is surprisingly slight: they give values closer to the true ones in about 40% of experiments (9). When outliers are present, or the least-squares calculations are done with wrong weights, the median estimates become much better than least-squares estimates. Other workers have obtained similar results in simulated experiments (17), and the method has been used without reported difficulties in many laboratories [for a diverse collection of examples, see (18-23)]. Furthermore, Monte Carlo trials of the procedure proposed for calculating confidence limits (15) show that, when modified in the light of the remarks above about bias, it provides results that are at least as reliable as those provided by the method of least squares and in some cases much more so (13).

Rather than repeat previously published results at length, I have chosen to tabulate some recent preliminary results in which the variances of the estimates of K_m/V and $1/V$ given by the least-squares and median methods are compared with those given by an alternative robust method that I shall discuss below. These results are given in Table 1, from which it is evident that the generaliza-

TABLE 1: Variances of parameter estimates given by three methods

Type of error	K_m/V			$1/V$		
	LS	M	BW	LS	M	BW
Relative errors, normal distribution	2.46	3.23	2.81	2.79	3.65	3.28
Relative errors, with outliers	2.63	1.86	1.97	2.87	1.93	1.92
Simple errors, normal distribution	14.19	11.06	15.75	9.13	5.62	9.05
Simple errors, with outliers	12.07	4.53	7.89	7.80	2.62	4.78

The Table shows the variances of estimates of K_m/V and $1/V$ obtained in Monte Carlo trials (1000 for each line of the Table) by the least-squares (LS), median (M) and biweight (BW) methods. In each experiment there were ten observations at s values spaced evenly from $0.2K_m$ to $2.0K_m$. The true values of K_m and V were 1.0. In all cases, the half-maximal v value (at $s=K_m$) had a coefficient of variation of 5% or (equivalently) a standard deviation of 0.025V. In experiments with "relative errors" each rate had the same co-efficient of variation; with "simple errors" each had the same standard deviation. In the least-squares and biweight calculation, in order to determine the effects of making a wrong weighting assumption, the errors were assumed to be relative regardless of whether they were or not. In experiments "with outliers" each error was drawn with probability 0.8 from one normal distribution with mean zero, or with probability 0.2 from a second normal dis-tribution with mean zero and standard deviation 4 times greater than that of the first. For reading convenience, all variances are multiplied by 1000 in the Table.

tions made previously in terms of the proportion of "successes" for each method (9) also apply to results expressed in terms of vari-ances, which offer a more convenient way of comparing more than two methods. The results shown form a small proportion of the total number obtained. The relative performances of the three methods are largely independent of the range of s/K_m values used, the arrangement of s/K_m values within the range, and the level of ex-perimental error (up to a coefficient of variation of about 20% for the half-maximal rate). They are also little affected by the

number of observations, provided this is 8 or more: as n is decreased below 8, the variances of the median estimates increase
very steeply.

Although only the Michaelis-Menten equation has been extensively discussed in the literature, the direct linear plot and the
corresponding median parameter estimates are in principle applicable to any two-parameter equation. Results (mostly unpublished)
similar to those observed with the Michaelis-Menten equation have
been obtained in several disparate cases, such as the straight
line (A. Cornish-Bowden, unpublished), the single-exponential decay
curve (24,25) and the two-parameter Adair equation (L. Wong, L.
Endrenyi and A. Cornish-Bowden, unpublished). There is therefore
good reason to believe that the direct linear plot may provide a
general robust approach to the fitting of two-parameter models.
Unfortunately, however, generalizations of the method would be
barely practicable for three-parameter models and quite impractic-
able for models with four or more parameters, because of the enormous numbers of intersections in three or more dimensions that
would have to be considered. As such models are of great and increasing importance in biochemistry and pharmacology, there is a
need for robust methods that are not restricted to two-parameter
models.

5. EXPERIMENTAL ERROR IN ENZYME KINETICS

Experimental studies of the type of error that occurs in
enzyme kinetic work (26-29) have provided little evidence of substantial deviations from normally distributed errors. This result
must be interpreted with caution, however, for several reasons.
First, all of the experiments cited were artificial, in the sense
that they were untypically concerned with error for its own sake.
Second, the size of sample required to establish non-normality
beyond reasonable doubt is much larger than the number of replicate
observations that it is practicable to make of an enzyme-catalysed
rate under constant conditions. Third, non-normality that is so
slight as to be barely detectable may be nonetheless severe enough
to vitiate the optimal properties of the method of least squares.

A more positive conclusion from the experiments cited is that
weighting is much more difficult and hazardous than was realized
before the experiments were done. Although equal weighting of v_i
values, with the implication of simple errors, has often been
recommended [*e.g.* (30)], it appears that it is exceptional for this
kind of weighting to be correct [though an example may be found in
the early study by Burk (4)]. On the whole, the experimental
results have been closer to relative errors, but several results
have suggested more complicated behaviour than implied by either
of the extreme assumptions.

Askelöf, Korsfeldt and Mannervik (28) have given one possible interpretation of complex error behaviour. They suggest that the variance of v may be given by an expression of the following kind:

$$\sigma^2(v) = Kv^\alpha \tag{9}$$

in which the exponent α is not constrained to be 0 (simple errors) or 2 (relative errors) but can have any value. I find it hard to imagine a physical model that would give rise to such a dependence and prefer to regard $\sigma^2(v)$ as the sum of independent "simple" and "relative" components:

$$\sigma^2(v) = \sigma_0^2 + v^2\sigma_2^2 \tag{10}$$

With either equation, an estimate of α or of σ_0^2/σ_2^2 permits appropriate weights for regression to be calculated. Mannervik, Jakobson and Warholm (31) have discussed methods of estimating α from the experimental data to be analysed, and it is likely that similar methods can be developed for σ_0^2/σ_2^2. The subject is, however, in its infancy, and much work needs to be done.

Although both Eqs. [9] and [10] permit a weighting assumption intermediate between simple errors and relative errors, e.g. with $\alpha=1$ in Eq. [9] or σ_0/σ_2 (which has the dimensions of v) within the range of experimentally observed v values in Eq. [10], the compromises implied by these two choices are very different, especially in the way they weight very small values of v. This is illustrated in Figure 2, where it may be seen that Eq. [10] resembles simple errors when v is small and relative errors when v is large, whereas the behaviour of Eq. [9] is the reverse. Thus, unless v is restricted to a narrow range, one should not expect the two compromises to yield essentially the same results. Moreover, people who believe in Eq. [10] may well feel that Eq. [9] embodies the worst features of both simple and relative errors and the good features of neither (and *vice versa*, of course).

6. WEIGHTS CALCULATED FROM THE LOCAL SCATTER

There is an alternative method of determining weights that is superficially very attractive, because it seems to avoid all of the complexities I have been discussing while remaining experimentally convenient. Provided that all observations are made in replicate, one can readily calculate the sample variance of the observations in each group of replicates and use its reciprocal as a weight for each of these observations (32). Unfortunately there is a fallacy in this approach: the variance we require for weighting is the local population variance, but the local sample variance is a highly erratic measure of this unless the number of replicate observations in each group is large. Monte Carlo studies of this method suggest that one requires *at least five* replicate observations in

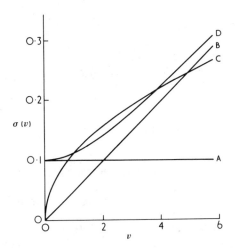

FIGURE 2: Dependence of $\sigma(v)$, the standard deviation of v, on v, under various assumptions. A: "Simple errors", *i.e.*, uniform standard deviation; B: "Relative errors", *i.e.*, uniform coefficient of variation; C: Calculated from Eq. [9] with $\alpha=1$; D: Calculated from Eq. [10] with $\sigma_0/\sigma_2=2$ (in the same units as v). The units of v and $\sigma(v)$ are arbitrary.

each group for it even to become acceptable, and even then it is less satisfactory than a weighting scheme based on a realistic assumption about the dependence of the variance on the measured quantity (26).

Although the method referred to above, suggested by Mannervik, Jakobson and Warholm (31) for estimating α for insertion into Eq. [9], is also based on estimates of local scatter, it is not subject to the same objections. First, it explicitly includes a requirement that each estimate of local variance be derived from at least five observations (not necessarily replicate observations). Second, the information is used not for the calculation of individual weights that can fluctuate wildly from group to group, but to calculate a single value of α that provides a reasonable summary of the total error behaviour.

7.   OTHER ROBUST METHODS

Probably the most widely known robust alternative to least-squares fitting is least-absolute-values fitting, in which one seeks to minimize the sum of absolute deviations from the fitted line, SA, which may be defined in the case of the Michaelis-Menten equation as follows:

$$SA = \sum_{i=1}^{n} w_i^{\frac{1}{2}} |v_i - Vs_i / (K_m + s_i)| \qquad [11]$$

Although such expressions are often given without weights (with the implication that $w_i = 1$ for all i), weights are appropriate for the same reason as in least-squares regression. They are shown in Eq. [11] as $w_i^{\frac{1}{2}}$ so that the same w_i apply to both SS and SA.

Least-absolute-values fitting is open to theoretical objections that suggest that it may sometimes give very poor results, so that it should be used cautiously, if at all (11). Nonetheless, in limited investigation of it as a method for fitting the Michaelis-Menten equation (9), it provided excellent protection against outliers, though poor protection against incorrect weighting (though not as poor as in the least-squares method). Over the whole range of conditions considered by Cornish-Bowden and Eisenthal (9), the median estimates defined by Eqs. [5] and [6] proved more uniformly robust and were preferred for that reason. Minimization of SA is in principle applicable to problems of more than two parameters, however, and so it might be an attractive choice if reliable method can be developed for estimating weights from internal evidence in the data. In any case, their appearance in Eq. [11] as $w_i^{\frac{1}{2}}$ rather than w_i indicates that less accurate information about their values will suffice for least-absolute-values fitting than is required for least squares.

Another robust procedure that has been strongly advocated in recent years is least-squares regression with weights that are iteratively adjusted in accordance with the magnitudes of the deviations found in the previous iteration (11). By assigning weights close to their ordinary least-squares values to observations with small or average residual deviations, and much smaller (and in extreme cases zero) weights to those with very large residuals, one may hope to combine robustness against outliers with efficiency approaching that of least squares when outliers are absent. There are numerous schemes for doing this, of which one recommended by Mosteller and Tukey (11) is called *biweight regression*. In the first iteration one uses the same weights w_i as in least squares. Subsequently one defines scaled deviations u_i in terms of the current fit as follows:

$$u_i = w_i^{\frac{1}{2}} (v_i - \hat{v}_i) / cS \qquad [12]$$

where $\hat{v}_i$ is the rate calculated from the best available parameter estimates (*i.e.*, those from the previous iteration), w_i is the ordinary weight as used earlier in this article, c is a constant and S is a robust measure of the scale of the deviations. In the discussion that follows I shall assume that c=6 and that S is the median absolute value of $w_i^{\frac{1}{2}} (v_i - \hat{v}_i)$. Other choices are possible,

however, and at present only limited information is available about
which choices are likely to be best. The next iteration is carried
out as in least squares but with modified weights W_i replacing w_i
in Eq. [3], defined as follows:

$$W_i = \begin{cases} w_i(1-u_i^2)^2 & \text{for } u_i^2 \leq 1 \\ 0 & \text{for } u_i^2 > 1 \end{cases}$$

[13]

This gives weights approximately the same as the ordinary weights
w_i to observations with small or moderate residuals, but small or
zero weights to observations with large residuals. The values of
the W_i are similarly adjusted in subsequent iterations until the
results are self-consistent, *i.e.*, until no further change takes
place.

The estimates whose variances are given in the columns headed
BW of Table 1 were obtained by this method, with c and S defined as
above and $w_i=1/v_i^2$ to incorporate the assumption of relative errors.
It may be seen that the method is substantially more efficient
than the median method when the least-squares assumptions are true,
and that it provides excellent protection against outliers (though
no more than with the median method). On the other hand it provides no protection against an incorrect weighting assumption.

8.   CONCLUSIONS

The direct linear plot and the median estimates derived from
it remain the most robust method currently available for fitting
data to the Michaelis-Menten equation. It compares acceptably with
least-squares fitting even when the least-squares assumptions are
true, and in addition provides excellent protection against failure
of these assumptions, specifically against the presence of outliers
and against an incorrect assumption about the dependence of the
variance of the rates on the rates themselves. Other robust methods,
such as least-absolute-values and biweight fitting, provide as good
or better protection against outliers, but they require correct
weights to give adequate results. Nonetheless, further investigation
of these methods is desirable because, unlike the direct linear
plot, they can, in principle, be applied to problems much more complex than the Michaelis-Menten equation. The immediate need is
for the development of reliable methods for assigning weights on
the basis of information contained within the data to be anlysed.

ACKNOWLEDGEMENTS

The most recent work referred to in this article was carried
out in the laboratory of Dr. L. Endrenyi at the University of
Toronto and supported by Research Grant No. 1851 from the North
Atlantic Treaty Organization.

REFERENCES

1. Dowd, J.E. and Riggs, D.S. (1965) J. Biol. Chem. 240, 863-869.
2. Lineweaver, H. and Burk, D. (1934) J. Amer. Chem. Soc. 56, 658-666.
3. Lineweaver, H., Burk, D. and Deming, W.E. (1934) J. Amer. Chem. Soc. 56, 225-230.
4. Burk, D. (1934) Ergebnisse der Enzymforschung, 3, 23-56.
5. Johansen, G. and Lumry, R. (1961) C.R. Trav. Lab. Carlsberg 32, 185-214.
6. Wilkinson, G.N. (1961) Biochem. J. 80, 324-332.
7. Watts, D.G. (1980) In "Kinetic Data Analysis: Design and Analysis of Enzyme and Pharmacokinetic Experiments" (L. Endrenyi, ed.) pp. 1 - 24 , Plenum, New York.
8. Eisenthal, R. and Cornish-Bowden, A. (1974) Biochem. J. 139, 715-720.
9. Cornish-Bowden, A. and Eisenthal, R. (1974) Biochem. J. 139, 721-730.
10. Cornish-Bowden, A. and Eisenthal, R. (1978) Biochim. Biophys. Acta, 523, 268-272.
11. Mosteller, F. and Tukey, J.W. (1977),   "Data Analysis and Regression", pp. 333-379, Addison-Wesley, Reading, Massachusetts.
12. Atkins, G.L. and Nimmo, I.A. (1980) In "Kinetic Data Analysis: Design and Analysis of Enzyme and Pharmacokinetic Experiments" (L. Endrenyi, ed.) pp. 121-135   , Plenum, New York.
13. Cornish-Bowden, A., Porter, W.R. and Trager, W.F. (1978) J. Theoret. Biol. 74, 163-175.
14. Cornish-Bowden, A. (1979),   "Fundamentals of Enzyme Kinetics", pp. 203-207, Butterworths, London and Boston.
15. Porter, W.R. and Trager, W.F. (1977) Biochem. J. 161, 293-302.
16. Henderson, P.J.F. (1978) In "Techniques in Protein and Enzyme Biochemistry", Part II (H.L. Kornberg, J.C. Metcalfe, D.H. Northcote, C.I. Pogson and K.F. Tipton, eds.), pp. 1-43, Elsevier/North-Holland, Amsterdam.
17. Atkins, G.L. and Nimmo, I.A. (1975) Biochem. J. 149, 775-777.
18. Ljones, T. and Flatmark, T. (1974) FEBS Lett. 49, 49-52.
19. Brooks, C.J.W. and Smith, A.G. (1975) J. Chromatog. 112, 499-511.
20. Debnam, E.S. and Levin, R.J. (1975) J. Physiol. (Lond.) 252, 681-700.
21. Simon, J.R., Atweh, S. and Kumar, M.J. (1976) J. Neurochem. 26, 909-922.
22. Browne, C.A., Campbell, I.D., Kiener, P.A., Phillips, D.C., Waley, S.G. and Wilson, I.A. (1976) J. Mol. Biol. 100, 319-343.
23. Lenk, W. (1976) Biochem. Pharmacol. 25, 997-1005.
24. Nimmo, I.A. and Atkins, G.L. (1979) Analyt. Biochem. 94, 270-273.
25. Endrenyi, L. and Tang, H.Y. (1980) Comput. Biomed. Res. 13, 430-436.

26. Storer, A.C., Darlison, M.G. and Cornish-Bowden, A. (1975)
    Biochem. J. 151, 361-367.
27. Siano, D.B., Zyskind, J.W. and Fromm, H.J. (1975) Arch. Biochem.
    Biophys. 170, 587-600.
28. Askelöf, P., Korsfeldt, M. and Mannervik, B. (1976) Eur. J.
    Biochem. 69, 61-67.
29. Nimmo, I.A. and Mabood, S.F. (1979) Analyt. Biochem. 94, 265-269.
30. Cleland, W.W. (1967) Adv. Enzymol. 29, 1-32.
31. Mannervik, B., Jakobson, I. and Warholm, M. (1979) Biochim.
    Biophys. Acta 567, 43-48.
32. Ottaway, J.H. (1973) Biochem. J. 134, 729-736.

ROBUST ALTERNATIVES TO LEAST-SQUARES

CURVE-FITTING

G.L. Atkins and I.A. Nimmo

University of Edinburgh Medical School
Department of Biochemistry
Teviot Place, Edinburgh   EH8 9AG
Scotland

ABSTRACT

The methods considered are a median method for fitting a straight line (which is mathematically identical to the direct linear plot), M-estimation and the jackknife technique. Their capacities to yield unbiased estimates of the parameters of the curve being fitted are discussed, and some of their limitations pointed out.

1.   INTRODUCTION

Biochemical experiments are often analyzed by fitting to the data an equation such as a straight line, the rectangular hyperbola of Michaelis and Menten, a single exponential or sums of exponentials. The objectives of the curve-fitting are generally: (i) to decide whether the equation really does fit the data satisfactorily; (ii) to estimate the values of the parameters characterizing the curve (*e.g.*, the slope and intercept of the straight line, or the K_m and V of the enzyme); and (iii) to calculate confidence limits for these parameters, so that values derived from different experiments can be compared statistically. All of these objectives can be achieved by conventional least-squares methods if: (i) the experimental errors in the data are normally distributed and their variances can be established; and (ii) the equation to be fitted is linear. (The term "linear" in this context means that the parameters to be estimated are linearly related to the dependent variable. For example, the slope b and intercept a of the straight line $y=a+b \cdot x$ are both linearly related to y, the dependent variable. On the other hand, in the Michaelis-Menten equation $v=V \cdot s/(K_m+s)$

121

where v is the initial velocity of the reaction at substrate concentration s, V is linearly related to v but K_m is not.)

In practice, of course, many equations of interest to biochemists (for instance, exponentials and rectangular hyperbolae)
are nonlinear, and as a result the least-squares confidence limits
of the parameters are approximations of largely unknown validity.
A second common difficulty is that the form of the distribution of
the errors cannot easily be worked out (in particular, the data may
contain outliers: that is, points with more error in them than
would be expected on the basis of a Gaussian or similar distribution). Similarly, it is often considered tedious or impractical to
find out how the error variance is related to the magnitude of the
dependent variable, and thus how the individual points should be
weighted (in least-squares they should be weighted according to the
reciprocals of their variances). In either or both of these situations the least-squares estimates of the parameters themselves as
well as of their confidence limits are likely to be biased (1).

A potential solution is to use a statistical method that makes
fewer assumptions than least squares about the nature of the underlying error. Ideally such a method should give unbiased estimates
of the parameters and of their confidence limits even when the data
contain outliers, and the goodness of fit of the model should be
testable. In this article we shall describe some of the alternatives that have been considered, including a (so-called) distribution-free method, more robust variants of least-squares, and the
jackknife technique. The distribution-free principle is quite different from that of least-squares, because the median rather than
the arithmetic mean is used as the index of "average". The more
robust versions of least-squares, for their part, work by giving
progressively less weight to outlying observations. Finally, the
jackknife technique is to leave out each point in turn from the
data set, fit the model to the truncated sets thus formed, and
average the values of the parameters that result. As will trans-
pire, most of the research into these methods has been concerned
with the estimation of unbiased parameters. Little attention has
been given to the question of confidence limits, and practically
none to model-testing.

2.   DISTRIBUTION-FREE CURVE-FITTING

2.1. Direct Linear Plot

The recent interest of biochemists in distribution-free methods
of curve-fitting stems largely from the publication of the direct
linear plot by Eisenthal and Cornish-Bowden in 1974 (2,3). This
plot is described by its originators as being in "parameter space":
that is, the axes refer to the parameters to be estimated (here,

the K_m and V of an enzymic reaction) and each observation of substrate concentration and initial velocity is represented by a
straight line rather than a point. However, an inspection of the
formulae involved shows that the direct linear plot is in fact
mathematically identical to the familiar Eadie-Hofstee (v *vs.* v/s)
plot, provided that the straight line in the latter is fitted by
the median method described below. Likewise, the revised version
of the direct linear plot, in which the axes are K_m/V and 1/V
rather than K_m and V (3,4), is equivalent to the plots of Lineweaver
and Burk (1/v *vs.* 1/s) and Hanes (s/v *vs.* s) (5). In this section,
we shall describe the median method just referred to, and then
examine the extent to which its principles can be applied to other
more complex curves.

2.2. Straight Line

The method has been described in rigorous terms by Sen (6),
and is illustrated in Figure 1. Assume that the line to be fitted
is:

$$y = a + b \cdot x \qquad\qquad [1]$$

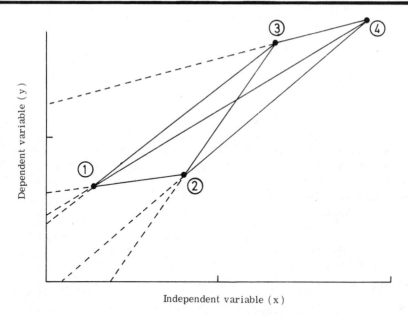

FIGURE 1: Fitting a straight line by the median method

Each pair of points is joined up, to give one estimate of the slope
($\Delta y/\Delta x$) and one of the intercept on the y axis. The 6 estimates
of each parameter are ranked, and the arithmetic means of the 3rd
and 4th taken as the best-fit values.

and, for the present, that there are no replicates. (As usual, y
is the dependent, error-containing variable, and x the fixed one.)
Each pair of points can be joined to give one estimate of the slope,
b, and one of the intercept, a; if there are n points there are
n(n-1)/2 such estimates. For example, for the i-th and j-th points:

$$b_{ij} = \frac{y_i - y_j}{x_i - x_j} \qquad [2]$$

and

$$a_{ij} = y_i - b_{ij} \cdot x_i = \frac{x_i y_j - x_j y_i}{x_i - x_j} \qquad . \qquad [3]$$

The estimates of b and a are then ranked in order of increasing
magnitude, and their medians (*not* means) taken as the best ones.
(If the number of estimates is even, the arithmetic mean of the
two median values is used.) Now consider what happens if replicates
are present. They cannot be paired with one another as the values
of the slope (infinity) and intercept (minus infinity) are nonsensical. Consequently, the total number of estimates to be ranked has
to be reduced, for example by $n \cdot q(q-1)/2$ if there are q replicates
of each of n observations (5).

The method is less dependent on assumptions about the nature
and distribution of random experimental error than, say, least
squares. It requires the errors in y to be uncorrelated with one
another and as likely to be positive as negative; and the values
of x to be in their correct rank order. There is no need to weight
the points, and the presence of outliers should make little difference to the median estimates (because the median is not affected
by how far on either side of it the other estimates are scattered).
Confidence limits can also be calculated for the slope, but only
if the errors in y come from a single distribution (1,6): an important limitation in practice, because it implies, *inter alia*, that
the standard deviation of y is constant.

In fact Sen (6) discussed the estimation of slopes but not
intercepts. However, one can transform the original equation for
the line (Eq. [1]) into:

$$\frac{y}{x} = b + a \cdot (1/x)$$

and then estimate the slope, a, of this regression; *e.g.*:

$$a_{ij} = \frac{(y_i/x_i) - (y_j/x_j)}{(1/x_i) - (1/x_j)} = \frac{x_i y_j - x_j y_i}{x_i - x_j}$$

which is the same result as before. The transformation would seem
to be valid in that the dependent (y/x) and independent $(1/x)$
variables still fulfill the requirements given above for a and b
to be unbiased. However the confidence limits of a and b cannot
both be valid, because if the errors in y all come from the same
distribution then those in y/x cannot (and *vice versa*).

It is instructive to consider at this juncture why the original
version of the direct linear plot (2) tends to underestimate K_m and
V, whereas the revised version (4) does not. In the former, v is
in effect plotted against v/s, so the slope of the line is given by:

$$b_{ij} = (-K_m)_{ij} = \frac{v_i - v_j}{(v_i/s_i)-(v_j/s_j)} = \frac{(v_i-v_j)}{v_i s_j - v_j s_i} \cdot s_i s_j \qquad [4]$$

whereas in the latter (which is equivalent to plotting $1/v$ against
$1/s$) it is:

$$b_{ij} = (K_m/V)_{ij} = \frac{(1/v_i)-(1/v_j)}{(1/s_i)-(1/s_j)} = \frac{s_i s_j (v_i-v_j)}{v_i v_j (s_i - s_j)} \qquad [5]$$

The denominator of Eq. [4], unlike that of Eq. [5], contains a
difference term in v, the error-containing variable, and not just
in s, the independent variable. Consequently in Eq. [4], increasing
error in v could change what should have been a substantial positive
denominator into a very small and finally a negative one, with the
result that the estimate of K_m calculated from it would grow very
large, and then flip "through infinity" to become large and negative. Put another way, the stipulation that the values of x be in
their correct rank order has not been met. There is a possible
solution, however, because the effect makes both $(K_m)_{ij}$ and V_{ij}
negative instead of infinitely large. It is simply to transpose
all such "double negative" values from the bottom to the top end
of the rank order (3,4).

The performance of the median method for fitting straight lines
has been examined in detail in the context of the direct linear
plot by Cornish-Bowden and his colleagues (2,7), and is described
elsewhere in this book (3). In short, it is somewhat less efficient
than least squares if the latter's assumptions hold, but the reverse
is true when least squares is used with the wrong weights or when
outliers are present in the data. Its confidence limits are only
approximate, for the theoretical reason given above, but are about
as good as those derived from least squares.

We are in the process of testing the median method, on two
sorts of simulated experimental data. The first has random error
in y but not x (the classical regression situation), whereas in
the second the error is in both variables: as, for example, when a

new analytical method is being compared with a reference one (8).
So far the median method has come out of the first test well, because it has always given unbiased estimates of a and b, even at
the highest error level employed (relative error of 10%). However,
whether or not the confidence limits are realistic remains to be
seen.

The analysis of the second set of data has confirmed that the
method underestimates b when both variables are subject to error
(Table 1). Nevertheless, it fared no worse (in terms of bias and
precision) than either classical least squares or, indeed, Deming's
least squares (9), which is supposed to allow for error in x and y.
Hence one must conclude that at present there is no satisfactory
statistical treatment of such method-comparison data.

TABLE 1: Slopes and intercepts estimated by using
          different linear regressions

Method	Intercept (a) ± SD	Slope (b) ± SD
Constant absolute error (AN)		
Linreg	217 (216) ± 31*	0.971 (0.967) ± 0.052*
Deming	179 (178) ± 31*	1.034 (1.036) ± 0.053*
Median	214 (214) ± 29*	0.978 (0.970) ± 0.048*
Constant relative error (RN)		
Linreg	220 (220) ± 52*	0.954 (0.975) ± 0.100*
Deming	165 (159) ± 58*	1.053 (1.076) ± 0.109*
Median	219 (222) ± 54*	0.963 (0.963) ± 0.105

Values are mean (median in parentheses) ± SD of 50 estimates.

The straight line was simulated with a=200 and b=1.000, and
comprised 51 points (400≤x≤800); cf. (8). Error types: AN: Gaussian
error of constant absolute magnitude, with SD of x=20 and of y=40;
RN: Gaussian error of constant relative magnitude, with coefficient
of variation of x=5% and of y=10%.

Mathematical methods: Linreg: unweighted least-squares, with
x the independent variable; Deming: Deming's method (9); Median:
the median method (see text).

*P < about 0.05 that there is no difference between the median
and its true value (22).

2.3. Application to Other Situations

The median method just described can in principle be generalized as follows (10). Suppose the equation to be fitted is $y=f(\theta,x)$, where θ is a set of p parameters. Set up p simultaneous equations with the parameters as the unknowns by taking p of the n data points:

$$f(\theta,x_1) - y_1 = 0$$

$$f(\theta,x_2) - y_2 = 0$$

$$- - - - - - - -$$

$$f(\theta,x_p) - y_p = 0$$

and solve them to get one set of estimates of the parameters. Repeat with another set of p data points, and continue until all N combinations of points have been used:

$$N = \frac{n!}{p!\,(n-p)!} \quad .$$

Rank all N estimates of each parameter and take the median as the best one. (Note that estimates calculated from replicates have to be discarded, as before.)

Some curves of biochemical interest that can in theory be fitted in this way include the following.

1.   The integrated Michaelis-Menten equation (11), the usual form of which may be written as:

$$V{\cdot}x = y - K_m{\cdot}\ln(1-y/s_0)$$

where y is the concentration of product at time x, s_0 is the initial concentration of substrate, and the parameters are V (θ_1) and K_m (θ_2). Putting $z=\ln(1-y/s_0)$, the simultaneous equations are:

$$\theta_1{\cdot}x_1 + \theta_2{\cdot}z_1 - y_1 = 0$$

$$\theta_1{\cdot}x_2 + \theta_2{\cdot}z_2 - y_2 = 0$$

which can easily be solved for θ_1 and θ_2.

2.   A single exponential (12,13), which may be written as:

$$y = \theta_1{\cdot}\exp(-\theta_2{\cdot}x) \quad .$$

Taking logarithms of both sides linearizes the equation to the straight-line form:

3.   A double exponential:

$$y = \theta_1 \cdot \exp(-\theta_2 \cdot x) + \theta_3 \cdot \exp(-\theta_4 \cdot x)$$

In this case, sets of four simultaneous equations have to be set up, and taking logarithms does not linearize them. Consequently, they have to be solved by an iterative procedure, which is bound to be extremely expensive in terms of computer time.

4.   The Hill equation, a Michaelis-Menten plus a linear term, the sum of two Michaelis-Menten terms, and some protein-ligand binding functions (10).

The ability of the median method to produce unbiased estimates of the parameters of some of these curves has been checked by analysing simulated experimental data. The parameters were unbiased in one instance, the single exponential (12,13), but biased in all the others (integrated Michaelis-Menten equation, protein-ligand binding function, and Michaelis-Menten plus linear terms (10)). The characteristic feature of the unbiased parameter seems to be that the denominator of the expression from which it is calculated contains different terms in the independent variable only: for example, $(x_i - x_j)$ in the case of the straight line (Eq.'s [2] and [3]) and $(1/s_i) - (1/s_j)$ in that of the Lineweaver-Burk plot (Eq. [5]). Biased parameters, in contrast, are derived from expressions whose denominators have difference terms in the dependent variable, such as $(v_i s_j - v_j s_i)$ for the Hanes plot (Eq. [4]), and $x_i \cdot \ln(1 - v_j/s_0) - x_j \cdot \ln(1 - v_i/s_0)$ for the integrated Michaelis-Menten equation. This generalization should help one decide which functions are likely to be fitted successfully by the median method. Closer inspection of the equations for the parameters should also help one decide whether any bias in those functions that are not fitted success- fully could be reduced or even eliminated by the modification which treats "double negative" estimates as large positive ones (see [4] and above). For example, this trick abolishes the downwards bias in the values of K_m and V derived from the integrated Michaelis-Menten equation as well as the Eadie-Hofstee plot (10).

3.   ROBUST VERSIONS OF LEAST SQUARES

In standard least-squares curve-fitting one minimizes $\Sigma w_i (y_i - y)^2$, where y_i and y are, respectively, the observed and cal- culated values of the i-th point, and w_i is the statistical weight given to it (*i.e.*, the reciprocal of its variance). The disadvan- tage with this formula is that if outliers are present they will bias the curve unduly in their direction. (Note that least-squares theory allows for some large deviations; outliers occur when the proportion is greater than expected on the basis of a Gaussian distribution.) In the more robust variants of least squares less weight is given to the points that are well away from a provisional fit of the curve. Their principle can be illustrated by considering how one estimates the location ("average") of a sample.

The least-squares estimate of average is the arithmetic mean, $\bar{x}$, which is derived by minimizing $\Sigma(x_i-\bar{x})^2$; $\bar{x}=\Sigma x/n$. A more robust estimate is obtained by "Winsorizing" (14). One ranks the observations $x_1,\ldots,x_n$ and then calculates the mean as $(g \cdot x_{g+1}+x_{g+1},\ldots,$ $+x_{n-h}+h \cdot x_{n-h})/n$, having "Winsorized" the g lowest and h highest points: that is, substituted for their actual values those of the points (g+1) and (h+1) in from each end. The difficulty is, of course, to choose suitable values for g and h, but it turns out that the answer depends on the proportion of outliers in the data (14). Thus if g=h=0, the Winsorized mean is identical to the arithmetic mean, and if $g=h=\frac{1}{2}(n+1)$ or $\frac{1}{2}n$ (depending on whether n is odd or even) it is equivalent to the median. Alternatively one can calculate a "trimmed" mean, by neglecting the first T% and the last T% of the points in the rank order (typically T=10-25%). A simple illustration of these possibilities is given in Table 2.

In the field of curve-fitting the general approach is rather similar to Winsorizing or trimming the residuals and then minimizing their sums of squares; different procedures treat the residuals in different ways. For example, Anscombe (15) has suggested one should minimize:

$$\Sigma_{(1)}d^2 + \Sigma_{(2)}k_1(2d-k_1) + \Sigma_{(3)}k_1(2k_2-k_1)$$

where $d=(|y_i-\hat{y}|)/sp$, sp is a scale parameter (sp>0), k_1 and k_2 are chosen numbers $(k_2>k_1>0)$, $\Sigma_{(1)}$ denotes summation of the points for which $d \leq k_1$, $\Sigma_{(2)}$ denotes summation of the point for which $k \leq d \leq k_2$,

---

TABLE 2:   Some ways of calculating "average"

Consider the following series of points:

Point No.	1	2	3	4	5	6	7
Value	10	20	30	40	60	80	110

Arithmetic mean=350/7=50.
Median=4th point=40.
Winsorized mean (with g=h=1): set point 1 to 20 and point 7 to 80, therefore, mean=330/7=47.
Approx. 25% trimmed mean: drop points 1 and 7, therefore, mean=230/5=46.

and $\Sigma_{(3)}$ denotes summation of the remaining points, for which $d>k_2$.
This is loosely equivalent to Winsorizing and trimming the residuals;
or to minimizing a weighted sum of squares, where the weights,
treated as though they were fixed, are:

$$w_i = 1 \qquad \text{if} \quad d \leq k_1$$

$$w_i = k_1/d \qquad \text{if} \quad k_1 < d \leq k_2$$

and $\qquad w_i = 0 \qquad \text{if} \quad d > k_2 \quad .$

Alternatively, if one wishes to be gentler and omits the third
summation $(\Sigma_{(3)})$, one is left with the formula for Huber's M-estimation (14,16).

So far as we are aware, only a few attempts have been made
to evaluate robust methods of curve-fitting in a biochemical context.
One of them was by Wahrendorf (17), who described how M-estimation
may be used to fit a binding isotherm to both real and artificial
data, and demonstrated that the results were relatively resistant
to outliers. (In terms of the above formulae, he set the robustness
constant, k_1, to either 0.5 or 1.5; k_2 was of course infinite.)
Cornish-Bowden has shown elsewhere in this book (3) how to fit the
Michaelis-Menten equation by biweight regression; he found that
this method also gave good protection against outliers, but not
against incorrect weighting.

The problem of deriving reliable confidence limits for parameters estimated by robust regression does not seem to have been
resolved yet. One suggestion (15) is that the jackknife technique
could be helpful here. This technique is therefore the subject of
the next section.

4. THE JACKKNIFE TECHNIQUE

The jackknife technique was devised by Quenouille in 1949 as
a way of reducing the bias in the estimate of a parameter. The
term itself was actually coined a decade later by Tukey, who also
suggested how the confidence limits of the estimate could be derived.
Miller (18) has written a lucid review of the whole topic, in which
he explains how to use the jackknife, and discusses its limitations.

The technique itself is quite simple in practice, the steps
involved in its first-order version being as follows (18). (i) The
parameter to be estimated, θ, is first found from the whole set of
data (of n points) in the usual way (e.g., a nonlinear least-squares
regression, or a robust alternative); let this estimate be $\hat{\theta}$.
(ii) The data are then divided into g groups, each of size h (n=g·h);
h can be unity, in which case g=n. (iii) The i-th group of data

is dropped, and θ re-estimated from the remaining (i-1) groups; call this estimate $\hat{\theta}_{-i}$. (iv) The process is repeated, leaving out each of the other (i-1) groups in turn. (v) Define $\tilde{\theta}_i = g\hat{\theta} - (g-1)\hat{\theta}_{-i}$; $\tilde{\theta}_i$ is chosen so that $\hat{\theta}$ is the weighted mean of $\tilde{\theta}_i$ and $\hat{\theta}_{-i}$, and was termed a "pseudo-value" by Tukey. (vi) The final estimate, $\tilde{\theta}$, is the mean of all g pseudo-values:

$$\tilde{\theta} = \Sigma\tilde{\theta}_i/g = g\hat{\theta} - [(g-1)/g]\,\Sigma\hat{\theta}_{-i} \qquad [6]$$

(vii) The confidence limits of $\tilde{\theta}$ are found by assuming the pseudo-values ($\tilde{\theta}_i$) follow Student's t-distribution with (g-1) degrees of freedom.

Theoretically, $\tilde{\theta}$ should be unbiased as long as the bias in $\hat{\theta}$ (the original estimate based on all n data points) is inversely proportional to n; or, in mathematical notation, if $E(\hat{\theta})=\theta+a_1/n+ 0(1/n^2)$, where $E(\hat{\theta})$ is the expectation of $\hat{\theta}$ and $0(1/n^2)$ means terms in n^2 and above can be ignored (18). Should the term in n^2 be appreciable, however, so that $E(\hat{\theta})=\theta+a_1/n+a_2/n^2+0(1/n^3)$, then the second-order jackknife ought to be used instead. In this case the final estimate, now termed $\tilde{\theta}^{(2)}$, is given by (18):

$$\tilde{\theta}^{(2)} = \frac{1}{2}\left[n^2\hat{\theta}-2(n-1)^2\left(\frac{1}{n}\,\Sigma\hat{\theta}_{-i}\right)+(n-2)^2\left\{\frac{2}{n(n-1)}\,\sum_{i<j}\hat{\theta}_{-ij}\right\}\right]$$

where $\hat{\theta}_{-ij}$ denotes the estimate obtained after the i-th and j-th groups of data have been removed.

The size (h) of the groups has to be decided by the investigator. The special case in which h=1 and g=h has the advantage that the groups are not chosen arbitrarily. Its drawback is that it may be impossibly expensive in terms of computer time, so that the groups have to be larger. For example, Cornish-Bowden and Wong (19) determined the initial velocity of an alcohol dehydrogenase reaction at six different concentrations of NAD^+ and six of ethanol, and then divided the data into six groups using a Latin square design.

The investigator also has to decide whether or not to transform the parameter θ when using the jackknife, to help reduce any bias in the results (18). For example, $\log s^2$ and $\tanh^{-1}r$ are preferable to the original parameters s^2 (variance) and r (correlation coefficient), because the former cannot be negative and the latter must lie between -1 and +1 (18). In the same way, it would seem sensible to transform K_m into $\log K_m$ when the Michaelis-Menten equation is being considered (19,20). The values of $(\tilde{K}_m)_i$ would then have to be positive, and Eq. [6] would become:

$$\log(\tilde{K}_m) = \Sigma\log(\tilde{K}_m)_i/g = g\cdot\log(\hat{K}_m) - [(g-1)/g]\Sigma\log(\hat{K}_m)_{-i} \quad .$$

The use of the jackknife in enzyme kinetics has been advocated by Cornish-Bowden and Wong (19), on the grounds that it may help overcome the computational problems associated with large, ill-conditioned matrices, and that it gives reasonable estimates of the precisions of the calculated parameters. They fitted the rate equation for their alcohol dehydrogenase data by nonlinear least-squares regression, and built the answers into a first-order jack-knife with a logarithmic transformation. In their view, the jack-knifed estimates of the rate constants and their precisions were realistic, whereas several of those given by least-squares alone were not. However, Duggleby (20) has been more sceptical, princi-pally because he analyzed simulated data for a Michaelis-Menten reaction and found that least-squares alone actually out-performed the jackknife.

We also have data that make us side with Duggleby. Essentially we simulated a Michaelis-Menten reaction for which $K_m=V=1$, and introduced into the initial velocities (v) errors that were either non-Gaussian but of constant SD (data sets AO), or Gaussian with a constant relative magnitude (data sets RN). The equation was fitted to the data by the revised version of the direct linear plot, in which the axes are scaled in K_m/V and $1/V$ (4), and by unweighted least squares (21). A first-order jackknife was also employed in conjunction with both these methods, the group size being chosen as unity; K_m was not transformed during the procedure, so the relevant calculations are summarized by Eq. [6] with K_m and V replacing θ and g=n=12.

The results (Table 3) show that overall the jackknife estimates of K_m were worse than the original ones, because they included far more "failures" (*i.e.*, final values which were less than 0.25 or greater than 4.0, or undetermined because the mathematical method failed to converge). The only successful application of the jack-knife was when the data with relative error in them were analyzed by unweighted nonlinear least-squares; in this instance both the accuracy and precision of K_m were improved.

It seems, therefore, that the case for the jackknife is (at best) not proven. The technique was introduced originally to reduce bias but, so far, it has failed to do this satisfactorily in the biochemical situation. Its other attribute, the provision of reli-able standard errors for the parameters being estimated, has yet to be tested properly; however, first impressions have been favour-able (19,20). It is apparent that, as Miller (18) has pointed out, the jackknife has still to be fully explored: for example, the technique is not a device for correcting outliers. The effect of outliers has not been investigated theoretically. Similarly, the effect of transforming in different ways the parameters to be esti-

TABLE 3: Reliability of K_m estimated with and
without first-order jackknife

| Direct linear plot | | Unweighted least squares | |
with jackknife	without jackknife	with jackknife	without jackknife
Absolute error with outliers (AO)			
1.242±0.686	1.119±0.548	0.940±0.482	1.137±0.513
1.061, 312	1.018, 4	0.853, 96*	1.061, 3*
Gaussian-distributed relative error (RN)			
1.027±0.248	1.036±0.178	1.005±0.157	1.044±0.237
1.003, 287	1.024, 0*	0.998, 0	1.010, 0

Values are mean ±SD (upper row) and median, number of failures
(lower row) for 500 simulated experiments. An experiment "failed"
if the mathematical method did not converge or the K_m did not lie
between 0.25 and 4.0; it was left out when the means &c. were
calculated.

Perfect data: $K_m=V=1$; s=0.2,0.4,0.6,0.8,1.0,1.25,1.5,1.75,2.0,
2.5,3.0,4.0; v calculated from Michaelis-Menten equation.

Error types: AO: Gaussian error in v of constant absolute
magnitude, SD=0.10, with 10% outliers, SD=0.20; RN: Gaussian error
in v of constant relative magnitude, coefficient of variation = 10%.

Mathematical methods: Revised direct linear plot (4) and unweighted nonlinear least-squares (23), both with and without first-
order jackknife (18).

*P< about 0.05 that there is no difference between the median K_m
and the true value of 1.00 (22).

mated is not known.

5. CONCLUSIONS

1. Most of the recent work on improving the traditional least-
squares method of curve-fitting has concentrated on finding ways of
making unbiased estimates of the characteristic parameters. Less
attention has been paid to the confidence limits of the parameters
or the design of the experiment, and practically none to testing

the goodness of fit of the curve. This emphasis, while understand-
able given the nature of the problems, is unfortunate in that the
consumer usually wants to know, first, whether a particular model
fits his data, and then whether its parameters are different from
those derived in another experiment.

2.   The distribution-free (median) method seems at present to be
suitable for fitting certain sorts of straight lines. Its main
advantage is that the data do not have to be weighted, as this is
usually done either by default or guesswork. It gives confidence
limits that are reasonable approximations, and is nearly as efficient
as least-squares when there are 8 or more data points (3). It would
be interesting to see whether the performance of the method could be
improved by taking a Winsorized or trimmed mean of the ranked estimates of the parameters, rather than the median.

3.   Robust methods have scarcely been tested with biochemical data,
but preliminary trials seem promising, at least as far as protection
against outliers is concerned. It is not clear how the robustness
coefficients or scale parameters should be chosen, partly because
little is known about the precise form of the errors expected in
the data (21). It might be possible to find reliable confidence
limits for the parameters by combining a robust method of regression
with the jackknife technique.

4.   The jackknife technique does not appear to be a way of reducing
the bias in parameter estimates. It may however be useful in the
derivation of confidence limits.

REFERENCES

1.   Cornish-Bowden, A., Porter, W.R. and Trager, W.F. (1978) J.
        Theor. Biol. 74, 163-175.
2.   Eisenthal, R. and Cornish-Bowden, A. (1974) Biochem. J. 139,
        715-720.
3.   Cornish-Bowden, A. (1980) In "Kinetic Data Analysis: Design
        and Analysis and Enzyme and Pharmacokinetic Experiments"
        (L. Endrenyi, ed.) pp. 105 - 119, Plenum, New York.
4.   Cornish-Bowden, A. and Eisenthal, R. (1978) Biochim. Biophys.
        Acta 523, 268-272.
5.   Porter, W.R. and Trager, W.F. (1977) Biochem. J. 161, 293-302.
6.   Sen, P.K. (1968) J. Amer. Stat. Ass. 63, 1379-1389.
7.   Cornish-Bowden, A. and Eisenthal, R. (1974) Biochem. J. 139,
        721-730.
8.   Wakkers, P.J.M., Hellendoorn, H.B.A., Op De Weegh, G.J. and
        Heerspink, W. (1975) Clin. Chim. Acta, 64, 173-184.
9.   Mandel, J. (1964) "The Statistical Analysis of Experimental
        Data", Chapter 12, Interscience, New York.
10. Atkins, G.L. manuscript in preparation.
11. Cornish-Bowden, A. (1975) Biochem. J. 149, 305-312.

12. Endrenyi, L. and Tang. H.-Y. (1980) Comput. Biomed. Res. 13, 430-436.
13. Nimmo, I.A. and Atkins, G.L. (1979) Anal. Biochem. 94, 270-273.
14. Huber, P.J. (1964) Ann. Math. Stat. 35, 73-101.
15. Anscombe, F.J. (1967) J. Roy. Stat. Soc. B29, 1-52.
16. Huber, P.J. (1973) Ann. Stat. 1, 799-821.
17. Wahrendorf, J. (1979) Int. J. Bio-Med. Comput. 10, 75-87.
18. Miller, R.G. (1974) Biometrika, 61, 1-15.
19. Cornish-Bowden, A. and Wong, J.T.-F. (1978) Biochem. J. 175, 969-976.
20. Duggleby, R.G. (1979) Biochem. J. 181, 255-256.
21. Nimmo, I.A. and Atkins, G.L. (1980) In "Kinetic Data Analysis: Design and Analysis of Enzyme and Pharmacokinetic Experiments" (L. Endrenyi, ed.) pp. 309 - 315, Plenum, New York.
22. Campbell, R.C. (1967) "Statistics for Biologists", p. 34, Cambridge Univ. Press, Cambridge.
23. Wilkinson, G.N. (1961) Biochem. J. 80, 324-332.

DESIGN OF EXPERIMENTS FOR ESTIMATING ENZYME

AND PHARMACOKINETIC PARAMETERS

Laszlo Endrenyi

University of Toronto
Department of Pharmacology and
Department of Preventive Medicine and Biostatistics
Toronto, Ontario   M5S 1A8, Canada

ABSTRACT

    Principles are described for optimally estimating precise
model parameters. D-optimized designs, which minimize the volumes
of joint confidence regions for the parameters, generally replicate
p sampling points. (p is the number of parameters.) The observations are repeated at equal frequency when the estimation of all
model parameters is essential, but often in unequal proportions
if only some of the parameters are of interest. Usually, one of
the design points yields the maximal experimentally attainable
response. In the presence of constant relative errors   another
point measures the smallest possible response. The principles of
optimization are applied to simple enzyme and pharmacokinetic models,
and designs yielding the most precise parameters are described.
These designs provide guidelines, and not rules, for experimentation. In practice, a few observations could test the validity of
the model while others could be obtained close to the proposed design points. The usefulness of the strategy is demonstrated by
the substantially reduced estimating efficiency (larger parameter
variance) of frequently applied experimental designs, especially
when the number of measurements is increased.

1.   INTRODUCTION

    Quantitative (including kinetic) investigations may have a
variety of goals (1-4). They may aim at *identifying* an underlying
biological, chemical, physical, *etc*. model by assessing the validity of a related, derived mathematical model. Once a model has
been established, the main purpose of the study may turn to *estimating* its parameters. Alternatively, an investigation may intend

137

to establish the basis for *predicting the response* under selected
conditions.

The effective design and execution of experiments depends on,
and therefore should be adjusted according to their main goal.
This report will consider the principles for designing experiments
for the precise estimation of model parameters and their application
to simple enzyme and pharmacokinetic models.

Optimal estimating designs will be discussed which will minimize the joint confidence intervals of the parameters. These proposed designs should serve not as rules but as guidelines for experimentation.

One reason for this somewhat relaxed but practical view lies
in the replicating nature of the designs. Under fairly general
conditions which apply to the kinetic models being considered, the
number of optimal design points equals the number of model parameters. Ideally, these optimal design points should be repeated
when the number of observations exceeds the number of parameters.

More pragmatically, it will be suggested on considering replicated designs (Section 2.2, Figure 1) that a few observations should
test the validity of the investigated model while the additional
readings should, approximately, follow the guidelines to be presented.

2.   PROPERTIES, APPLICATIONS AND LIMITATIONS OF OPTIMAL DESIGNS

2.1. Optimality

The designs aim at yielding the altogether most precisely estimated parameters of an investigated model. This purpose can be
approached by various design criteria (5). Most frequently applied
among these is the so-called *D-optimization* criterion which selects
experimental conditions minimizing the determinant of the variance-
covariance matrix.

This criterion yields optimal parameters in the sense of minimizing the volume of their joint confidence regions. [Strictly
speaking, the statement applies to linear models or to linear approximations for nonlinear model parameters; see *e.g.* (5,6).] Corre-
spondingly, the criterion selects experimental conditions maximizing
information about the parameters. D-optimization also maximizes,
in a Bayesian interpretation, the posterior probabilities of the
parameters (7). Furthermore, it yields the maximal gain of information (8) about the parameters [ref. (9), pp. 476-477].

The designs obtained by D-optimization are independent of the selection or transformation of the model parameters.

One of the alternatives to D-optimization involves *sensitivity* analysis which evaluates the maximum variation, or sensitivity, of the model function with respect to the various parameters as measured by the corresponding derivatives. For instance, Kanyár (10) calculated optimal designs by this approach for hyperbolic and one-exponential models and Bogumil (11) described principles for designing tracer kinetic experiments on the basis of sensitivity analysis. The results often agree with those of D-optimization. However, this is not always true. For instance, the designs calculated by the two methods are different in the case of multi-exponential models (10,12).

This report will consider the application of the D-optimization criterion to the question of designing experiments which aim at the precise estimation of enzyme and pharmacokinetic parameters.

2.2. Replicated Observations for Parameter Estimation

The importance of distinguishing between the various purposes of experimentation was emphasized in the Introduction. When we assume that the main goal of the experiments to be performed is the evaluation of model parameters, then we imply that the model itself has already been defined and its validity substantially confirmed.

In such cases, it appears to be intuitively reasonable to obtain readings, repeatedly, at only a few design points which contain the most information about the parameters. For instance, the slope, and with it the intercept, of a *straight line* is evaluated most precisely if replicated measurements are made at the two end-points of the observation range (5,9) since this design provides the highest "leverage" for the estimation.

Indeed, designs most precisely estimating linear or nonlinear parameters seem to be *repeating* p design points, where p is the number of parameters (13-15). This is the situation also in the case of designs optimally estimating enzyme and pharmacokinetic parameters which will be considered in this report.

How useful is this suggestion of replicating p-point designs? Two opinions on kinetic parameter estimation seem to have emerged recently. Duggleby (16,17) suggests that such designs should be followed because they have the additional advantage of enabling the nonparametric estimation of parameters. On the other hand, Endrenyi and Dingle (18) believe that replicating designs will likely not be accepted by experimental kineticists since, in practice, rarely is complete confidence placed on the supposed form of the model. Therefore, they suggest that while the majority of the readings should be made near the optimal design points, other observations could aim at confirming the validity of the model. Obviously, I advocate this approach which will be illustrated in Figure 1.

2.3. Assumptions and Limitations

The D-optimization design criterion is based on the method of least squares and shares its assumptions and limitations. The assumptions have been repeatedly reviewed (*e.g.*, [6,9,19,20]).

We shall be particularly concerned with one of them: the requirement that either all observations should have the same error variance, or that adjustments for differing variances should be made by applying appropriate weights to the readings. The question of weighting in kinetic experiments has been discussed in the literature (21-26) and in this volume (4,27-29).

Optimal experimental designs will depend, then, on the *error behaviour* of the studies. We shall consider two limiting error conditions: (i) the case of constant error variance, and (ii) the equivalent conditions of constant relative or percentage error, or constant coefficient of variation. Optimal designs will be considered for these two error behaviours.

The construction of optimal designs implies two additional assumptions: (i) as already discussed, the form of the model is presumably known, and (ii) preliminary information is available about the magnitudes of the parameters. Such prior knowledge is usually available from early experiments or from the literature. As a rule, these guessed values deviate from the true magnitudes of the parameters or from their final, precisely estimated values on the basis of a careful design. These deviations may give rise to designs which are not quite optimal for the problem under investigation. Generally, however, the discrepancies would not yield substantial deviations from optimality.

3.   PRINCIPLES OF OPTIMAL DESIGN

3.1. Design Criteria

As already mentioned in the Introduction, the estimation of parameters follows, in principle, the identification of a model. The model (f) relates a response (y) or responses observed at n design points to some independent variable(s) $[\underline{x}'=(x_1,x_2,\ldots,x_n)]$ and p model parameters $[\underline{P}'=(P_1,P_2,\ldots,P_p)]$:

$$y = f(\underline{x},\underline{P})$$

[1]

(Vectors and matrices are underlined. A superscript $^{-1}$ refers to an inverse and ' to a transpose.)

As seen earlier, the optimal designs aim at minimizing the volume of the joint confidence region of the parameters. In the linear least-squares approximation, this is proportional to the determinant of the parameter variance-covariance matrix which can be calculated from

$$\underline{V} = (\underline{F}'\underline{WF})^{-1}\sigma^2 \qquad . \qquad [2]$$

Here σ^2 is the true variance, $\underline{F}$ is a $(p\times n)$-dimensional *design matrix* defined by its elements

$$F_{ij} = \frac{\partial f(x_i,\underline{P})}{\partial P_j} \qquad , \qquad [3]$$

and $\underline{W}$ is an $(n\times n)$-dimensional diagonal weighting matrix, the elements of which are inversely proportional to the variances of the observations (when these are assumed to be uncorrelated):

$$W_{ii} \propto 1/\sigma_i^2 \qquad , \qquad i=1,2,\ldots,n,$$

$$W_{ij} = 0 \qquad , \qquad i\neq j \qquad . \qquad [4]$$

The *design criterion*, to be applied for the precise estimation of the parameters, minimizes the determinant of their variance-covariance matrix:

$$D_V = \text{Min}\{\text{Det}(\underline{V})\}$$

$$= \text{Min}\{\text{Det}(\underline{F}'\underline{WF})^{-1}\}\sigma^2 \qquad . \qquad [5]$$

Equivalently, the *D-optimization* criterion maximizes the determinant of the *information matrix* which is proportional to the inverse of $\underline{V}$:

$$D = \text{Max}\{\text{Det}(\underline{F}'\underline{WF})\} \qquad . \qquad [6]$$

3.2. Effect of Error Variance

In the presence of constant variance, all weights (W_{ii}) are the same. Consequently, the weighting matrix $(\underline{W})$ has no effect on the optimization, and the design condition reduces to

$$D = \text{Max}\{\text{Det}(\underline{F}'\underline{F})\} \qquad . \qquad [7]$$

This criterion was first applied to nonlinear models in the pioneering study of Box and Lucas (30). St. John and Draper reviewed its application and that of the more general D-optimality (Eq. [6]) to regressions (31).

   With approximately constant relative errors, the standard deviation of an observation is proportional to the true response:

$$\sigma_i \propto f(x_i, \underline{P}) \qquad .$$
[8]

Therefore, the corresponding weight is inversely proportional to
the square of the response:

$$W_{ii} \propto [f(x_i, \underline{P})]^{-2} \qquad .$$
[9]

By substituting these weights, the design criterion [5] can be
solved.

   Alternatively, we can consider a transformation of the response
which yields a reduced design matrix $(\underline{G})$ such that (18,32)

$$\underline{G}'\underline{G} = \underline{F}'\underline{W}\underline{F} \qquad .$$
[10]

For instance, with constant relative errors, by substituting the
logarithm of the response into the simplified design criterion [7],
the same results are obtained as when the weights [9] are inserted
into the full, more complicated criterion [5] (18,32).

4.   DESIGNS FOR TRACER- AND PHARMACOKINETIC EXPERIMENTS

4.1. Response: Sums of Exponentials

   The response in many time-dependent kinetic systems can be
described by a sum of exponentials:

$$y = \Sigma_j A_j e^{-B_j t} \qquad .$$
[11]

The interpretation, for instance in terms of compartments, usually
involves a combination of first-order processes.

   Again, the optimal designs depend on the form of the response
equation. However, they do not vary with mechanistic variants of a
given equation form. It will be found that in the presented systems,
one of the design points yields always the maximal experimentally
feasible response. With constant relative errors, another sampling
point will measure the smallest attainable response.

4.2. Designs for One-Exponential Disappearance

   In the simplest case, the time (t) course of first-order disappearance of substance R in the process

$$R \rightarrow S$$
[12]

is given by

$$y = A \cdot e^{-Bt} \qquad .$$
[13]

With *constant variance*, the most precise parameters are obtained if about half of the measurements are obtained as soon as possible (t_{min}) after starting the experiment, giving rise to a response of y_{max}. The second half of the observations should be around (18,30)

$$t_2 = t_{min} + 1/B$$

$$= t_{min} + 1.44\ t_{1/2} \qquad , \qquad\qquad\qquad [14a]$$

where the response is

$$y_2 = y_{max}/e \qquad , \qquad\qquad\qquad\qquad [14b]$$

and $t_{1/2}$ the half-life.

Optimally designed experiments following this suggestion are illustrated in Figure 1. For highest efficiency, observations should be made repeatedly at the earliest possible time and later at a time given by Eq. [14] (Fig. 1A). This repetitive 2-point design yields, indeed, precise parameters but relies on assuming the validity of the one-exponential model.

More practically, as suggested in Section 2.2, a few readings taken at various times could be allocated to confirming the model, while more frequent measurements around the suggested optimal design points could aim at the precise evaluation of the parameters (Fig. 1B).

If, in a non-replicating design, measurements are spaced at equal time intervals then the last reading should optimally be (9) at

$$t_2 = t_{min} + 1.69/B$$

$$= t_{min} + 2.43\ t_{1/2} \qquad . \qquad\qquad\qquad [15]$$

This result has been derived for an asymptotically large sample size.

In the presence of *constant relative errors* we can consider the logarithm of the response:

$$\log y = \log A - Bt \qquad , \qquad\qquad\qquad [16]$$

and base the optimization of the design, according to Eq. [10], on the resulting reduced design matrix, $\underline{G}$. In this case, for optimal efficiency, half of the observations should be made as early as possible (18):

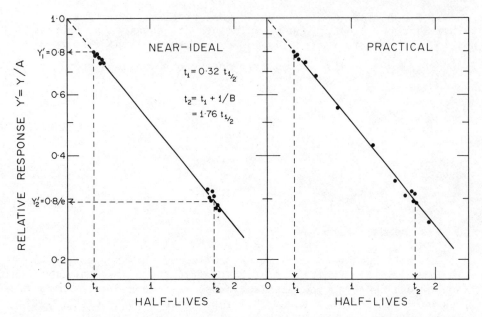

FIGURE 1: Experimental designs for estimating the parameters of the one-exponential disappearance model when the variance of observations is constant. It is assumed that, due to experimental limitations, meaningful measurements can be obtained only starting at 0.32 half-lives after the commencement of the experiment. This means that the highest attainable relative response is $Y'=y/A=0.80$.
(A) For optimal precision of the parameters, replicated observations should be obtained at the earliest possible time ($t_1=t_{min}$) after starting the experiment, and $1/B=1.44t_{1/2}$ later ($t_2=t_{min}+1/B$). In the near-ideal design shown in the Figure, readings are obtained at great frequency near the optimal design points.
(B) In the "practical" design, measurements made over a wide range assess the validity of the model. However, for precise parameter estimation, additional readings are concentrated again close to the optimal sampling points.

$$t_1 = t_{min}$$
$$y_1 = y_{min}$$

[17]

and the other half as late as possible:

$$t_2 = t_{max}$$
$$y_2 = y_{max}$$

[18]

This result is not surprising since the transformed Eq. [16] is linear in the parameters and, consequently, the optimal design is the one given in Section 2.2 for straight lines: Readings should be as far from each other as possible.

Deviations from the optimal design result in larger parameter variances and standard errors. These can be characterized by the *efficiency* of the non-optimal relative to the corresponding optimal design. The relative efficiencies are ratios of generalized parameter variances for the two cases, and can be calculated as percentages in terms of the determinants (D) for their information matrices (15):

$$\% \text{ Efficiency} = 100(D_{\text{non-opt}}/D_{\text{opt}})^{1/p} \quad . \qquad [19]$$

Here p is again the number of parameters.

Figure 2 illustrates percent efficiencies of frequently applied experimental designs. These contain observations which range from a time of zero to a value indicated on the horizontal axis and are distanced in geometrically (logarithmically) increasing intervals. High efficiencies are noted when the majority of readings is near the optimal design points ($t_2=1.44$ in Fig. 2A and $t_2=5.00$ in Fig. 2B). Otherwise, however, the efficiency of these "usual" designs can be quite low, especially when the number of readings is increased.

This suggests that the "practical" design strategy, described in Section 2.2 and illustrated in Fig. 1B, is effective: The decline in estimation efficiency is generally moderate when a few observations are allocated to verifying the model. In contrast, the efficiency is reduced quite substantially if additional readings are distributed throughout the range of observations. Therefore, these readings would be concentrated around the optimal design points (*cf*. Fig. 1B).

4.3. Designs for One-Exponential Accumulation

The time course for the accumulation of substance S in the first-order process [12] is described by

$$y = A(1 - e^{-Bt}) \quad . \qquad [20]$$

The optimal design involves again two observation times replicated at equal frequency. One of these is the largest practically reachable value:

$$t_2 = t_{\text{max}}$$

$$y_2 = y_{\text{max}} \quad . \qquad [21]$$

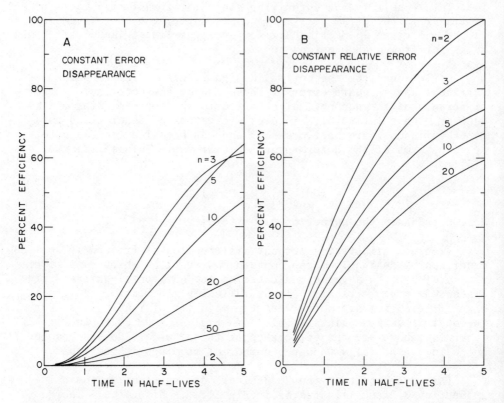

FIGURE 2: Percent efficiencies for estimating the parameters of
the one-exponential disappearance model by frequently applied experimental designs. The efficiencies compare the joint variance of the
parameters estimated by the non-optimal designs with that obtained
from the corresponding optimal design.

The parameter estimating efficiencies of the frequently used
designs are generally lower than those of the corresponding optimal
designs. The efficiencies tend to decrease, and consequently the
estimated variances increase, when number of observations is higher.
Therefore, not more than a few measurements should be made away from
the optimal design points.
(A) Constant variance; (B) Constant relative error (constant coefficient of variation).

In the non-optimal designs, n=2,3,5,10,20 or 50 observations
are spaced in geometrically (logarithmically) increasing distances
between times of zero and the value indicated on the horizontal
axis. Optimal designs with the same number of observations follow
the recommendations presented in the text; for constant relative
errors, the upper limit of measurement is assumed to be, for illustration, 5 half-lives.

The other design point is, in the presence of constant variance
(18,30):

$$t_1 = 1/B - t_2(A - y_2)/y_2 \; ,$$ [22]

and with constant relative errors:

$$t_1 = t_{min}$$ [23]

the shortest attainable time (18).

If it is possible to extend measurements quite close to the
maximal response (A) then, from Eq. [22], the first design point
becomes

$$t_1 = 1/B = 1.44t_{1/2} \qquad ,$$ [24a]

with a response of

$$y_1 = (1 - 1/e)A = 0.632A \; .$$ [24b]

Figure 3 shows percent efficiencies again for the frequently
used, "usual" designs applied now to studies of first-order accumulation. Again, efficiencies can be quite low, especially when the
number of observations is increased. Consequently, the "practical"
design strategy which allocates additional measurements (say, in
excess of n=10) near the optimal design points, can be recommended
again.

For non-replicating measurements of constant variance spaced
equally between times of 0 and t_{max}, the last reading should be
(9) at

$$t_{max} = 7.18/B$$
$$= 10.3t_{1/2} \qquad .$$ [25]

4.4. Experimental Designs for the Precise Estimation of One Model
     Parameter

If the goal of an investigation involves the precise evaluation of only some (say, the first p_1), and not all of the model
parameters then the design criterion [6] is modified. The variance-
covariance matrix [2] is partitioned into a $p_1 \times p_1$-dimensional matrix
(V_{11}) referring to the parameters of interest, and complementary
matrices of $p_i \times p_j$ dimensions:

$$V = \begin{pmatrix} V_{11} & V_{12} \\ V_{21} & V_{22} \end{pmatrix}$$ [26]

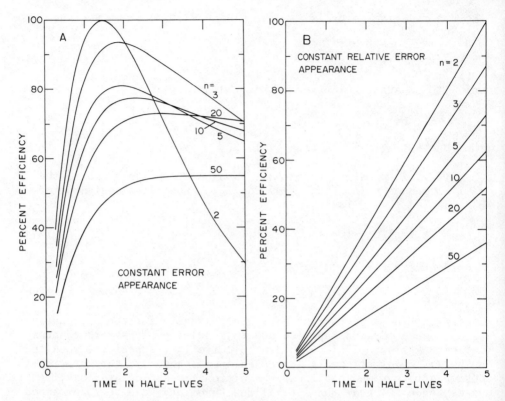

FIGURE 3: Percent efficiencies for estimating the parameters of
the one-exponential accumulation (appearance) model by frequently
applied experimental designs. The interpretation of the diagrams
is similar to that of Figure 2.
(A) Constant variance; (B) Constant relative error (constant coefficient of variation).

with

$$p = p_1 + p_2 \qquad .$$

The design criterion requires now maximizing a modified determinant
(33,34):

$$D_1 = \text{Max}\{\text{Det}(\underline{V}_{11} - \underline{V}_{12}\underline{V}_{22}^{-1}\underline{V}_{21})\} \qquad . \qquad\qquad [27]$$

For estimating either parameter of the one-exponential *disappearance model* (Eq. [13]), observations should be obtained again as
early as possible:

$$t_1 = t_{min} \tag{28}$$

$$y_1 = y_{max} \qquad .$$

With *constant variance*, the second design point is slightly beyond the one given in Eq. [14], at (35)

$$t_2 = t_{min} + 1.28/B \tag{29}$$

with

$$y_2 = 0.278y_{max} \qquad .$$

In contrast to designs aiming at the evaluation of all model parameters in which all to sampling points are replicated at equal frequency, designs intending to estimate precisely a subset of parameters may involve *unequal allocations* of the design points. For instance, if only the evaluation of the exponent B is important then only an approximate fraction

$$f_1 = 0.218 \tag{30a}$$

of the readings should be made at the first design point (t_1) while the remaining proportion

$$f_2 = 0.782 \tag{30b}$$

is the allocation for the second sampling point (t_2).

The picture can be even more complicated. For instance, when the estimation of only the coefficient A is judged to be essential then the approximate proportion of taking the first reading (Eq. [28]) depends on the experimental constraint of the initial observation (35):

$$f_1 = (0.78t^* + 1)/(0.59t^* + 1) \tag{31}$$

where

$$t^* = Bt_{min} \qquad . \tag{32}$$

With *constant relative errors*, the optimal design is identical to the one given for the evaluation of both parameters (Eqs. [17,18]): repeated measurements should be made as early and as late as possible:

$$t_1 = t_{min} \qquad , \tag{17}$$

$$t_2 = t_{max} \qquad . \tag{18}$$

When only the exponent (B) is important then, as before, the two readings should be obtained at about equal frequency, *i.e.*,

$$f_1 = f_2 = 0.5 \qquad . \qquad [33]$$

On the other hand, if only the coefficient (A) is required then the allocations depend on both experimental constraints:

$$f_1 = 1 + \left(1 + \frac{Y_{max} \cdot \ell n \, Y_{max}}{Y_{min} \cdot \ell n \, Y_{min}} \right) \qquad [34]$$

with

$$Y = y/A \qquad . \qquad [35]$$

Similar guidelines (35) are available for estimating only one parameter of the one-exponential *accumulation model* (Eq. [20]). Only the design points will be presented. For constant variance, the design points depend on the experimental limitation for the longest attainable time (t_{max}) evoking the highest possible response (y_{max} or Y_{max}). This is one of the optimal points:

$$t_2 = t_{max} \qquad , \qquad [36]$$

while the other point can be read from Figure 4. The resulting values of t_1, which have a limiting maximum of $1.108t_{1/2}$, are lower than the corresponding times given for the estimation of both parameters (Eqs. [22] and [24a]).

In the presence of constant relative errors, as before (Eq. [23]), measurements should be taken as early and as late as possible:

$$t_1 = t_{min}$$
$$\qquad\qquad\qquad\qquad\qquad\qquad\qquad [23]$$
$$t_2 = t_{max} \qquad .$$

When only the exponent is estimated, these observations should be obtained again at approximately equal frequency (Eq. [33]).

4.5. Designs for First-Order Consecutive Processes

The simplest example of first-order consecutive processes is

$$R \rightarrow S \rightarrow T \qquad . \qquad [37]$$

It describes simple unimolecular (or quasi-unimolecular) chemical reactions, the decay of atomic nuclei as well as various physiological and pharmacological transfers and transformations. These include the absorption and elimination of a drug, or the formation and excretion of a metabolite.

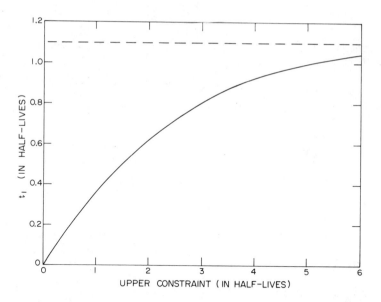

FIGURE 4: Relationship between the first optimal design point and the maximal experimentally attainable time when one parameter of the one-exponential accumulation model is precisely estimated in the presence of constant variance. All times are expressed as multiples of the half-life.

For these processes, the time course for the amount or concentration (y) of the intermediate product (S) is given by

$$y = \frac{Ak_1}{k_1 - k_2} (e^{-k_2 t} - e^{-k_1 t})$$ [38]

where k_1 and k_2 are the rate constants for the two consecutive reactions and A is a scaling parameter.

General guidelines for designing experiments, which rely on this model, have not been presented. However, numerical results indicate (30,36) that, with constant variance, replicate observations are optimally obtained shortly before and after maximum response, and at a time following the second design point by about 1/B. These results are sensible. They are also in line with those given for one-exponential disappearance (Eq. [14]).

4.6. Designs for the Two-Exponential Model

The model

$$y = A_1 e^{-B_1 t} + A_2 e^{-B_2 t} \tag{39}$$

has 4 parameters. Consequently, the optimal experimental designs involve 4 replicated sampling points.

In the presence of *constant variance*, the design points are related to the ratio of exponents,

$$B = B_1/B_2 \quad . \tag{40a}$$

The first observation should again be made as soon as possible after starting the experiment. Ideally,

$$t_1 = 0 \quad . \tag{41a}$$

The other 3 optimal sampling points can be calculated (12) from

$$\log T_2 = -0.20 - 0.90 \log B$$

$$\log T_3 = 0.30 - 0.74 \log B \tag{41b}$$

$$T_4 = T_2 + T_3 + 1.04 \quad ,$$

where

$$T_i = B_2 t_i \quad . \tag{41c}$$

The expressions are optimized for values of B ranging from 1.5 to 1,000. The logarithms have a base of 10.

For *constant relative errors*, two of the design points are again at the boundaries of the observation region. Ideally then,

$$t_1 = 0 \tag{42a}$$

$$t_4 = \infty \quad .$$

The two middle sampling points are related to the ratio of the exponents (Eq. [40a]) and the coefficients,

$$A = A_1/A_2 \quad , \tag{40b}$$

according to

$$\log T_2 = 1.160 + 0.044 \log A - 1.016 \log B \tag{42b}$$

$$\log T_3 = 0.294 + 0.214 \log A - 1.102 \log B \quad .$$

These expressions are based on A-ratios ranging from 1 to 1,000, and B-ratios from 3 to 100.

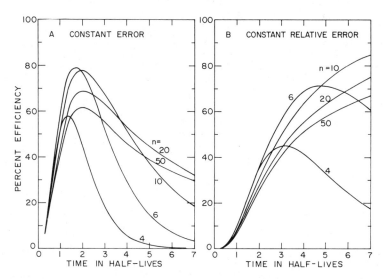

FIGURE 5:   Percent efficiencies for estimating the parameters of
the two-exponential model by frequently applied experimental designs.
The interpretation of the diagrams is similar to that of Figure 2,
except that the range of experimentation extends now from 0 to 7
half-lives in terms of B_2.   A ratio of A=4 is assumed for the coefficients, and B=4 for the exponents.
(A) Constant variance; (B) Constant relative error (constant coefficient of variation).

Figure 5 illustrates that frequently applied experimental
designs in which measurements cover the entire range of observations,
yield reduced efficiencies, *i.e.*, larger parameter variances. As
before, the reduction is most pronounced when the number of readings
is increased.

Figure 6 shows confidence contours for contrasts of the coefficients and exponents. In each case, the replicating optimal
design (marked by o) results in the smallest joint (or, here, even
separate) parameter errors. This is anticipated since D-optimization is expected to yield the smallest possible hypervolume of the
joint confidence region for the parameters (Section 2.1). Experiments in which the same number of observations is spaced either
uniformly (u) or in geometrically (g) increasing intervals, have
much larger parameter errors. These can become so large that evaluation of one or more of the parameters becomes impossible. For
instance, it may not be possible to estimate meaningfully either
the exponent B_1 (Fig. 6B, geometric spacing), or B_2 (Fig. 6D, uniform spacing), or both of these parameters (Fig. 6B, uniform spacing). This illustrates the importance and effectiveness of careful
experimental designs.

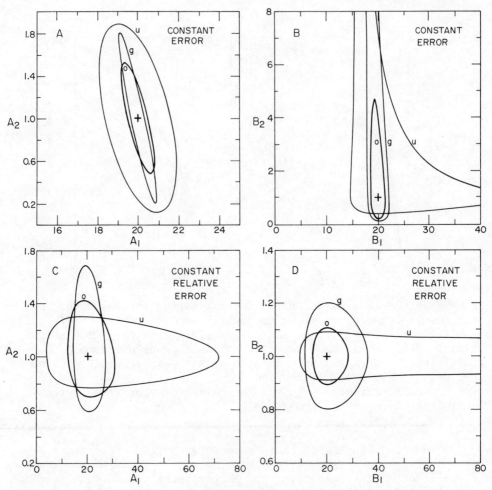

FIGURE 6: Confidence contours of the two-exponential model parameters estimated by three differently designed experiments.

Each contour is a section of a 5-dimensional sum-of-squares
(SS) surface involving SS and the 4 parameters. Here, the SS and,
consequently, the contours correspond, in the linear approximation,
to 95% confidence regions. The probability levels are likely not
quite correct but the shapes of the contours are. For these SS
levels, the rather high standard deviations of $\sigma=2.5$ for constant
variances, and $\sigma=0.25y$ for constant relative errors were chosen.

Parameter ratios of A=B=20 are assumed. If, in particular,
$A_1=B_1=1$, then $A_2=B_2=20$. The corresponding true parameter values
are marked by +. The range of experimentation is assumed to extend
from 0 to 5 half-lives in terms of B_2. In each design, 40 observations are included.

5. DESIGNS FOR DOSE-RESPONSE/ENZYME KINETIC EXPERIMENTS

5.1. Response Curves

When plotted against the logarithm of concentration or dose, a representation favoured by pharmacologists, many responses show a sigmoidal shape. They include the reaction velocity observed in steady-state enzyme kinetic experiments, or the bound or free concentration of a ligand in binding studies including, for instance, radioimmunoassays, or various drug responses.

In the simplest case, the response is described by a two-parameter, rectangular hyperbola:

$$y = Ax/(Bx+1) \qquad\qquad\qquad [43a]$$

or, equivalently, with the parameters used in enzyme kinetics, the asymptotic reaction velocity (V) and the Michaelis constant (K, omitting the subscript):

$$y = Vs/(s+K) \qquad . \qquad\qquad [43b]$$

(Here y and s correspond to the reaction velocity and (free) substrate or ligand concentration, respectively.)

When biochemical or pharmacological experimental data can not be characterized by the hyperbola, they can often be reasonably described by the essentially empirical Hill equation:

$$y = Vs^H/(s^H + K^H) \qquad . \qquad\qquad [44]$$

The Hill coefficient (here denoted by H) is a measure of deviations from hyperbolicity (37-39).

(A) Constant variance, contrast of coefficients; (B) Constant variance, contrast of exponents; (C) Constant relative error, contrast of coefficients; (D) Constant relative error, contrast of exponents.

o: optimal design; u: uniformly, g: geometrically (logarithmically) spaced observations. Optimally designed experiments have comparatively low parameter errors. By contrast, uniform spacing can result in very large estimated errors.

In more complicated systems, the dependence of the response on free concentration can be described by rational functions which are ratios of polynomials:

$$y = \frac{a_\ell x^\ell + a_{\ell-1} x^{\ell-1} + \cdots + a_1 x + a_0}{b_m x^m + b_{m-1} x^{m-1} + \cdots + b_1 x + b_0} \qquad [45]$$

where some of the coefficients may equal zero.

In binding studies, this expression takes the more restricted form of the Adair equation (40);

$$y = \frac{P_{max}}{m} \cdot \frac{m a_m x^m + (m-1) a_{m-1} x^{m-1} + \cdots + a_1 x}{a_m x^m + a_{m-1} x^{m-1} + \cdots + a_1 x + 1} \qquad . \qquad [46]$$

The hyperbola [43a] or [43b] can be considered to be a special case of expressions [45] and [46].

Again (excepting only the design given in Eqs. [58]), one of the repeated sampling points will measure the maximal response. In the presence of constant relative errors, another point will yield the smallest possible response.

5.2. Designs for Hyperbolic Kinetic Experiments

If the experimental variance is approximately constant then the optimal design requires that about half of the observations should be obtained at the highest experimentally attainable concentration (s_{max}) giving rise to the largest possible reaction velocity (y_{max}). The other half of the readings should yield (10,16,17,32)

$$y_2 = y_{max}/2 \qquad . \qquad [47]$$

The corresponding concentration is

$$s_2' = s_{max}'/(2 + s_{max}') \qquad [48]$$

with $s' = s/K$

With constant relative errors, about half of the observations should be obtained at the highest possible concentration (s_{max}), and the other half at the smallest practically reasonable concentration (s_{min}) (10,17,32).

Results of computer simulated experiments agree with the theoretical predictions (41).

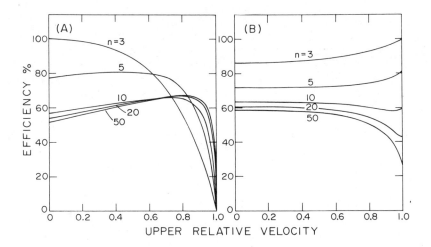

FIGURE 7:   Percent efficiencies for estimating the parameters of a
hyperbola in kinetic experiments in which the concentration is
spaced uniformly.   The interpretation of the diagrams is similar
to that of Figure 2.
(A) Constant variance; (B) Constant relative error (constant coefficient of variation).

If the standard deviation of the measurements is proportional
to y^2 then the second optimal design point is (17)

$$y_2 = y_{max}/3 \qquad . \qquad [49]$$

Figure 7 shows relative efficiencies for frequently applied
designs of kinetic experiments.   In these "practical" designs, the
concentrations are spaced uniformly between the limits of zero and
s_{max} which yields the maximal velocity indicated on the horizontal
axis.   The corresponding optimal designs have the same number of
observations and are restricted to the same concentration range.

With the uniform spacing of the data illustrated in Figure 7,
the relative efficiencies are fairly constant up to constrained
maximal velocities of about 0.8.   Increasing the number of measurements reduces the efficiency, but not below 50-60%.

This is in contrast with efficiencies calculated for geometrically (logarithmically) increasing spacing of the concentrations (32). In this case, the efficiencies are generally lower than those seen with uniform, arithmetic spacing, and are substantially reduced when the number of readings is increased. These contrasts suggest that among non-optimal designs, uniformly spaced concentrations yield more precise parameters than logarithmically spaced ones.

Duggleby (16) presented optimal designs for inhibition models, assuming constant variance. For competitive inhibition, the first of three sampling points is, as before,

$$y_1 = y_{max} \tag{50a}$$

obtained with

$$s_1 = s_{max} \tag{50b}$$
$$i_1 = 0 \qquad ,$$

where i is the concentration of the inhibitor. The second and third optimal response is, corresponding to Eq. [47],

$$y_2 = y_3 = y_{max}/2 \tag{51a}$$

based either on

$$i_2 = 0 \tag{51b}$$

and s_2 calculated from Eq. [48], or on

$$s_3 = s_{max} \tag{51c}$$

with i_3 appropriately adjusted (16).

Optimal designs for noncompetitive and uncompetitive inhibition have also been given (16).

5.3. Designs for Hyperbolic Binding Experiments

Binding studies evaluate the relationship between ligand concentrations which are, in an equilibrium, bound (b) to a macromolecule, and unbound or free (f). In the simplest case, this relationship is described, in analogy to Eq. [43], by a hyperbola:

$$b = Pf/(f+K) \qquad , \tag{52}$$

where P and K are the parameters to be estimated.

Often, however, either f or b is measured, and the other concentration is obtained by subtraction from their known sum, the total ligand concentration:

$$c = b + f \qquad . \qquad [53]$$

Consequently, for statistical design and analysis, the proper independent variable is c.

By expressing explicitly, for instance, the free ligand concentration,

$$f = \tfrac{1}{2}\{(c-K-P) + [(c-K-P) + 4Kc]\}^{\tfrac{1}{2}} \qquad [54]$$

is obtained (41). Optimal designs can be evaluated from this relationship and the corresponding one for b (32).

The behaviour of these expressions, and that of the related optimal designs, is characterized by an intrinsic parameter, the ratio K/P. It is an important parameter in pharmacological practice since $1/(1+K/P)$ equals the fraction of ligand being bound at limitingly small concentrations. For instance, with $K/P=0.05$, at small concentrations, $100/1.05=95\%$ of a drug is bound and only 5% is free and available for therapeutic action.

With constant variance, the optimal sampling points are, for measuring either b or f:

$$c_1 = \infty$$

$$c_2 = K + P \qquad . \qquad [55]$$

When observations are restricted to

$$c_1 = c_{max}$$
$$b_1 = b_{max} \qquad , \qquad . \qquad [56a]$$

then the second design point also decreases. The second, lower optimal sampling point can be obtained from Figure 8A. With weakly binding substances (large K/P), the lower optimal binding approaches one-half of the constrained, limiting binding:

$$b_2/b_1 = 0.5 \qquad . \qquad [56b]$$

Reasonably, this corresponds to the design condition described for kinetic studies (Eq. [47]). For more strongly bound substances, the ratio of the two optimal bindings is higher (Fig. 8A).

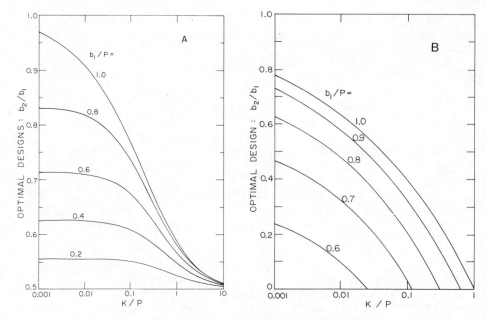

FIGURE 8: Optimal design of hyperbolic binding experiments when the highest attainable concentration and binding is, in practice, limited. b_1/P is the maximally feasible fractional binding; the second optimal binding (b_2) to be measured parallels the magnitude of this constraint. It is related to the K/P-ratio which is a measure of the limiting fractional binding: low values indicate strong binding.
(A) Constant variance for measuring either bound or free substrate concentrations; (B) Constant relative error (constant coefficient of variation) for measuring bound ligand concentrations.

The optimal design for measuring bound concentrations with constant relative errors is (32)

$$c_1 = \infty$$

and

$$c_2 = 0 \qquad \text{if } K \geq P \qquad\qquad\qquad [57a]$$

or

$$c_2 = P - K \qquad \text{if } K \leq P \qquad .$$

The corresponding measured bound concentrations are

$$b_1 = P$$

and

$$b_2 = 0 \qquad \text{if } K \geq P \qquad\qquad\qquad [57b]$$

or

$$b_2 = P - K^{\frac{1}{2}} \qquad \text{if } K \leq P \qquad .$$

Again, the condition for weak binding is identical to the one given for kinetic experiments: Measurements should be made at the highest and lowest possible concentrations. In the presence of such an upper constraint, the second optimal bound concentration can be read from Figure 8B. When the upper experimental limitation for b_1/P becomes lower, the kinetic design conditions become applicable over a wider range of binding strength (K/P).

When the relative error for measuring free concentrations is approximately constant, then the optimal design is

$$c_1 = 0$$

and

$$c_2 = K + P \qquad\qquad [58a]$$

with corresponding free ligand concentrations of

$$f_1 = 0$$

and

$$f_2 = K^{\frac{1}{2}}(K+P)^{\frac{1}{2}} \qquad\qquad [58b]$$

This is the only case reported in this survey, in which the optimal design does not require the measurement of the maximal experimentally attainable response.

Figure 9 presents the efficiencies of binding experiments with "practical" designs. In these, the (total) ligand concentration is spaced uniformly between limits of zero and a maximal value which gives rise to the fractional binding shown on the horizontal axis.

As before, the "practical" experiments have reduced efficiencies, especially when the number of observations is increased. However, in similarity to kinetic studies (Figure 7), equal spacing of the concentrations leads only to moderate loss of efficiency. By contrast, geometrically increasing distances between concentrations result in very substantial reduction of the efficiency (32).

Computer simulated experiments are in agreement with the theoretical conclusions (41).

5.4. Designs for Nonhyperbolic Responses

Very little is available about designing experiments involving nonhyperbolic responses.

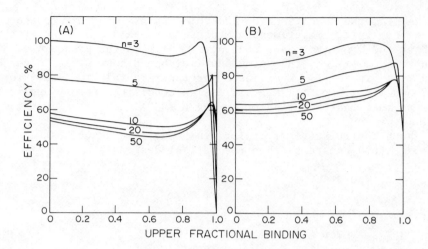

FIGURE 9: Percent efficiencies for estimating the parameters of a
hyperbola in binding experiments in which the concentration is
spaced uniformly. The interpretation of the diagrams is similar
to that of Figure 2. Here K/P=0.1, corresponding to fairly strong
binding.
(A) Constant variance for measuring either bound or free ligand
concentrations; (B) Constant relative error (constant coefficient
of variation) for measuring bound substrate concentrations.

        For the *Hill equation* [44], two of the 3 design points are
identical with those given for the hyperbola. Thus, the first
group of observations should be made again at the highest possible
concentration which results in the maximal practical response.
With constant variance, the second group of sampling points yields
one-half of the maximal response. Ideally, then,

$$y_1 = V$$

$$y_2 = V/2 \qquad . \qquad\qquad\qquad\qquad [59a]$$

The third group of design points should be obtained at

$$s_3^H = 3.59K$$

yielding

$$y_3 = 0.782V \qquad . \qquad\qquad\qquad\qquad [59b]$$

H is again the Hill coefficient.

With constant relative errors, the second group of measurements should be made, as before, at the smallest possible concentration. Ideally, the corresponding response is

$$y_2 = 0 \qquad . \qquad\qquad [60a]$$

The third group of readings could be obtained, symmetrically either at

$$s_3^H = 0.214K$$

or at

$$s_3^H = 4.68K \qquad\qquad [60b]$$

corresponding to

$$y_3 = 0.176V$$

or

$$y_3 = 0.824V \qquad . \qquad\qquad [60c]$$

Optimization results are not available for higher-degree rate or binding equations. Simulated experiments were performed for the 4th-degree Adair equation [46] by spacing observations over various ranges of the predicted response (4). It was suggested that they should cover at least 50% of the range. The highest and lowest attainable concentrations and responses appeared to be again of great importance. Sensitivity calculations for the hemoglobin-oxygen binding system which is described again by a 4th-degree Adair equation, emphasized particularly the low-concentration measurements (43).

In practice, the described principles have been applied only to the design of experiments for obtaining kinetic progress curves (44,45). They were found to estimate effectively the parameters of fairly complex systems.

6.   GENERAL DESIGN STRATEGIES

The designs described for simple kinetic systems provide guidelines for experimentation. Even if such general information is not available, the efficient design can be evaluated in an investigated system. After all, the D-optimizing criterion [5] can be always applied for designing experiments whenever an investigator is prepared to assume:
(i)   a model;
(ii) initial, crude values of the parameters;

(iii) properties of the experimental error (distribution, serial correlation and, especially, the relationship between error variance and predicted response).

All optimizing algorithms can be applied to the evaluation of the D-criterion. Nevertheless, computer programs, written specifically for this purpose, facilitate the task (16,46).

A useful strategy for estimating parameters suggests their sequential experimental design. As seen above, D-optimization, together with other design procedures, requires preliminary estimates of the parameters. Consequently, it would be reasonable to consider a cycle (2,15,27) in which a set of experiments is performed on the basis of such early information. The parameters are evaluated, the design is modified and a new set of measurements is obtained. The cycle is continued until no substantial further gain in parameter precision is recorded.

The sequential design strategy is particularly advantageous when the measurements are time-consuming and/or expensive. This is the situation in many engineering and industrial investigations. However, enzyme, tracer or pharmacokinetic observations are usually taken in larger groups at a time, possibly including complete sub-experiments (for instance, a series of drug kinetic measurements obtained in a given subject). Consequently, for these studies, the guidelines for designing experiments, described in the present survey, could be useful.

7. CONCLUSIONS

1.   Model parameters can be evaluated precisely in suitably de-signed experiments. The principles of such designs are presented. The criterion of D-optimization generally yields *replications* of p sampling points. If all p model parameters are of interest then the points should be measured at *equal frequency*.

2.   These theoretical designs should serve as guidelines for ex-perimentation, and not as rules. In a practical design, some ob-servations would test the validity of the model. However, addition-al readings obtained close to the recommended sampling points could aim at the estimation of precise parameters.

3.   Generally, one of the critical sampling points measures the *maximal experimentally attainable response*. With constant relative errors, another design point usually yields the smallest possible response.

4.   Time-dependent kinetic, such as tracer and pharmacokinetic models are described by sums of exponentials. With constant vari-ance, the parameters of the *one-exponential disappearance* model are

estimated most precisely by obtaining observations at 1/B following the first reading. Similarly, in studies of accumulation, measurements should be made at 1/B or, depending on the upper constraint, earlier (Eq. [22]).

5. When *one of the parameters* is of particular interest then also the design points are often not equally important. For instance, for estimating, in the presence of constant variance, the exponent of the one-exponential disappearance model, the second group of measurements is obtained, at 1.28/B following the first group, with a relative frequency of 0.78. For the evaluation of the coefficient, the allocation of the same design point is given by Eq. [31].

6. For *two-exponential models*, the 4 optimal design points are related to the ratio of exponents and, with constant relative errors, also to the ratio of coefficients (Eqs. [41] and [42]).

7. In *hyperbolic kinetic experiments*, performed in the presence of constant variance, the second optimal response (velocity) is one-half of the maximal experimentally attainable one.

8. In *hyperbolic binding* studies, in which either bound (b) or free (f) ligand concentrations are measured with constant variance, the second optimal b-value is higher than one-half of the maximal experimentally feasible one (Fig. 8A). With constant relative errors, the second optimal b to be measured is obtained from Figure 8B. The same error condition involved in obtaining f requires observations at the lowest possible concentration and, ideally, at values given by Eq. [58].

9. Deviations from optimality result in the *loss of estimating efficiency*. Such reduced efficiencies are illustrated for various frequently applied designs. Generally, they decline with increasing number of observations. This confirms the usefulness of the practical design strategy given above in paragraph 2.

ACKNOWLEDGEMENTS

Financial support from the Medical Research Council of Canada is gratefully acknowledged.

REFERENCES

1. Box, G.E.P. and Hunter, W.G. (1962) Technometrics, 4, 301-318.
2. Endrenyi, L. (1974) In "Mathematical Models of Metabolic Regulation" (T. Keleti and S. Lakatos, eds.) Akademiai, Budapest, pp. 11-30.
3. Bozler, G., Heinzel, G., Koss, F.W. and Wolf, M. (1977) Arzneim.-Forsch. (Drug Res.) 27 (I), 897-900.

4. Mannervik, B. (1980) In "Kinetic Data Analysis: Design and Analysis of Enzyme and Pharmacokinetic Experiments" (L. Endrenyi, ed.) pp. 235-270 , Plenum, New York.

5. Fedorov, V.V. (1972) "Theory of Optimal Experiments", Academic Press, New York-London.

6. Bard, Y. (1974) "Nonlinear Parameter Estimation", Academic Press, New York-San Francisco-London.

7. Box, G.E.P. and Hunter, W.G. (1965) IBM Scientific Computing Symposium in Statist., 113-137.

8. Kullback, S. and Liebler, R.A. (1951) Ann. Math. Stat. $\underline{22}$, 79-86.

9. Beck, S.V. and Arnold, K.J. (1977) "Parameter Estimation in Engineering and Science", Wiley, New York-London-Sidney-Toronto.

10. Kanyár, B. (1978) Acta Biochim. Biophys. Acad. Sci. Hung. $\underline{13}$, 153-160.

11. Bogumil, R.J. (1980) Federation Proc. $\underline{39}$, 97-103.

12. Endrenyi, L. and McDougall, J. (1979) Proc. Can. Fed. Biol. $\underline{22}$, 91.

13. Atkinson, A.C. and Hunter, W.G. (1968) Technometrics, $\underline{10}$, 271-289.

14. Box, M.J. (1968) J. Roy. Statist. Soc. $\underline{B30}$, 290-302.

15. Box, M.J. (1970) Technometrics, $\underline{12}$, 569-589.

16. Duggleby, R.G. (1979) J. Theor. Biol. $\underline{81}$, 671-684.

17. Duggleby, R.G. (1980) In "Kinetic Data Analysis: Design and Analysis of Enzyme and Pharmacokinetic Experiments" (L. Endrenyi, ed.) pp. 169-179 , Plenum, New York.

18. Endrenyi, L. and Dingle, B.H. (1980) In "Pharmacokinetics During Drug Development - Data Analysis and Evaluation Techniques" (G. Bozler and J.M. van Rossum, eds.) Gustav Fischer Verlag, Stuttgart, in press.

19. Draper, N.R. and Smith, H. (1967) "Applied Regression Analysis", Wiley, New York-London-Sydney.

20. Watts, D.G. (1980) In "Kinetic Data Analysis: Design and Analysis of Enzyme and Pharmacokinetic Experimenst", (L. Endrenyi, ed.) pp. 1-24, Plenum, New York.

21. Siano, D.B., Zyskind, J.W. and Fromm, H.J. (1975) Arch. Biochem. Biophys. $\underline{170}$, 587-600.

22. Storer, A.C., Darlison, M.G. and Cornish-Bowden, A. (1975) Biochem. J. $\underline{151}$, 361-367.

23. Wagner, J.G. (1975) "Fundamentals of Clinical Pharmacokinetics" Drug Intelligence Publ., Hamilton, Ill., pp. 280-290.

24. Clements, J.A. and Prescott, L.F. (1976) J. Pharm. Pharmacol. $\underline{28}$, 707-709.

25. Mannervik, B., Jakobson, I. and Warholm, M. (1979) Biochim. Biophys. Acta, $\underline{567}$, 43-48.

26. Nimmo, I.A. and Mabood, S.F. (1979) Anal. Biochem. $\underline{94}$, 265-269.

27. Cornish-Bowden, A. (1980) In "Kinetic Data Analysis: Design and Analysis of Enzyme and Pharmacokinetic Experiments", (L. Endrenyi, ed.) pp. 105-119 , Plenum, New York.

28. Endrenyi, L. and Kwong, F.H.F. (1980) In "Kinetic Data Analysis:
     Design and Analysis of Enzyme and Pharmacokinetic Experi-
     ments" (L. Endrenyi, ed.) pp. 89 - 103 , Plenum, New York.
29. Nimmo, I.A. and Atkins, G.L. (1980) In "Kinetic Data Analysis:
     Design and Analysis of Enzyme and Pharmacokinetic Experi-
     ments" (L. Endrenyi, ed.) pp. 309 - 315 , Plenum, New York.
30. Box, G.E.P. and Lucas, H.L. (1959) Biometrika, 46, 77-90.
31. St. John, R.C. and Draper, N.R. (1975) Technometrics, 17,
     15-23.
32. Endrenyi, L. and Chan, F.Y., J. Theor. Biol., submitted for
     publication.
33. Atkinson, A.C., Hunter, W.G. and Henson, T.L. (1969) Can. J.
     Chem. Eng. 47, 76-80.
34. Box, M.J. (1971) Biometrika, 58, 149-153.
35. Dingle, B.H. (1975) M.Sc. Thesis, University of Toronto.
36. Süverkrup, R. (1980) In "Pharmacokinetics During Drug Develop-
     ment - Data Analysis and Evaluation Techniques" (G. Bozler
     and J.M. van Rossum, eds.) Gustav Fischer Verlag, Stuttgart,
     in press.
37. Cornish-Bowden, A. and Koshland, D.E., Jr. (1975) J. Mol. Biol.
     95, 201-212.
38. Endrenyi, L. and Wong, J.T.F. (1971) Can. J. Biochem. 49,
     581-598.
39. Bardsley, W.G. and Waight, R.D. (1978) J. Theor. Biol. 72,
     321-372.
40. Adair, G.S. (1925) J. Biol. Chem. 63, 529-545.
41. Endrenyi, L. and Kwong, F.H.F. (1972) In "Analysis and Simula-
     tion of Biochemical Systems" (H.C. Hemker and B. Hess,
     eds.) pp. 219-237, North-Holland, Amsterdam.
42. Cornish-Bowden, A. and Koshland, D.E., Jr. (1970) Biochemistry,
     9, 3325-3336.
43. Endrenyi, L. and Kwong, F.H.F. (1973) Acta Biol. Med. Germ.
     31, 495-505.
44. Markus, M. and Plesser, Th. (1976) Biochem. Soc. Trans. 4,
     361-364.
45. Markus, M. and Plesser, Th. (1980) In "Kinetic Data Analysis:
     Design and Analysis of Enzyme and Pharmacokinetic Experi-
     ments" (L. Endrenyi, ed.) pp. 317-339 , Plenum, New York.
46. Metzler, C.M., Elfring, G.L. and McEwen, A.J. (1974) "A Users'
     Manual for NONLIN and Associated Programs", Upjohn Co.,
     Kalamazoo, Mich.
47. Draper, N.R. and Hunter, W.G. (1967) Biometrika, 54, 147-153.

EXPERIMENTAL DESIGNS FOR THE DISTRIBUTION-FREE ANALYSIS OF ENZYME KINETIC DATA

Ronald G. Duggleby[*]

Australian National University
John Curtin School of Medical Research
Department of Biochemistry
Canberra, ACT, Australia

ABSTRACT

The statistical analysis of experimental data always involves making certain assumptions about the distribution of experimental errors. Methods based on the least-squares principle are appropriate when these errors follow a normal distribution but several studies have shown that, for enzyme kinetic measurements, such a distribution may not be common. Under these circumstances, distribution-free methods of analysis should be employed. In this paper a distribution-free method, which can be applied to a wide variety of data analysis problems, is proposed. The method is based on a special type of experimental design in which replicate measurements are made under as many experimental conditions as there are parameters to be estimated. This design offers great simplicity in the execution of the experiment and in the analysis of the results. In addition, the design can be made optimal in the sense that the overall variance of the parameter estimates is minimized. Formulae are given for choosing the optimal designs for selected cases and a worked example of the design and analysis of a competitive inhibition experiment is presented.

1. INTRODUCTION

The purpose of an enzyme kinetic experiment is to obtain some information about the relationship between the measured reaction rate and experimental variables such as the concentrations of

─────────────
*Present address:
University of Queensland, Department of Biochemistry, St. Lucia, Queensland, Australia 4067.

substrates, activators and inhibitors, pH, temperature, ionic
strength, and so on. This relationship is usually expressed as a
mathematical model or equation which conveys the general form of
the relationship and a set of parameter values which determine the
specific numerical relationship. For example, the Michaelis-Menten
equation

$$v = VA/(K_a + A) \tag{1}$$

describes the form of the relationship between the reaction velocity (v) and the substrate concentration (A) as a rectangular
hyperbola passing through the origin. Assigning specific numerical
values to the parameters V and K_a then calibrates the axes.

The information which is being sought in an enzyme kinetic
experiment is most commonly of two types (1). In model discrimination, the objective is to determine which of several possible
model types best approximates the kinetic properties of the enzyme
which is under study. An experiment to determine whether a com-
pound acts as a competitive or as a non-competitive inhibitor with
respect to a particular substrate typifies such studies. In model
fitting, also known as parameter estimation or as model optimization, the form of the model is known or at least it is tentatively
assumed. The objective of the study is to determine numerical
values for the parameters associated with the model.

For both model discrimination and parameter estimation some
sort of statistical analysis of the data is necessary. While
much of the early work concentrated on the relative merits of
different linear transformations [e.g. Dowd and Riggs (2)] such
studies became of largely historical interest with the introduction
of nonlinear regression techniques into the analysis of enzyme
kinetic data (3,4). The value of nonlinear regression has been
challenged in the past few years with the realization that the
errors associated with kinetic measurements may not vary in a
simple manner with the experimental variables (5,6) and may not
even be normally distributed (7,8). Appropriate weighting factors
may be employed to correct for differences in the variance, but
if the errors are not normally distributed then one of the basic
assumptions of least-squares analysis is violated. In practice,
the data are often insufficient to calculate accurate weighting
factors, and only rarely would the error distribution be defined.
These uncertainties regarding the error structure affect both
the location of the "best" estimates of the parameters, and the
calculation of meaningful standard errors.

Statisticians have recognized that classical statistical
methods depend on assumptions about the distribution of experimental errors which are difficult to verify in practice. This

has led to the development of nonparametric or distribution-free methods which dispense with most of these assumptions. Cornish-Bowden and his associates (7,9-11) have introduced nonparametric methods into the analysis of enzyme kinetic measurements which conform to Eq. [1], but extensions of the technique to more complex models have not yet been described.

In view of the uncertainties concerning the error structure of the data, it is perhaps surprising that very little consideration has been given to the design of experiments in such a way as to minimize the influence of such uncertainties. In the present report a type of experimental design is proposed which largely avoids the need for untested assumptions in order to estimate values for the parameters. The designs suggested have the additional advantage that they can be chosen to be optimal, in the sense that they minimize the overall variance of the parameter estimates. Finally, the designs greatly simplify both the execution and the analysis of the experiment. However, they are specifically tailored to the problem of parameter estimation and are incapable of providing information on whether a particular model or equation is an adequate description of the system under study. It is necessary that preliminary experiments be conducted to establish the appropriate model.

2. THEORY

2.1. The Replicate Design

Consider the simple model described by Eq. [1]. If a measurement of v is available at each of two values of A, then unique values of V and K_a may be calculated by solving the pair of simultaneous equations:

$$v_1 = VA_1/(K_a + A_1) \tag{2}$$

$$v_2 = VA_2/(K_a + A_2) \tag{3}$$

The usual situation is that v is measured at each of several values of A, and Cornish-Bowden and Eisenthal (7) have devised a nonparametric analysis for this type of data. V and K_a are calculated for each possible pair of measurements and the median values of these sets of V and K_a are used as estimates of the true values for these parameters. Porter and Trager (10) have extended this analysis to place useful confidence limits on these parameter estimates.

It is rarely pointed out that if one is willing to assume that the data obey Eq. [1] then the only reason for performing experiments at more than two values of A is to improve upon, and provide estimates of, the reliability of the parameter values.

An alternative way in which these objectives may be achieved, is to perform multiple measurements of v at each of two values of A. These multiple measurements may then be used to estimate the "true" value of v at each value of A and then the parameters may be calculated algebraically using Eq.'s [2] and [3]. If the error structure is not known (which is usually the case) then the median value of the velocities may be used to estimate the population value. In those rare instances where the data are sufficiently extensive to define the error distribution then the appropriate measure of central tendency may be used. If the data are normally distributed then the mean value of the velocities should be used, in which case the parameter estimates will be least-squares estimates. It is worth emphasizing that, with this experimental design, it is necessary to make only minimal assumptions about the error in v in order to determine V and K_a.

The above considerations may be generalized and applied to any model whether linear or nonlinear in the parameters. Provided that the dependent variable is measured at as many points in variable space as there are parameters in the model then the values of the parameters may be determined algebraically. For example, if data are to be fitted to the equation for a Ping-Pong mechanism (a three-parameter model) then velocities must be measured under three appropriate sets of experimental conditions with each characterized by the concentrations of two substrates. Values for the three parameters are calculated from the median velocities, while the variability between the replicate measurements at each of the design points may be used to assess the reliability of the parameter estimates, as will be shown later.

2.2. Selection of Design Points

Given that experiments are to be performed as outlined above, there remains the problem of determining where in variable space the measurements are to be made. For the Michaelis-Menten equation, the above considerations indicate that v should be measured several times at two concentrations of A, but they do not suggest what these concentrations should be.

Normally the aim would be to obtain the best estimates of both V and K_a which suggests that we should design the experiment such that the area of the joint confidence region (at a chosen probability level) is minimized. A convenient approximation to this area is the determinant of the variance-covariance matrix, $viz.$ $\sigma^2 (X^T X)^{-1}$ where σ^2 is the variance, X is the matrix of partial derivatives evaluated at each experimental point, and X^T is the transpose of X. This criterion, which was proposed by Box and Lucas (12) and which will be referred to as the Box-Lucas design or criterion, has been recognized to be applicable to enzyme kinetic studies (1,13-16).

2.3. Designs for the Michaelis-Menten Equation

We turn now to consider which experiments must be performed in order to satisfy the Box-Lucas criterion for a particular model and will use the Michaelis-Menten equation as an example. This model, like all nonlinear models, has partial derivatives which are functions of the parameters and estimates of the parameter values are a prerequisite for designing the experiment. This paradoxical situation is rarely a problem in practice as approximate values for the parameters will normally be available.

The optimal design for estimating V and K_a is achieved by making multiple measurements of v, half of which are performed at infinite substrate concentration ($A_1 = \infty$) and half at a substrate concentration equal to the Michaelis constant ($A_2 = K_a$). In practice, A_1 will always be finite and due to technical or economic considerations, may often be comparable in magnitude to K_a. In any event, A_1 should be held as high as possible ($A_1 = A_m$, the maximum concentration at which A may be used) and A_2 calculated from Eq. [4].

$$A_2 = A_m K_a / (A_m + 2K_a) \qquad [4]$$

It is interesting that the expected value of v at A_2 is half of that expected at A_1.

In formulating the design presented above, it has been assumed that the variance of v is constant. If this is not the case then this design will not be optimal and two other simple cases may be considered. If the error is proportional to v then the optimal design occurs when $A_1 = A_m$ and A_2 is kept as low as possible, consistent with the sensitivity of the analytical method used to measure the velocity. A safe compromise between these two extremes is to assume that the errors are proportional to $v^{\frac{1}{2}}$ [see Reich (1)] which leads to the design $A_1 = A_m$, $A_2 = A_m K_a / (2A_m + 3K_a)$. Note that in this case, $v_2 = v_1/3$. Hereafter, constant variance will be assumed and the designs discussed will be optimal only when this condition is met.

Cleland (17) has argued that the fundamental parameters for an enzyme-catalyzed reaction are V and the ratio V/K_a. The design which is optimal for estimating this pair of parameters is the same as that for estimating V and K_a, *viz*. $A_1 = A_m$, while A_2 is calculated from Eq. [4].

2.4. Complex Models

In the previous section, the simplest of enzyme kinetic models, the Michaelis-Menten equation, was considered but the principles which have been outlined may be applied directly to models invol-

ving more than one reactant or more than two parameters. To
illustrate this point, consider the model for linear competitive
inhibition. Since there are three parameters and two independent
variables, the design points are described by six variable values:
viz. the concentrations of substrate and inhibitor at three design
points. Two of the design points are identical to those for the
Michaelis-Menten equation [$A_1 = A_m$, $I_1 = 0$; $A_2 = A_m K_a/(A_m + 2K_a)$,
$I_2 = 0$]. The optimum position for the third point is at $A_3 = A_m$,
$I_3 = (A_3 + K_a)K_{is}/K_a$ (*i.e.*, at an inhibitor concentration which
gives half the rate of A_1, I_1). Inhibitor concentrations, like
substrate concentrations, are subject to constraints such that
there is a maximum practical concentration of inhibitor (I_m)
which may be used. If I_3 calculated as above is greater than I_m,
the third design point should be at A_m, I_m. Finally, if this choice
for the design point predicts a reaction velocity which is greater
than $V/2$, then A_3 should be reduced so that $v = V/2$. In this case,
$A_3 = K_a (1 + I_3/K_{is})$. While these conditions yield the optimal
design, the overall variance of the parameter estimates is relatively insensitive to small departures from the optimum. Thus,
the concentrations of substrate and inhibitor need not be set
exactly at the design points described above; any convenient concentration in the vicinity of the design point will be sufficient.
Note the practical simplicity of this design. Rather than having
to prepare 20-30 reaction mixtures, each of different composition,
it is only necessary to make three reaction mixtures (albeit
repetitively) in order to perform the experiment.

Designs for other complex models will not be presented here
but may be found by using formulae presented elsewhere [Duggleby
(18)], or by using a computer program which has been designed for
this purpose and is available from the author.

2.5. Analyzing the Data

If the principles outlined above are followed, then a data
set will consist of several replicate measurements of v, obtained
under p experimental conditions where p is the number of parameters in the model which is to be fitted to the data. Thus, if
we are fitting Eq. [1] then an equal number of replicate measurements are made at each of two substrate concentrations. Each set
of replicates is used to estimate the true velocity (using the
median, the mean or whatever measure of central tendency is the
most appropriate) to yield two simultaneous equations (Eq.'s [2]
and [3]). These may be solved by using any convenient algebraic
technique such as that outlined by Cornish-Bowden and Eisenthal
(7). For more complex models, the solution of sets of nonlinear
simultaneous equations involves some tedious algebra, and linear
transformation followed by multiple linear regression is the
method of choice. This is the method used in the nonlinear
regression computer programs outlined by Cleland (4) to obtain

initial estimates of the parameters. Since this method will produce an exact solution of the p simultaneous equations, these
initial estimates will also be the best estimates of the parameters.
This will be true irrespective of differences in the variance of the
the velocities at different design points. This procedure is valid
as the v values are taken to be without any associated error and
any transformation on v will not introduce any bias. The simplicity
of this analysis is in marked contrast to the complexities of nonlinear regression.

In principle, the precision of the parameter values should
be determined by using distribution-free methods such as that
suggested by Porter and Trager (10). However, is is usually much
simpler to calculate asymptotic standard errors using the matrix-
inversion method (4) and these values will be acceptable for most
purposes. Naturally, standard errors calculated in this way will
be less reliable if the data are not normally distributed but it
should be remembered that for nonlinear models these statistics
are inexact under ideal circumstances and caution should always
be employed in their interpretation. A worked example for data
conforming to the model for linear competitive inhibition is shown
in Table 1. Note that the residual variance is calculated from
the individual determinations of v rather than from the median
values. This is necessary as the fit passes through the median v
values and if these were used the residual variance, and hence
the standard errors of the parameters, would be undefined. In
the example shown, the variances at each design point are approximately equal and could be pooled, but this situation will not
generally hold and it may be useful to apply weighting factors in
calculating $(\underline{X}^T\underline{X})^{-1}$. Since several replicate measurements will
be available for each set of experimental conditions, such
weighting factors may be calculated with moderate accuracy.

2.6. The Number of Replicates

In the preceding discussion no mention has been made of the
number of replicates (r) which should be performed at each design
point. Naturally the more replicates there are, the better will
be the estimate of the true velocity. When the experimental error
is small then fewer replicates will be necessary than when it is
large, but five replicates should be regarded as a minimum in
order to obtain even an approximate estimate of the variance (5).
The standard errors of the parameters are proportional to $(1/r)^{\frac{1}{2}}$
and this may be used as a guide for choosing the number of replicate measurements to be performed. In fact if an estimate of
the experimental variance is available then the standard errors
of the parameters may be predicted prior to conducting the experiment and r may be chosen accordingly.

TABLE 1:   Analysis of data conforming to the model for linear
           competitive inhibition*

Assumed values	$V = 10.0$	$K_a = 0.5$ mM	$K_i = 2.3$ mM
Design points	$A = 3.5$ mM $I = 0.0$ mM	$A = 0.389$ mM $I = 0.0$ mM	$A = 3.5$ mM $I = 18.4$ mM
Measured velocities	7.63 7.72 7.97 8.08 8.21	3.45 3.72 3.79 3.93 4.17	3.10 3.35 3.45 3.66 3.67
Median velocity	7.97	3.79	3.45
Variance	0.05907	0.07052	0.05623
		pooled variance = 0.06194	
$(X^T X)^{-1}$	0.569 0.0957 0.156	0.0957 0.0262 0.0648	0.156 0.0648 0.386
Parameter values	$V = 9.24$	$K_a = 0.560$ mM	$K_i = 1.94$ mM
Standard errors	0.19	0.040 mM	0.15 mM

An inhibitor (I) was assumed to act as a competitive inhibitor
with respect to the substrate (A). An experiment to determine
the parameter values was designed as described in the text for
$A_m = 3.5$ mM and $I_m = 20$ mM, assuming the indicated values for the
parameters. Five replicate determinations of v at each design
point were simulated to give the data listed in the table. The
median values of v were taken as estimates of the true values
which were then used to calculate the parameter values. The
variance of each set of replicates was calculated and these were
pooled. The matrix X, which consisted of the three partial deri-
vatives ($\partial v/\partial V$, $\partial v/\partial K_a$ and $\partial v/\partial K_i$) evaluated at each of the three
design points, was multiplied by its own transform and the resul-
ting matrix inverted to yield the $(X^T X)^{-1}$ matrix. Standard
errors of the parameters were calculated as the square root of
the product of the pooled variance and the diagonal elements of
the $(X^T X)^{-1}$ matrix.

---

* Reproduced from the Journal of Theoretical Biology (ref. 18) with
  the permission of the Publisher.

3. DISCUSSION

A crucial factor in the analysis of enzyme kinetic data is
the error associated with measuring the steady-state velocity.
This error is unknown and it is necessary to make some assumptions
regarding these errors in order to perform the analysis. Generally speaking, the assumptions made are governed by mathematical
rather than experimental considerations and this can lead to the
use of inappropriate methods for the determination of kinetic parameters. While this would appear to be a problem of data analysis,
the root of the problem lies in the manner in which enzyme kinetic
experiments are designed. The designs commonly employed usually
follow the principles outlined by Cleland (4) which involve varying
the concentrations of substrates and inhibitors over ranges which
span the kinetic parameters associated with these compounds. Such
designs are widely used and they are useful for model discrimination
as well as obtaining approximate values for the kinetic parameters.
However, if the form of the model is known then these designs are
often quite inappropriate as they lead to difficulties in analysis
as outlined above.

In this paper, an experimental design for parameter estimation is suggested which involves performing replicate measurements under as many experimental conditions as there are parameters
to be estimated. The results are analyzed by taking the median of
the replicates as an estimate the true value of v, and then solving
a set of simultaneous equations to obtain unique values for the
parameters.

The equations usually encountered in enzyme kinetics are nonlinear in the parameters and it is necessary to assume values for
the parameters in order to construct the best possible design.
If these estimated values differ from the true values, then the
design will be suboptimal, but will still be better than a standard kinetic design. Even standard designs must assume tentative
values for the parameters as a guide to the appropriate concentrations of substrates and inhibitors to employ. If the assumed
values for the parameters are grossly in error than this will
become apparent when the results are analyzed. The experimenter
may then decide whether more data are needed and will be able to
design a better experiment in the light of the newly-acquired
information.

The main advantage of the proposed designs is the fact that
they circumvent any difficulties which arise from uncertainties
in the error structure of the data. Like the method proposed by
Cornish-Bowden and Eisenthal (7) the analysis of experimental
results obtained using the designs proposed here is nonparametric
as it avoids unsubstantiated assumptions regarding the error
distribution of the observations. The procedure described here

has the additional advantage that it is readily applicable to any kinetic equation, no matter how complex.

4. CONCLUSIONS

It is now accepted that the experimental errors associated with enzyme kinetic measurements rarely exhibit constant variance. Rather, the variance depends on the experimental variables in a complex manner. In several cases which have been subjected to close scrutiny, there is good evidence that the errors are not normally distributed. For such systems, estimation of the kinetic parameters by simple least-squares analysis is inappropriate and distribution-free methods of analysis should be used. However, such methods have been described for relatively simple kinetic models only.

In this paper it has been shown that, by using the proper experimental design, the kinetic parameters may be estimated unambiguously while making minimal assumptions regarding the error structure of the data. A design criterion is presented for selecting experimental conditions which are optimal in the sense that they minimize the overall variance of the parameter estimates. The design involves replicate measurements of velocity under as many experimental conditions as there are kinetic parameters to be estimated. Such a design greatly simplifies the execution of the experiment and the analysis of the results.

REFERENCES

1. Reich, J.G. (1970) FEBS Lett. 9, 245-251.
2. Dowd, J.E. and Riggs, D.S. (1965) J. Biol. Chem. 240, 863-869.
3. Wilkinson, G.N. (1961) Biochem. J. 80, 324-332.
4. Cleland, W.W. (1967) Adv. Enzymol. 29, 1-32.
5. Storer, A.C., Darlison, M.G. and Cornish-Bowden, A. (1975) Biochem. J. 151, 361-367.
6. Askelöf, P., Korsfeldt, M. and Mannervik, B. (1976) Eur. J. Biochem. 69, 61-67.
7. Cornish-Bowden, A. and Eisenthal, R. (1974) Biochem. J. 139, 721-730.
8. Nimmo, I.A. and Mabood, S.F. (1979) Anal. Biochem. 94, 265-269.
9. Eisenthal, R. and Cornish-Bowden, A. (1974) Biochem. J. 139, 715-720.
10. Porter, W.R. and Trager, W.F. (1977) Biochem. J. 161, 293-302.
11. Cornish-Bowden, A., Porter, W.R. and Trager, W.F. (1978) J. Theor. Biol. 74, 163-175.
12. Box, G.E.P. and Lucas, H.L. (1959) Biometrika, 46, 77-90.

13. Bártfai, T. and Mannervik, B. (1972) FEBS Lett. 26, 252-256.
14. Markus, M. and Plesser, T. (1976) Biochem. Soc. Trans. 4, 361-364.
15. Kanyár, B. (1978) Acta Biochim. Biophys. Acad. Sci. Hung. 13, 153-160.
16. Endrenyi, L. (1980) In "Kinetic Data Analysis: Design and Analysis of Enzyme and Pharmacokinetic Experiments" (L. Endrenyi, ed.), pp. 137 - 167, Plenum, New York.
17. Cleland, W.W. (1975) Accounts Chem. Res. 8, 145-151.
18. Duggleby, R.G. (1979) J. Theor. Biol. 81, 671-684.

A PRIORI IDENTIFIABILITY ANALYSIS IN

PHARMACOKINETIC EXPERIMENT DESIGN

Claudio Cobelli

Laboratorio per Ricerche di Dinamica dei Sistemi
   e di Bioingegneria
Consiglio Nazionale delle Ricerche
Corso Stati Uniti 4
35100 Padova, Italy

ABSTRACT

   Mathematical modelling and dynamic identification experiments
are increasingly employed in quantitative pharmacokinetic studies.
This paper addresses the so-called identifiability problem which
has to be faced *a priori, i.e.*, once a certain pharmacokinetic
model structure has been postulated and the input-output experiment
planned, but prior to its performing. More precisely, identifiability
analysis addresses the question of whether it is possible to obtain
solutions for the unknown parameters of the chosen model structure
from data collected *via* those input-output tests which can be carried
out. The prerequisite value of identifiability analysis for the
design of a well-posed pharmacokinetic experiment is emphasized.
A precise set of identifiability definitions are given with reference
to a general pharmacokinetic experiment design model. Three classes
of pharmacokinetic models are discussed in some detail, namely the
nonlinear saturable models, the linear or linearizable models and
the linear compartmental models. Available methods for testing in
practice identifiability for these three classes of experiment design models are reviewed, compared and exemplified. Connections
between the identifiability property of a given model and the possibility of reconstructing/predicting system variables of interest
not directly accessible to measurement, which is one of the purposes
for which pharmacokinetic models are often built, are stressed.

1.   INTRODUCTION

   Mathematical models are increasingly employed in pharmacokinetic studies both at the organ and whole-organism level with the

purpose of evaluating parameters not directly accessible to measurements relating to the uptake, distribution and elimination of a drug, and of predicting its time courses and levels in non-accessible sites (1,2). For this kind of models, in contrast to the so-called dose-response models, usually one or more physiologically isomorphic structures are, at least tentatively, postulated which try to reflect through the incorporation of all available *a priori* knowledge of the system, consistent with purpose and level of the study, the various biochemical processes explicitly, *i.e.*, in a parametric form. In this context, lumped-parameter dynamic models, that is models described by ordinary linear and nonlinear differential equations, are widely employed in pharmacokinetic studies and will be considered in this paper.

However, it is worth noting that under certain circumstances, a lumped-parameter model may not be adequate, for instance, if the assumptions of homogeneous distribution or perfect mixing are not valid. In these cases, therefore, distributed models should be considered [see (3) for recent examples of lumped *vs*. distributed pharmacokinetic modelling approaches].

The type of models we shall deal with may be easily grasped by looking at Figure 1 where some hypothetical pharmacokinetic model structures are diagrammed. Once, for a given system, one or more well-posed model structures have been, at least tentatively, postulated on the basis of validated *a priori* knowledge, and thus the set of unknown parameters of the interest has been clearly delineated (*e.g.*, $V_1, k_{12}, k_{21}, k_{01}$ and k_{02} in the example of Figure 1a; $V_1, k_{12}, k_{01}, V_m, K_m$ in the system of Figure 1b), the input-output identification experiment has to be designed. The following problems/questions are relevant for obtaining accurate estimates for parameters of interest:

1.   Which are the accessible input ports into which test-input signals can be introduced, and which are the accessible output ports from which output signals can be measured? For example, if the three models of Figure 1 refer to an intact organism study, it could happen that in cases **a**, b and c, only 1 is accessible for the input-output experiment (*e.g.*, an intravenous injection of a drug dose and the measurement of the plasma concentration), and that in case d, whilst pools 1 and 3 are both accessible for a test input (*e.g.*, oral and intravenous doses, respectively), only pool 3 is accessible for output measurement.

2.   Which inputs ports must be probed and what types of test-input signals should be employed (*e.g.*, pulse dose or an **infusion**)?

3.   Which output ports must be measured, how long and at which times should the samples be collected and with what error?

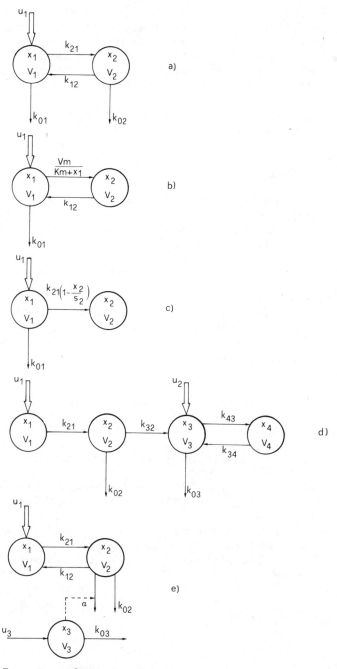

FIGURE 1:   Examples of pharmacokinetic models.   a) and d) are linear
    models; b) and c) are nonlinear saturable models and e) is a
    linear model with control.   See text for the dynamical equa-
    tions.

A *priori* identifiability is strictly linked to the experimental design problems outlined above. More precisely, it addresses questions which have to be answered *a priori*, *i.e.*, prior to performing the experiment, of whether it is possible to obtain solutions for the unknown parameters of the chosen model structure from data collected *via* those input-output tests which can be carried out.

The importance of identifiability analysis lies in the fact that, in a specific pharmacokinetic study, a clear distinction can be made *a priori* among those input-output experiments which cannot succeed and those which might. Consequently, parameter unidentifiability for a given model-experiment guarantees failure, whilst parameter identifiability is a necessary condition for the well-posedness of the experimental parameter estimation problem. The identifiability issue is of particular relevance in the context of biological systems, especially at the whole organism level, for which usually severe ethical and practical limitations exist, *e.g.*, on the number of accessible input-output ports for the identification experiments and on the type of test-input signals that can be applied.

The notion of *a priori* or structural identifiability was first formalized in a system theoretic framework in ref. (4). Since then, the identifiability literature and methods for testing have appeared in increasing number. This subject area has been recently reviewed and critically examined with special emphasis on application in biology and medicine (5). In the present work, after carefully presenting the identifiability problem in a formal context, emphasis will be placed on practical methods for testing the identifiability of dynamical models, with special attention to the class of linear compartmental models, which is of particular relevance in pharmacokinetics, and also to a class of nonlinear saturable systems.

2.   A KINETIC EXPERIMENTAL DESIGN MODEL AND THE IDENTIFIABILITY PROBLEM

In order to define the *a priori* identifiability problem it is convenient to resort to a formal description of a pharmacokinetic experiment. The following model will serve the purpose:

$$\dot{\underline{x}}(t) = \underline{f}[\underline{x}(t,\underline{p}), \underline{u}(t), t, \underline{p}], \quad \underline{x}_0 = \underline{x}(t_0,\underline{p}) \tag{1}$$

$$\underline{y}(t) = \underline{g}[\underline{x}(t,\underline{p}), \underline{p}], \quad\quad t_0 \leq t \leq T \tag{2}$$

$$\underline{h}[\underline{x}(t,\underline{p}), \underline{u}(t), \underline{p}] \geq 0 \tag{3}$$

where $\dot{x} \equiv d\underline{x}/dt$; $\underline{x} = [x_1, x_2, \ldots, x_n]^T$ (T denotes transpose) is the vector of state variables (*e.g.*, masses, concentrations); $\underline{u} = [u_1, u_2, \ldots, u_r]^T$

denotes the input vector which usually includes both the endogenous inputs (*e.g.*, production or secretion rates of materials) and the exogenous (test) inputs (*e.g.*, a pulse dose or an infusion); $\underline{y}=[y_1,y_2,\ldots,y_m]^T$ is the vector of m outputs or measurement variables (*e.g.*, concentrations in the plasma pool); $[t_0,T]$ denotes the observation interval; $\underline{p}=[p_1,p_2,\ldots,p_p]^T$ is the vector of unknown constant parameters (*e.g.*, transport parameters, enzyme-substrate interaction parameters); $\underline{f}$ defines the nonlinear known input-state coupling between $\underline{u}$ and $\underline{x}$, parameterized by $\underline{p}$, *i.e.*, the model structure; $\underline{g}$ defines the (often linear) known output-state couplings, also parameterized by $\underline{p}$, and $\underline{h}$ incorporates all additional and independent algebraic equality or inequality constraints relating $\underline{x}$, $\underline{u}$, $\underline{p}$ or any combination of these, known *a priori*.

The pharmacokinetic experimental design model [1]-[3], originally introduced as the constrained structure (5,6), is a fairly comprehensive one and allows us to state the identifiability problem in rather general terms. More precisely, given Eq.'s [1]-[3], *i.e.*, a well-defined model structure which exhibits P unknown parameters and the feasible identification experiments designed for parameter estimation, the problem of whether all the p_i can be estimated uniquely or not from the input-output data has to be solved prior to actually performing the experiment. Three situations are possible in regard to identifiability of the unknown parameter vector, which will be termed, respectively, unidentifiability, system identifiability and parameter identifiability according to the following definitions (5,6):

1.   The single parameter p_i of the constrained structure [1]-[3] is said to be *unidentifiable* in the interval $[t_0,T]$ if there exists an infinite number of solutions for p_i from these relationships. If one or more p_i is unidentifiable, then the model is said to be *system unidentifiable*.

2.   The single parameter p_i of the constrained structure [1]-[3] is said to be *identifiable* in the interval $[t_0,T]$ if there exists a finite number of solutions (greater than zero) for p_i from these relationships. If all p_i are identifiable, the model is said to be *system identifiable*.

3.   The single parameter p_i of the constrained structure [1]-[3] is said to be *uniquely identifiable* in the interval $[t_0,T]$ if there exists a unique solution for p_i from these relationships. If all p_i are uniquely identifiable, the model is said to be *parameter identifiable*.

It should be noted that if the model is unidentifiable, then there will always exist identifiable combinations of parameters which may have one or a finite number of solutions.

It is also worth remarking on the theoretical framework into which the *a priori* identifiability problem has been posed. More precisely it is assumed that the model structure describes perfectly the real system, *e.g.*, its drug kinetics, and that ideal (noise-free) measurements of the time course of the accessible variables, *e.g.*, plasma concentrations of the drug, can be made. In fact, the aim of identifiability analysis is to ensure that the unknown parameters are estimable from ideal input-output data, this being an obvious necessary prerequisite for any model used in a well-designed kinetic experiment, if the resulting real data are to be used to obtain meaningful values of the parameters of interest.

One example may serve to gain some confidence with the problem. Let us consider the linear pharmacokinetic model of Figure 2, and assume that only pool 1 (*e.g.*, plasma) is accessible for the input-output experiment (u_1, y_1), say a rapid injection of a known amount D of drug and the measurement of its concentration y_1. The experimental design model [1]-[3] particularizes to

$$\dot{x}_1(t) = -(k_{21}+k_{31}+k_{01})x_1(t)+k_{12}x_2(t)+k_{13}x_3(t)+u_1(t), \qquad [4]$$
$$x_1(0)=0$$

$$\dot{x}_2(t) = k_{21}x_1(t) - k_{12}x_2(t) , \qquad\qquad x_2(0)=0 \qquad [5]$$

$$\dot{x}_3(t) = k_{31}x_1(t) - (k_{13}+k_{03})x_3(t) , \qquad x_3(0)=0 \qquad [6]$$

$$y_1(t) = \frac{1}{V_1} x_1(t) \qquad\qquad\qquad\qquad\qquad [7]$$

$$k_{21},k_{31},k_{01},k_{12},k_{13},k_{03},V_1 \geq 0 \qquad\qquad\qquad [8]$$

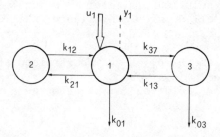

FIGURE 2:   A three-compartment linear model.

where $x_i(t)$ is the amount of drug in the i-th compartment, $u_1(t)$ is given by $D \cdot \delta(t)$ where $\delta(t)$ is the Dirac delta function, V_1 is the distribution volume of pool 1 and the k_{ij} are transport rate parameters. In this case, the unknown parameter vector $\underline{p}$ is given by $\underline{p} = [k_{12}, k_{21}, k_{01}, k_{31}, k_{13}, k_{03}, V_1]^T$ and the identifiability problem addresses the question of whether it is possible to estimate $\underline{p}$ from the relationships [4]-[8]. No error is assumed in the model struc-ture, *i.e.*, the model describes perfectly the drug kinetics, and ideal (noise-free) measurements are considered, whilst the real, *e.g.*, N discrete-time measurements $z_1(t_i)$ are related to the ideal (continuous time) data $y(t)$ by

$$z_1(t_i) = y_1(t_i) + e_1(t_i) \quad , \qquad i=1,2,\ldots,N \qquad [9]$$

where $e_1(t_i)$ is the measurement error at time t_i.

3.   SOME USEFUL CLASSES OF EXPERIMENTAL DESIGN MODELS

In this section we shall briefly review some classes of experi-mental design models which are of particular relevance to pharmaco-kinetic studies and for which methods for testing identifiability will be presented and discussed in Section 4.

3.1. Nonlinear Constrained Kinetic Models

The first class of models is the nonlinear constrained structure [1]-[3] already described in the previous section:

$$\underline{\dot{x}}(t) = \underline{f}[\underline{x}(t,\underline{p}), \underline{u}(t), t, \underline{p}], \qquad \underline{x}_0 = \underline{x}(t_0,\underline{p}) \qquad [10]$$

$$\underline{y}(t) = \underline{g}[\underline{x}(t,\underline{p}), \underline{p}] \quad , \qquad t_0 \le t \le T \qquad [11]$$

$$\underline{h}[\underline{x}(t,\underline{p}), \underline{u}(t), \underline{p}] \ge 0 \qquad . \qquad [12]$$

This experimental design model is a rather general one and allows the quantitative description of several nonlinear mechanisms involved in drug kinetics.

Very often, the observation equation [11] is linear in the state variables, that is, the measurement equation can be written:

$$\underline{y}(t) = \underline{C}(\underline{p}) \cdot \underline{x}(t,\underline{p}) \qquad [13]$$

where $\underline{C}$ is an m×n matrix.

An important category of nonlinear models which is widely used for analysing pharmacokinetic data and which is a particularization of [10], [12], [13], is described by

$$\dot{\underline{x}}(t) = \underline{A}(\underline{x}(t,\underline{p}), \underline{p}) \cdot \underline{x}(t) + \underline{B}(\underline{p}) \cdot u(t) \qquad\qquad [14]$$

$$\underline{y}(t) = \underline{C}(p) \cdot \underline{x}(t) \qquad\qquad [15]$$

$$\underline{h}[\underline{x}(t,\underline{p}), \underline{u}(t), \underline{p}] \geq 0 \qquad\qquad [16]$$

where $\underline{A}$ is an $n \times n$ matrix and $\underline{B}$ is an $n \times r$ matrix.

The well-known pharmacokinetic models exhibiting Michaelis-Menten or Langmuir type saturation fall, for instance, into this category. Examples of this class are the nonlinear models of Figures 1b and 1c which can be described through relationships [14]-[16], respectively, as

$$\dot{x}_1(t) = -\left[k_{01} + \frac{V_m}{K_m + x_1(t)}\right] x_1(t) + k_{12}x_2(t) + u_1(t) \quad , \qquad [17]$$

$$x_1(0) = 0$$

$$\dot{x}_2(t) = \frac{V_m}{K_m + x_1(t)} x_1(t) - k_{12}x_2(t) \quad , \qquad x_2(0) = 0 \qquad [18]$$

$$y_1(t) = \frac{1}{V_1} x_1(t) \qquad\qquad [19]$$

$$k_{01}, V_m, K_m, k_{12}, V_1 \geq 0 \qquad\qquad [20]$$

and as:

$$\dot{x}_1(t) = -\{k_{01} + k_{21}[1 - x_2(t)/s_2]\}x_1(t) + u_1(t) \quad , \qquad [21]$$

$$x_1(0) = 0$$

$$\dot{x}_2(t) = k_{21}[1 - x_2(t)/s_2]x_1(t) \quad , \qquad x_2(0) = 0 \qquad [22]$$

$$y_1(t) = \frac{1}{V_1} x_1(t) \qquad\qquad [23]$$

$$k_{01}, k_{21}, s_2, V_1 \geq 0 \qquad\qquad [24]$$

3.2. Linear Constrained Kinetic Models

In many instances the adoption of a linear dynamical model may
be judged adequate. One obvious case is when the intrinsic dynamics
of the system is essentially linear; then, linear models are valid
even when the applied test signals result in relatively large per-
turbations of system variables. Linear kinetic models are often
employed in pharmacokinetic studies. In these cases, a linear constrained structure is appropriate:

$$\underline{\dot{x}}(t) = \underline{A}(\underline{p}) \cdot \underline{x}(t) + \underline{B}(\underline{p}) \cdot \underline{u}(t) \quad , \qquad \underline{x}_0 = \underline{x}(t_0, \underline{p}) \qquad [25]$$

$$\underline{y}(t) = \underline{C}(\underline{p}) \cdot \underline{x}(t) \quad , \qquad t_0 \leq t \leq T \qquad [26]$$

$$\underline{h}[\underline{x}(t,\underline{p}), \underline{u}(t), \underline{p}] \geq 0 \qquad [27]$$

where $\underline{A}$, $\underline{B}$, $\underline{C}$ are constant coefficient matrices of suitable dimensions.

Representative examples would be the ones of Figures 1a, d and
e and Figure 2. The dynamical equations of the models of Figures 1a,
d and Figure 2 are of the same type and have been already reported
for the model of Figure 2 in Eqs. [4-8] ; the model of Figure 1e can
be written as

$$\dot{x}_1(t) = -k_{21}x_1(t) + k_{12}x_2(t) + u_1(t) \quad , \qquad x_1(0)=0 \qquad [28]$$

$$\dot{x}_2(t) = k_{21}x_1(t) - k_{12}x_2(t) + \alpha x_3(t) - k_{02}x_2(t), \qquad [29]$$

$$x_2(0)=0$$

$$\dot{x}_3(t) = -k_{03}x_3(t) + u_3(t) \quad , \qquad x_3(0)=x_{30} \qquad [30]$$

$$y_1(t) = \frac{1}{V_1} x_1(t) \qquad [31]$$

$$k_{21}, k_{12}, k_{02}, k_{03}, V_1 \geq 0 \qquad . \qquad [32]$$

It is worth noting that in this last example a control effect
($\alpha x_3(t)$) by a factor x_3 on the drug elimination from pool 2 has
been postulated.

There are also other cases in which a linear dynamic experimental design model of the type [25]-[27] is adequate, that is, when data

are obtained from a tracer or some other type of linearizing experiment. If a tracer experiment about some constant steady-state
operating point $(x_{1\infty}\cdots x_{n\infty}, u_{1\infty}\cdots u_{r\infty}, y_{1\infty}\cdots y_{m\infty})$ of the system [10]-
[12] is correctly performed, a linear constant-coefficient structure
of the type [25]-[27] can be easily proved to be appropriate (7,8),
where now $\underline{x}$, $\underline{u}$, $\underline{y}$ are deviations of the system variables from their
steady-state values, and the entries of $\underline{A}$, $\underline{B}$, $\underline{C}$ are partial derivatives of $\underline{f}$ and $\underline{g}$ with respect to $\underline{p}$ evaluated at the steady state.
These models are widely used especially in conjunction with radio-
active tracer experiments. As an example, consider a linearizing
experiment on the model of Figure 1b, where the steady-state operating point is assumed to be zero, $i.e.$, there is no endogenous
source of the test substance introduced as a tracer. The resulting
linearized model can be easily derived:

$$\dot{\Delta x}_1(t) = p_1 \Delta x_1(t) + p_2 \Delta x_2(t) + u_1(t) \quad , \quad \Delta x_1(0) = 0 \qquad [33]$$

$$\dot{\Delta x}_2(t) = p_3 \Delta x_1(t) + p_4 \Delta x_2(t) \quad , \qquad\qquad \Delta x_2(0) = 0 \qquad [34]$$

$$\Delta y_1(t) = \frac{1}{V_1} \Delta x_1(t) \qquad\qquad\qquad\qquad [35]$$

where Δ means deviation (sufficiently small) from the steady state
and $p_1 = -k_{01} - V_m/K_m$; $p_2 = k_{12}$; $p_3 = V_m/K_m$ and $p_4 = -k_{12}$.

It is worth noting here that, in general, a tracer experiment
will not yield sufficient information to estimate all the unknown
parameters of the original nonlinear structure ($e.g.$, k_{01}, k_{12}, V_m,
K_m, V_1 in the above example) and more than one linearizing experiment could be necessary.

3.3. Compartmental Models

A special class of models which is of particular importance
in pharmacokinetic experiment design as well as in other fields,
$e.g.$, endocrinology and metabolism, are the so-called compartmental
models (9,10). These models are variously defined in the literature,
for instance some authors (8) practically define as compartmental
the lumped parameter dynamic models described by Eq.'s [1]-[3],
whilst other authors (9) tend to distinguish between compartmental
systems without ($e.g.$, the examples of Figure 1a and d) and with
control ($e.g.$, the example of Figure 1e). I shall concentrate here
on a particular class of linear compartmental models which have
been referred to as exclusively compartmental (3), $i.e.$, the flux
of material from one compartment to another depends in a linear
manner solely upon the mass or concentration of material in the

source compartment. Representative examples of this class are the already considered linear models of Figure 1a, d and of Figure 2. This class of models can be obviously described by Eq.'s [25]-[27], but it constitutes a specially appealing subclass which is worth considering separately due to its structural properties. For this class of models matrix $\underline{A}$ has certain special features, $i.e.$:

$$a_{ij} \geq 0, \qquad\qquad\qquad i \neq j \qquad\qquad [36]$$

$$a_{ii} \leq 0 \quad \text{and} \quad |a_{ii}| \geq \sum_{\substack{j=1 \\ j \neq i}}^{n} a_{ji} , \quad i = j \qquad [37]$$

which are a direct result of the "exclusively compartmental" assumption with all the k_{ij} as nonnegative parameters. These models have nice structural and asymptotic properties (11), which have a relevant effect on their identifiability properties.

4. METHODS FOR TESTING FOR IDENTIFIABILITY

The problem is to assess on the only basis of the knowledge of the assumed model structure and of the chosen experiment configuration if the model is unidentifiable, $i.e.$, there is an infinite number of solutions, or if it is system identifiable, $i.e.$, if there is a finite number of solutions (in this case it would be desirable to know how many and possibly what they are) or if it is parameter identifiable, $i.e.$, there is one and only one solution. Emphasis will be placed on reviewing practical methods for testing identifiability both for linear and nonlinear systems (5). Explicit identifiability results which are available for the special class of linear exclusively compartmental models (see Section 3.3), will also be reviewed and some topological easy-to-test necessary conditions for testing for identifiability of this class will be presented and exemplified. It is worth noting that as we are concerned with structural properties only, the presented methods are independent of the numerical values of the parameters.

4.1. Linear Models

4.1.1. Markov Parameter Matrix and Transfer Function Matrix. In regard to linear dynamical models [25]-[27], two approaches will be discussed, the first based on the Markov parameters and the second one based on the transfer matrix. Other less practical methods have been reviewed (5). We shall assume in the following zero-initial conditions, $i.e.$, $\underline{x}_0 = 0$, which is a rather common situation in pharmacokinetic studies; in any case, a non-zero initial condition in the model can always be incorporated into the dynamic equations as an appended impulsive input. The tests we shall discuss, can accommodate only equality constraints, *i.e.*,

$$\underline{h}[\underline{x}(t,\underline{p}), \underline{u}(t), \underline{p}] = 0 ;$$

available inequality constraints in some special cases may be useful in distinguishing among multiple solutions for $\underline{p}$, in a structural sense, after a test has been applied.

The Markov parameter matrix approach makes references to the so-called Markov parameter matrix:

$$\underline{M}(\underline{p}) = \begin{bmatrix} \underline{C}(\underline{p}) \cdot \underline{B}(\underline{p}) \\ \underline{C}(\underline{p}) \cdot \underline{A}(\underline{p}) \cdot \underline{B}(\underline{p}) \\ \vdots \\ \underline{C}(\underline{p}) \cdot \underline{A}^{2n-1}(\underline{p}) \cdot \underline{B}(\underline{p}) \end{bmatrix} \qquad [38]$$

where equality constraints of the type [27] have been incorporated into the matrices $\underline{A}$, $\underline{B}$, $\underline{C}$. Then the model is system identifiable if and only if:

$$\text{rank} \begin{bmatrix} \dfrac{\partial M_1}{\partial p_1} & \cdots & \dfrac{\partial M_1}{\partial p_P} \\ \vdots & & \vdots \\ \dfrac{\partial M_{2n}}{\partial p_1} & \cdots & \dfrac{\partial M_{2n}}{\partial p_P} \end{bmatrix} = P \qquad [39]$$

where $\underline{M}_1 = \underline{CB}$; $\underline{M}_2 = \underline{CAB}, \ldots, \underline{M}_{2n} = \underline{CA}^{2n-1}\underline{B}$, for all $\underline{p}$ in the admissible parameter space. Otherwise the model is unidentifiable.

In general, if the model is system identifiable, to test if some or all the unknown parameters are uniquely identifiable, the linear and nonlinear relationships contained in [38] must be solved for the p_i's along with the available parameter constraints. The system and parameter identifiability tests based on the Markov parameters are usually cumbersome. A generally simpler approach is based on the analysis of the $r \times m$ transfer function matrix:

$$\underline{H}(s,\underline{p}) = [H_{ij}(s,\underline{p})] = \begin{bmatrix} \dfrac{Ly_i(t,\underline{p})}{Lu_j(t)} \end{bmatrix} = \underline{C}(\underline{p})[s\underline{I}-\underline{A}]^{-1}\underline{B}(\underline{p}) \qquad [40]$$

where each element H_{ij} of $\underline{H}$ is the Laplace transform of the response in the measurement variable at port i, $y_i(t,\underline{p})$, to a unit impulse test input (Dirac delta function) at port j, $u_j(t)=\delta(t)$, and $\underline{I}$ is the identity matrix. Thus, each element $H_{ij}(s,p)$ reflects an experiment performed on the system between input port j and output port i.

The transfer function matrix approach makes reference to the coefficients of the numerator and denominator polynomials of each of the mr elements $H_{ij}(s,p)$ of the transfer function matrix, $\beta_1^{ij}(p),\ldots,\beta_{n-1}^{ij}(p)$ and $\alpha_1^{ij}(p),\ldots,\alpha_n^{ij}(p)$, respectively, The (2n-1)rm×P Jacobian matrix is formed:

$$
\underline{G}(\underline{p}) =
\begin{bmatrix}
\dfrac{\partial \beta_1^{11}}{\partial p_1} & \cdots & \dfrac{\partial \beta_1^{11}}{\partial p_P} \\[2ex]
\vdots & & \vdots \\[1ex]
\dfrac{\partial \alpha_m^{11}}{\partial p_1} & \cdots & \dfrac{\partial \alpha_n^{11}}{\partial p_P} \\[2ex]
\vdots & & \vdots \\[1ex]
\dfrac{\partial \beta_1^{rm}}{\partial p_1} & \cdots & \dfrac{\partial \beta_1^{rm}}{\partial p_P} \\[2ex]
\vdots & & \vdots \\[1ex]
\dfrac{\partial \alpha_n^{rm}}{\partial p_1} & \cdots & \dfrac{\partial \alpha_n^{rm}}{\partial p_P}
\end{bmatrix}
\qquad [41]
$$

Then the model is system identifiable if and only if

$$\text{rank } \underline{G}(\underline{p}) = P \qquad\qquad [42]$$

for all $\underline{p}$ in the parameter space. The test based on Eq. [42] is usually simpler than the one based on Eq. [39].

Again, possible unique identifiability of certain parameters can be in general tested only by solving for the p_i's the nonlinear relationships which define the α's and β's along with all parameter constraints on the p_i's:

$$\beta_1^{11}(\underline{p}) = c_1^{11}$$

$$\vdots \qquad \vdots$$

$$\alpha_n^{11}(\underline{p}) = c_{2n-1}^{11}$$

$$\vdots \qquad \vdots$$

$$\beta_1^{rm}(\underline{p}) = c_1^{rm}$$

$$\vdots \qquad \vdots \qquad\qquad\qquad\qquad [43]$$

$$\alpha_n^{rm}(\underline{p}) = c_{2n-1}^{rm}$$

$$h_1(\underline{x}(t,\underline{p}), \underline{u}(t), \underline{p}] = 0$$

$$\vdots$$

$$h_\ell(\underline{x}(t,\underline{p}), \underline{u}(t), \underline{p}] = 0$$

where the c_k^{ij} are measurable constants.

In case some of the p_i's are not uniquely identifiable, the number of possible distinct solutions can be evauated from Eq,'s [43]. The analysis of [43] will reveal for a model which is unidentifiable or system identifiable, combinations of parameters which can be identified uniquely. The following example reported also in (5) will serve to illustrate the practical application of the two tests.

4.1.2. Example 1. Consider the following two inputs-two outputs experiment design model:

$$\dot{x}_1(t) = p_1 x_1(t) + u_1(t) \qquad\qquad , \qquad\qquad x_1(0)=0 \qquad [44]$$

$$\dot{x}_2(t) = p_2 x_1(t) + p_3 x_2 + u_2(t), \qquad\qquad x_2(0)=0 \qquad [45]$$

$$y_1 = x_1 \qquad\qquad\qquad\qquad\qquad\qquad\qquad\qquad\qquad [46]$$

$$y_2 = x_2 \qquad\qquad\qquad\qquad\qquad\qquad\qquad\qquad\qquad [47]$$

The Markov parameter test, Eq. [39], is applied as follows. Matrices $\underline{A}$, $\underline{B}$ and $\underline{C}$ are given by, according to Eq.s' [25]-[27],

$$\underline{A} = \begin{bmatrix} P_1 & 0 \\ P_2 & P_3 \end{bmatrix} \quad ; \quad \underline{B} = \begin{bmatrix} 1 & 0 \\ 0 & 1 \end{bmatrix} \quad ; \quad \underline{C} = \begin{bmatrix} 1 & 0 \\ 0 & 1 \end{bmatrix} \qquad [48]$$

and the Markov parameters are:

$$\underline{M}_1 = \underline{CB} = \begin{bmatrix} 1 & 0 \\ 0 & 1 \end{bmatrix} \quad ; \quad \underline{M}_2 = \underline{CAB} = \begin{bmatrix} P_1 & 0 \\ P_2 & P_3 \end{bmatrix} \quad ; \qquad [49]$$

$$\underline{M}_3 = \underline{CA^2B} = \begin{bmatrix} P_1^2 & 0 \\ P_1P_2+P_2P_3 & P_3^2 \end{bmatrix} \quad ;$$

$$\underline{M}_4 = \underline{CA^3B} = \begin{bmatrix} P_1^3 & 0 \\ P_1^2P_2+P_1P_2P_3+P_3^3 & P_3^3 \end{bmatrix}$$

The test of Eq. [39] can be now applied: deleting all zero rows and columns for convenience we have:

$$\text{rank} \begin{bmatrix} 1 & 0 & 1 & 0 & 1 \\ 2P_1 & P_2 & P_1+P_3 & P_2 & 2P_3 \\ 3P_1^2 & 2P_1P_2+P_2P_3 & P_1^2+P_1P_3 & P_1P_2+3P_3^2 & 3P_3^2 \end{bmatrix} = 3 \qquad [50]$$

because each of the rows is linearly independent. Therefore the model is system identifiable. Inspection of the Markov parameter matrix, Eq. [38], $\underline{M}=[\underline{M}_1 \ \underline{M}_2 \ \underline{M}_3 \ \underline{M}_4]^T$ indicates that the model is also parameter identifiable; in fact $\underline{M}_2$ provides three linearly independent equations in p_1, p_2 and p_3.

The test based on the transfer function coefficients, Eq. [42], requires first the computation of the transfer function matrix, Eq. [40], which is given by:

$$\underline{H}(s,\underline{p}) = \begin{bmatrix} H_{11}(s,\underline{p}) & H_{12}(s,\underline{p}) \\ H_{21}(s,\underline{p}) & H_{22}(s,\underline{p}) \end{bmatrix} = \begin{bmatrix} \dfrac{1}{s-P_1} & 0 \\ \dfrac{P_2}{s^2-(P_1+P_3)s+P_1P_3} & \dfrac{1}{s-P_3} \end{bmatrix} \qquad [51]$$

Now the test, Eq. [42], can be applied: deleting all zero rows and columns we have

$$\underline{G}(\underline{p}) = \begin{bmatrix} -1 & 0 & 0 \\ 0 & 1 & 0 \\ p_3 & 0 & p_1 \\ -1 & 0 & -1 \\ 0 & 0 & -1 \end{bmatrix} \qquad [52]$$

and it can be easily shown that

rank $\underline{G}(\underline{p})$ = 3 .

The model is therefore system identifiable. In order to test for parameter identifiability, Eq.'s [43] must be analyzed. From Eq. [51] we have, in particular:

$$-p_1 = c_1^{11}$$

$$p_2 = c_1^{21} \qquad\qquad [53]$$

$$-p_3 = c_1^{22}$$

and thus it is clear that the model is also parameter identifiable. In order to note some specific properties/features of identifiability analysis which may not be so apparent and which are very important for practical applications, let us consider the following two case studies on linear (compartmental) models.

4.1.3. Case Study 1. Consider a proposed (12) digitoxin kinetic model (Figure 3), described by

$$\dot{x}_1(t) = -(k_{01}+k_{21}+k_{31})x_1(t) + u_1(t) \quad , \qquad x_1(0)=0 \qquad [54]$$

$$\dot{x}_2(t) = k_{21}x_1(t) \qquad\qquad\qquad\qquad , \qquad x_2(0)=0 \qquad [55]$$

$$\dot{x}_3(t) = k_{31}x_1(t) - (k_{03}+k_{43})x_3(t) \quad , \qquad x_3(0)=0 \qquad [56]$$

$$\dot{x}_4(t) = k_{43}x_3(t) \qquad\qquad\qquad\qquad , \qquad x_4(0)=0 \qquad [57]$$

$$y_1 = \frac{1}{V} x_2(t) \qquad\qquad\qquad\qquad\qquad\qquad\qquad [58]$$

$$y_2 = \frac{1}{V} x_4(t) \qquad\qquad\qquad\qquad\qquad\qquad\qquad [59]$$

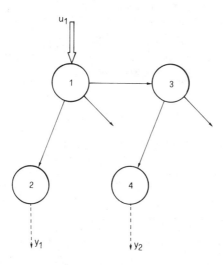

FIGURE 3:   Digitoxin kinetics model (12).

where x_1 represents digitoxin in the body, x_2 is urinary digitoxin, x_3 is digoxin in the body, x_4 is urinary digoxin, u_1 is an injection (of unknown amount) of digitoxin, y_1 and y_2 are the measurable digitoxin and digoxin urinary concentrations and V is the known urinary volume.

The model exhibits five unknown parameters, $i.e.$, $\underline{p}=[k_{01},k_{21},k_{31},k_{03},k_{43}]^T$. It may be proved, for instance by re-sorting to the test based on the transfer function coefficients, Eq. [42] (the one based on the Markov parameters requires computations up to $\underline{CA^7B}$), that the model is unidentifiable. It is, in general, a good strategy to inspect carefully equations of the type [43]. In this case, for instance, from $H_{11}(s,\underline{p})$ and $H_{21}(s,\underline{p})$, we have:

$$k_{21}/V = c_1^{11}$$

$$k_{01}+k_{21}+k_{31} = c_2^{11}$$

$$k_{31}k_{43}/V = c_1^{21} \tag{60}$$

$$(k_{01}+k_{21}+k_{31})(k_{43}+k_{03}) = c_2^{21}$$

$$k_{01}+k_{21}+k_{31}+k_{43}+k_{03} = c_3^{21}$$

and therefore the set of uniquely identifiable parameter combina-
tions can be determined: $k_{21}, k_{01} + k_{31}, k_{31} k_{43}$ and $k_{43} + k_{03}$. Thus,
identifiability analysis reveals that if no other *a priori* know-
ledge is available, the planned experiment will certainly be unsuc-
cessful in quantifying the individual transport rate parameters,
but it could be successful in quantifying the above set of parameter combinations.

This example allows us also to illustrate another important
point, *i.e.*, the effect of using independent additional *a priori*
knowledge on the identifiability properties. An independent physiologically based constraint for identifying the proposed model has
been considered (12):

$$k_{21} + k_{01} = \alpha \qquad\qquad\qquad\qquad [61]$$

where α is an independently measured constant.

It may be easily shown from Eq.'s [60] and [61] that the new
model [54]-[59] and [61] is now parameter identifiable, *i.e.*, all
the individual transport rate parameters can be now estimated uniquely. Therefore it is very important that identifiability analysis
explicitly includes all available *a priori* information, and this
was in fact the original motivation for introducing the constrained
structure and the identifiability definitions (5,6) which have been
reviewed in Section 2.

4.1.4. <u>Case Study 2</u>. Consider the linear compartmental model
diagrammed in Figure 4 and described by

$$\dot{x}_1(t) = -(k_{21} + k_{31}) x_1(t) + k_{12} x_2(t) + k_{13} x_3(t) + u_1(t) \quad,$$
$$x_1(0) = 0 \qquad [62]$$

$$\dot{x}_2(t) = k_{21} x_1(t) - (k_{12} + k_{02}) x_2(t) \quad, \qquad x_2(0) = 0 \qquad [63]$$

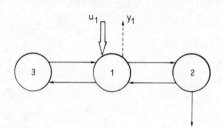

FIGURE 4:   A system identifiable three-compartment linear model.

$$\dot{x}_3(t) = k_{31}x_1(t) - k_{13}x_3(t) \qquad , \qquad x_3(0)=0 \qquad [64]$$

$$y_1 = \frac{1}{V_1} x_1 \qquad \qquad . \qquad\qquad\qquad [65]$$

The model exhibits six unknowns, *i.e.*, $\underline{p}=[k_{13},k_{31},k_{21},k_{12},k_{02},V_1]^T$. It may be proved, for instance through [41]-[43], that the model is system identifiable but not parameter identifiable. More precisely, V_1 can be estimated uniquely but there are two solutions for each of the k_{ij}, both of which are physiologically feasible; this result was first reported in (13) and full details on the two sets of solutions in a transfer function language may be found in (5). In these cases, additional information about the system or its parameters is needed in order to distinguish between these two solutions. A good example is provided by (14), where a model like the one of Figure 4 has been employed for studying unconjugated bilirubin kinetics. In this case, one of the two solutions was rejected because the corresponding predicted time pattern of the substance in compartment 2 (liver), in compartment 3 (extravascular tissues) and the liver-to-bile efflux were both in magnitude and shape against experimental evidence. In general, however, in other or more complicated circumstances it could be impossible or very difficult to distinguish between two or more solutions.

This example has thus shown that system identifiability, *i.e.*, the existence of a finite number of solutions for the unknown parameters, may occur in practical situations. The following remark is of particular importance in regard to a model which is only system identifiable, especially when the model is used for predictive purposes. It has been shown (5,6) that a system identifiable model does not necessarily guarantee that the time courses of system variables (not parameters) of interest inaccessible to direct measurement (*e.g.*, tissue concentrations) can be unambigously reconstructed, *i.e.*, the predictions of these variables are not necessarily unique. If the model is not parameter identifiable, several ambiguities may arise both in magnitude and waveform [see (5,6) for a simple example on a glucose-insulin model] and thus predicting state variables *via* only system identifiable models must be evaluated very carefully. This issue is of particular relevance in pharmacokinetic studies, as often the purpose for which models are built is for predicting drug concentrations in compartments which are inaccessible to direct measurements.

4.2. Compartmental Models

For the special class of linear exclusively compartmental models introduced in Section 3, the same methods for testing for identifiability presented for the general linear case obviously still apply, and in fact have already been applied in the two case studies. In this section some explicit identifiability results and some necessary topological conditions for linear compartmental models will be reviewed which have been obtained due to the peculiar structure and properties of this class of models.

4.2.1. Identifiability of Catenary and Mamillary Systems.

Among the explicit identifiability results available for specially constrained compartmental systems, the ones related to catenary and mamillary systems seem to be of particular relevance in this context as catenary and mamillary compartmental systems are often applied in pharmacokinetic studies. The basic structures of a catenary and a mamillary system are diagrammed in Figures 5 and 6, respectively.

The following results for single input-single output experiments have been proved (15), where also other more general classes of constrained compartmental systems have been investigated:

1.   A catenary compartmental system, closed (no leak) or almost closed (only one leak), is parameter identifiable if the single input/single output experiment is performed in an extremal compartment (compartment 1 or n in Figure 5), but it is only system identifiable if the experiment is performed in a single intermediate compartment, and the number of different solutions increases with the distance of the experiment from either end.

2.   A mamillary compartmental system, closed or almost closed, is always system identifiable and the number of different solutions is (n-1)! if the experiment is performed in the central compartment (compartment 1 in Figure 6) and (n-2)! if it is performed in a peripheral one.

3.   Combinations of catenary and mamillary systems (radial- and tree-type compartmental systems) are always system identifiable by means of a single input/single output experiment performed in any compartment and the number of different solutions can be evaluated.

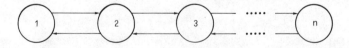

FIGURE 5:   The basic structure of a catenary compartmental model.

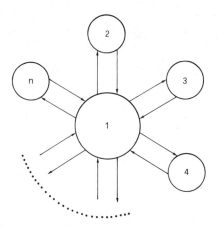

FIGURE 6:   The basic structure of a mamillary compartmental model.

4.2.2. Simplified Identifiability Analysis for Linear Compartmental
Models. The methods for testing for system and parameter identifiability requires, in general, a considerable amount of symbolic
computation. For the class of linear compartmental models, two
set of results are available which can speed up the analysis and
also avoid unnecessary computations.

        For writing the symbolic expressions of the nonlinear relation-
ships [43] of the α's and β's as a function of the unknown parameter
vector $\underline{p}$, it is usually necessary to invert (symbolically) a matrix
of dimension n, that is of dimension equal to the number of compartments. To speed up this procedure, a new technique has been proposed
which avoids the inversion and allows a fast-writing of the symbolic
expressions [43] directly from the diagram of the compartmental
structure and experiment configuration. The details of this fast-
writing procedure and practical examples may be found in (16).

        Identifiability analysis of linear compartmental models can be
greatly simplified by using a set of necessary topological conditions
to system and parameter identifiability (17-19), which can be easily
applied directly to the compartmental diagram without symbolically
writing the Markov parameters or the transfer function coefficients.
This set of necessary conditions is an easy and quick test for model
unidentifiability and its use is therefore suggested as a preliminary
test in identifiability analysis. For presenting these conditions
in a simple way, the following definitions are given (see also Figure
7):

Compartmental diagram: a set of compartments and directed arcs corresponding to the flows, where experiment inputs and outputs are
shown.

Path: a succession of arcs such that the compartment entered by an arc is the same from which the next arc is starting and every compartment is crossed only once; the *length* of the path is the number of its intercompartmental arcs.

Input-and output-connectability: a compartment is input- and/or output-connectable if there exists a path from a compartment directly entered by an input to the considered one and/or a path from it to a measured compartment; a system is input-output connectable if all its compartments are input and output connectable.

hk subsystem: the subsystem formed by all compartments connectable to input k and to output h; the hk subsystems are arbitrarily ordered from 1 to r×m.

Closed subsystem: a set of compartments such that no arcs connect a compartment of the set either to compartments outside it or to the environment; closed subsystems may reduce to one compartment (closed compartment) and may be contained in a larger closed subsystem.

Common cascade part (between two hk and lm subsystems): a set of compartments such that each of its compartments is influenced by compartments outside it only through one compartment f and influences compartments outside the common cascade part only through one compartment g. Compartment f may coincide with input compartments k and m of both subsystems; similarly g may coincide with the output compartments h and l of both subsystems.

By resorting to the above notions, the following two necessary conditions for identifiability have been proposed:

1.   the system has to be input and output connectable;
2.   the number n_u of unknown parameters shall not exceed the number n_e:

$$n_e = n - n' + \sum_{\substack{h=1,m \\ k=1,r}} w_{hk} - \sum_{\substack{h=1,m \\ k=1,r}} z_{hk} \qquad [66]$$

where: n is the number of compartments of the system;

   n' is the number of closed subsystems (if a closed subsystem, with measurement output, is contained in a larger subsystem, the latter is not taken into account; for instance in Figure 7, n'=2 corresponding to subsystems {6} and {4,5});

   w_{hk} refers to the hk subsystem and corresponds to the number n_{hk} of their compartments minus the length of the short-

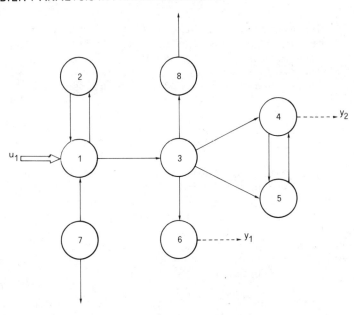

FIGURE 7:   Definitions of Section 4 are exemplified with reference
            to an 8-compartment diagram.

            Input 1 in compartment 1;
            Output 1 in compartment 6;
            Output 2 in compartment 4;
            Path from 1 to 6: k_{31}, k_{63} (length=2);

            Path from 1 to 4: $\begin{cases} k_{31}, k_{43} & \text{(length=2)}; \\ k_{31}, k_{53}, k_{45} & \text{(length=3)}; \end{cases}$

            Input-connectable compartments: 1,2,3,4,5,6,8;
            Output-connectable compartments: 1,2,3,4,5,6,7;
            Input-output connectable subsystem: {1,2,3,4,5,6};
            hk subsystems: h=1,k=1: {1,2,3,6};
                             h=2,k=1: {1,2,3,4,5};
            Closed subsystems: {6}, {4,5};
            Common cascade parts between 1,1 and 2,1;
            Subsystems: {1,2,3}; $f \equiv 1$; $g \equiv 3$.

est hk path; if the length of the shortest path is zero
(input and output in the same j-th compartment) and if
the output coefficient c_{hj} is known, then w_{hk} is computed as above minus one;

z_{hk} refers to the possible cascade parts, common between
the hk subsystem and any of the previously considered
ones (following the chosen order); for each common
cascade part the corresponding addendum of z_{hk} is the

difference between the number of compartments of the
common cascade part minus the length of the shortest
path between f and g, minus one.

If necessary conditions 1) and 2) are not satisfied, the model
is unidentifiable; if 1) and 2) are satisfied, the test based for
instance on the transfer function coefficients coupled with the
fast-writing procedure previously described has to be applied in
order to investigate identifiability.

As an illustration of the test, let us consider the example of
Figure 7. In this case it is immediately evident that the system
is not identifiable, as it is not input-output connectable. However
if compartments 7 and 8 and their respective outgoing arcs are
neglected, then the remaining system is input-output connectable
and the second necessary condition may be tested. Let us assume
that the output coefficients are known, then the system exhibits
the following set of unknown parameters: $p=[k_{12},k_{21},k_{31},k_{43},k_{45},$
$k_{53},k_{54},k_{63},k_{03}]^T$ (k_{03} instead of k_{83}), then $n_u=9$. For the computation of n_e we have: $n=6$; $n'=2$; $w_{11}=4-2=2$; $w_{21}=5-2=3$; $z_{11}\equiv0$;
$z_{21}=3-1-1=1$; and finally $n_e=6-2+2+3=8$. Therefore the model is not
identifiable.

As a further illustration of this quick test for unidentifiability we can apply it to the already discussed models of Figures 2
and 3. The model of Figure 2 has $n_u=7$ and is input-output connect-
able; for n_e we have: $n=3$; $n'=0$, $w_{11}=3$; $z_{11}\equiv0$; thus $n_e=6<n_u$ and the
model is unidentifiable. The digitoxin kinetics model (Figure 3)
exhibits five unknown parameters, $i.e.$, $n_u=5$ and is input-output
connectable; for n_e we have: $n=4$; $n'=2$; $w_{11}=2-1=1$; $w_{21}=3-2=1$;
$z_{11}\equiv0$; $z_{21}=0$, thus $n_e=4<n_u=5$ and the model is unidentifiable.

4.3. Nonlinear Models

4.3.1. A Test for System/Parameter Identifiability. For nonlinear
experiment design models [10]-[12], a set of conditions for identifiability has been recently become available (20), which are based
on the analysis of the power series expansion of the output function y evaluated at time t_0, as a function of the unknown parameters.
More precisely, a sufficient condition for the model to be system
or parameter identifiable is that the set of equations

$$\frac{d^k y(x(t_0,p),p)}{dt^k} = c_k(t_0) \qquad k=0,1,2,\ldots \qquad [67]$$

has a finite or a unique solution for p.

For the important class of models [14]-[16] described in Section 3, this set of sufficient conditions is also necessary, and thus important classes of saturable pharmacokinetic systems may be tested *a priori* for identifiability *via* this method. Let us consider a simple example in order to familiarize us with the test.

4.3.2. Example 2. Consider a pharmacokinetic model described by:

$$\dot{x}_1(t) = -\frac{V_m x_1(t)}{K_m + x_1(t)} + D \cdot \delta(t) \quad , \qquad x_1(0) = 0 \qquad [68]$$

$$y_1(t) = \frac{1}{V_1} x_1(t) \qquad\qquad\qquad\qquad\qquad\qquad [69]$$

where $D \cdot \delta(t)$ refers to a known bolus dose of drug applied at time 0.

The model may also be written in the form:

$$\dot{x}_1(t) = -\frac{V_m x_1(t)}{K_m + x_1(t)} \quad , \qquad x_1(0_+) = D \qquad [70]$$

$$y_1(t) = \frac{1}{V_1} x_1(t) \qquad . \qquad\qquad\qquad\qquad\qquad [71]$$

The unknown parameter vector is given by $\underline{p} = [V_m, K_m, V_1]^T$. The test [67] assumes that $y_1(0_+)$ and its successive time derivatives are (ideally) available (like the c_i^{ij} in [43]) and gives

$$y_1(0_+) = \frac{D}{V_1} = c_0$$

$$\dot{y}_1(0_+) = -\frac{V_m x_1(0_+)}{K_m + x_1(0_+)} \cdot \frac{1}{V_1} = c_1$$

$$\ddot{y}_1(0_+) = -\frac{V_m K_m \dot{y}_1(0_+)}{[K_m + x_1(0_+)]^2} = c_2 \qquad\qquad [72]$$

$$\dddot{y}_1(0_+) = \frac{\ddot{y}_1(0_+)}{\dot{y}_1(0_+)} \left[\ddot{x}_1(0_+) - \frac{2\dot{x}_1^2(0_+)}{K_m + x_1(0_+)} \right] \cdot \frac{1}{V_1} = c_3$$

It may be verified from Eq.'s [72] that V_m, K_m and V_1 can be uniquely identified and thus the model is parameter identifiable.

An interesting point was made (21) concerning identifiability properties of the model of Figure 1c and of a linearized version of it. The nonlinear model may be proved to be parameter identifiable, *i.e.*, k_{01}, k_{21}, s_2 and V_1 may be estimated uniquely, but a linearized version of it obtained under negligible saturable state feedback, *i.e.*, with $x_2(t) \ll s_2$, is only system identifiable. Other applications of the test [67] may be found in (21) and also in (22,24) where several nonlinear glucose-insulin models were tested for identifiability.

5. CONCLUSIONS

In this paper the *a priori* identifiability problem for pharmacokinetic experiment design models has been analyzed. *A priori* identifiability addresses the question, before actually performing an experiment, whether, given only the structure of a proposed model and the planned experimental design, it is possible to obtain unique estimates of all model parameters on the basis of those input-output tests which can be carried out. The prerequisite value of the identifiability concept in the design of a well-posed pharmacokinetic identification experiment is obvious as a clear distinction can be made *a priori* among those experiments that cannot succeed and those that might, *i.e.*, parameter unidentifiability for a given model-experiment guarantees failure, whilst parameter identifiability is a necessary condition for the success of the design problem for pharmacokinetic experiments.

A precise set of identifiability definitions in terms of unidentifiability, system identifiability and parameter identifiability has been given with reference to a general experiment design model which incorporates all *a priori* available structural information. Three classes of pharmacokinetic models and input-output experiments have been discussed in some detail: nonlinear (saturable) models, linear (or linearized) models and the classical linear compartmental models.

Available practical methods for testing identifiability have been reviewed, compared and exemplified. Necessary and sufficient (algebraic) conditions both for linear and nonlinear models have been given; for the class of linear compartmental models some *ad hoc* procedures that may expedite the test have been presented. It has been also remarked that prediction of system variables of interest which are not accessible to direct measurement (a common use of pharmacokinetic models) depends intimately on the identifiability properties. It is worth concluding that *a priori* parameter identifiability, although necessary for the well-posedness of the parameter estimation problem and very useful in delineating *a priori* all possible alternative experimental configurations, is

not sufficient to ensure that parameter estimation can be successfully achieved from the actual experimental data or even more, that
the so-called model validation problem has been automatically
solved. In fact, the theoretical framework in which *a priori* identifiability analysis is carried out assumes that the model structure
is perfect and that noise-free measurements are available for estimating the unknown parameters; therefore an *a priori* identifiable
model can be rejected for several reasons, *e.g.*, the accuracy with
which its parameters can be estimated from the actual data is very
poor due either to a falsely postulated structure or to a structure
which is too complex for the available data set or simply because
of the paucity and bad quality of the collected data. However, in
the overall model validation process which relates to the ability
of determining from input-output data a model which represents an
adequate description of the system, and which requires the application of a whole battery of both theory-based (*e.g.*, sensitivity)
and data-based tests (*e.g.*, *a posteriori* identifiability, *i.e.*,
accuracy, goodness-of-fit *vs.* number of parameters, whiteness of
residuals) so as to reveal theoretical and statistical deficiences
of a given model [see (23,24) for the application of some of the
above tests to several glucose-insulin models], *a priori* parameter
identifiability represents a necessary preliminary theory-based
step.

REFERENCES

1. Wagner, J.G. (1975) "Fundamentals of Clinical Pharmacokinetics",
       Drug Intelligence Publ., Hamilton, Ill.
2. Gibaldi, M. and Perrier, D. (1975) "Pharmacokinetics",Dekker,
       New York.
3. Carson, E., Cobelli, C. and Finkelstein, L., Am. J. Physiol.
       (to appear).
4. Bellman, R. and Åström, K.J. (1970) Math. Biosci. 7, 329-339.
5. Cobelli, C. and DiStefano, J.J. III, Am. J. Physiol. (to appear).
6. DiStefano, J.J. III and Cobelli, C., IEEE Trans. Autom. Contr.
       (to appear).
7. DiStefano, J.J. III (1976) Am. J. Physiol. 230, 476-485.
8. Berman, M. (1971) In "Advances in Biological and Medical
       Physics" (J.S. Laughlin, ed.) Proc. 2nd Intl. Conf. on
       Medical Physics, pp. 279-296, Boston, Mass.
9. Jacquez, J.A. (1972) "Compartmental Analysis in Biology and
       Medicine", Elsevier, Amsterdam.
10. Rescigno, A. and Segre, G. (1966) "Drug and Tracer Kinetics",
       Blaisdell, Waltham, Mass.
11. Hearon, J.Z. (1963) Ann. N.Y. Acad. Sci. 108, 36-68.
12. Jeliffe, R.W., Buell, J., Kalaba, R., Shridar.R. and Rockwell,
       R. (1970) Math. Biosci. 6, 387-403.
13. Skinner, S.M., Clark, R.E., Baker, N. and Shipley, R.A. (1959)
       Am. J. Physiol. 196, 238-244.

14. Berk, P.D., Hower, R.B., Bloomer, J.R. and Berlin, N.I. (1969)
     J. Clin. Invest. $\underline{48}$, 2176-2190.
15. Cobelli, C., Lepschy, A. and Romanin Jacur, G. (1979) Math.
     Biosci. $\underline{47}$, 173-195.
16. Bossi, A., Cobelli, C., Colussi, L. and Romanin Jacur, G. (1979)
     Math. Biosci. $\underline{43}$, 187-198.
17. Cobelli, C., Lepschy, A. and Romanin Jacur, G. (1979) In
     "Theoretical Systems Ecology" (E. Halfon, ed.) pp. 237-258,
     Academic Press, New York.
18. Cobelli, C., Lepschy, A. and Romanin Jacur, G. (1979) Math.
     Biosci, $\underline{44}$, 1-18.
19. Cobelli, C., Lepschy, A. and Romanin Jacur, G. (1980) Fed.
     Proc. $\underline{39}$, 85-90.
20. Pohjanpalo, H. (1978) Math. Biosci. $\underline{41}$, 21-34.
21. Brown, R.F. (1979) In "Preprints 5th IFAC Symposium on Identi-
     fication and System Parameter Estimation" (R. Isermann, ed.)
     Vol. 1, pp. 727-734, Pergamon, Oxford.
22. Bergman, R.N., Bortolan, G., Cobelli, C. and Toffolo, G.
     (1979) In "Preprints 5th IFAC Symposium on Identification
     and System Parameter Estimation" (R. Isermann, ed.) Vol. 2,
     pp. 883-890, Pergamon, Oxford.
23. Bergman, R.N., Ider, Y.Z., Bowden, C.R. and Cobelli, C. (1979)
     Am. J. Physiol. $\underline{236}$, E 667-E 677.
24. Toffolo, G., Bergman, R.N., Bowden, C.R. and Cobelli, C.
     Diabetes (to appear).

A FREQUENCY RESPONSE METHOD FOR PHARMACOKINETIC

MODEL IDENTIFICATION

Victor F. Smolen

Ayerst Laboratories, Inc.
Rouses Point, N.Y.   12979
U.S.A.

ABSTRACT

A mathematical model must be constructed in order for either
body fluid measurements of drug concentration or pharmacological
response intensity data to be useful for computation of the time
course of systemic drug bioavailability. For linear compartment
models the time course of either drug concentration or pharmacological response is represented by a sum of exponentials. The usual
method for establishing such models involves the use of some form
of nonlinear least-squares estimation technique. For systems of
order higher than first, such techniques are often cumbersome, time
consuming, and in some cases inadequate. A general frequency response technique commonly used in systems engineering applications,
pulse testing, has been successfully applied to pharmacological data
for model identification. The method is general and can be applied
equally well to blood or urine data. Theoretical descriptions of
both the nonlinear least-squares estimation technique in the time
domain and the frequency domain estimation technique are presented
here. The frequency response data obtained by pulse testing are
fitted with accurate model parameters by using a nonlinear least-
squares estimation technique in the frequency domain. However, the
number of parameters which must be determined, is decreased when
modelling in the frequency domain. This results in decreased computation time. For example, the least-squares time domain technique
required 25 and 190 seconds of computation time for second- and
third-order models, respectively, while the frequency domain technique required 8 and 10 seconds for the same models.

1. INTRODUCTION

A requirement for pharmacokinetic analysis is the construction
of an accurate model. Various modelling techniques have been used.
The popular method of estimating model parameters by logarithmic
peeling (1) is adequate in certain cases. However, greater accuracy
may be required for many applications. The method of modelling by
a nonlinear least-squares estimation technique (2-5) produces more
accurate estimates of the model parameters. However, this technique
is expensive to implement for systems higher than second order because of the computation time involved. A widely used technique
in engineering analysis, pulse testing (6-8), has been used to obtain pharmacokinetic models (9). This method is illustrated in
this report. The method is more generally applicable than the nonlinear least-squares parameter estimation technique. For example,
it can be used with any general input of drug, that is, the i.v.
input does not have to be a bolus injection (an impulse input) or
a sustained constant rate input (zero-order infusion). All that
is required is that the shape of the input pulse be known. Pulse
testing also provides accurate estimates of the model parameters
and is relatively inexpensive to implement even for higher order
system.

In this work, models are produced for bolus i.v. inputs of drugs
by using both the pulse testing and nonlinear least-squares estimation methods. A theoretical description of both methods is presented, as is a comparison of the models produced and the computation time required. The pharmacokinetic response illustrating
the modelling procedures is the change in pupil diameter in rabbits
(9,10) produced by chlorpromazine. However, the method is general
and can be applied equally well to blood or urine data for other
drugs.

2. THEORETICAL APPROACH

2.1. Estimation by Least Squares and Frequency Response

In order for pharmacological response intensity data to be
useful for the pharmacokinetic analysis and computation of the time
course of systemic drug bioavailability, a mathematical model must
be constructed to describe the time course of variation of the
relative biophasic drug levels in the system being considered.
The relative biophasic drug levels are first obtained by transforming the pharmacological response intensity data using a dose-
effect curve.

In the time domain, the time course of variation of dose (D)-
normalized relative biophasic drug levels is represented by a sum
of exponentials for linear compartment models (4,9,11-18):

$$f(I)/D = \sum_{i=1}^{n} A_i e^{-m_i t} \quad . \tag{1}$$

The sum of exponentials given in Eq. [1] is obtained as the mathematical solution of equations describing the rates of change of relative biophasic drug level for linear compartment models. It needs to be pointed out that though the expression results from linear theory, the expression itself is nonlinear in its parameters A_i and m_i. Therefore, to identify the model from experimental data, a nonlinear least-squares estimation technique must be implemented (19,20). The logarithmic peeling technique is used to provide initial estimates of the parameters as a starting point for this iterative procedure.

Frequency response analysis is widely used in several fields of engineering as a powerful method of model identification. The corresponding expression to Eq. [1] in the Laplace transform domain is

$$\frac{f(I)(s)}{D} = \sum_{i=1}^{n} \frac{A_i}{s+m_i} = G(s) \quad . \tag{2}$$

The ratio of output to input in the Laplace domain is referred to as the *transfer function*, $G(s)$ (6,8,12,16). A simple, convenient and widely used technique called *pulse testing* is easily implemented to estimate the A_i and m_i in Eq. [2]. This procedure results in a *Bode diagram* which provides estimates of A_i and m_i. These estimates are further refined to obtain the best model. This is done by using a nonlinear least-squares procedure to fit the frequency response of the system, as obtained from pulse testing, to the expression on the right-hand side of Eq. [2]. In spite of the fact that the frequency response method seems to involve two steps much like the nonlinear least-squares estimation technique in the time domain, the frequency response method requires an order of magnitude less computation time on the computer. Also, the pulse testing technique is more generally applicable. Therefore, it will often be advantageous to use the pulse testing procedure in the frequency domain instead of the nonlinear least-squares estimation in the time domain.

In the time domain nonlinear least-squares analysis, estimates of the parameters of Eq. [1] can be obtained by a graphical method often referred to as logarithmic peeling (5). If these estimates are not adequate then the more sophisticated nonlinear least-sqaures estimation technique is required to produce the best fit for the observed data. Since the right-hand side of Eq. [1] is a function of the individual A_i and m_i, it can be written as

$$f(I)/D = F(A_1, A_2, \ldots, A_n, m_1, m_2, \ldots, m_n, t) \tag{3}$$

where F represents the functional relationship explicitly shown in Eq. [1]. One criterion that is frequently used to obtain the best model for any given data is to minimize the sum of squares of errors between the observed data and the mathematical model (21), that is minimize

$$\varepsilon^2 = \sum_{j=1}^{N} w_j \{ [f(I)/D]_j - F_j(A_1, A_2, \ldots, A_n, m_1, m_2, \ldots, m_n, t_j) \}^2 \tag{4}$$

where w_j represents appropriate weights, perhaps unity, and $(f(I)/D)_j$ represents the j-th observation made at time t_j. It is assumed that there are a total of N observations. It should be pointed out here that N must be greater than 2n (the total number of unknowns in the original equation being estimated) to obtain a unique solution.

Posed in the form of Eq. [4], the nonlinear estimation problem becomes simply an optimization problem where the best values of A_i and m_i are sought. The optimization techniques that would be useful in this application fall under two broad classes: (i) derivative-free methods, and (ii) derivative methods. A general derivative-free method of minimization of a nonlinear objective function, like Eq. [4], makes use of regular patterns of search involving simplexes (21,22). A technique very similar to the one using simplexes, called iterative guessing, was developed to model biological data (2,3). The major difference between the simplex technique and the iterative guessing technique is that the simplex method uses triangles and tetrahedrons for its search patterns, while the iterative guessing method utilizes squares and cubes. Hazelrig *et al.* (3) were among the first to use this technique to model biological data.

For a two-dimensional problem, the iterative guessing method involves treating the initial guess as the center of a square and considering eight points around this initial point such that the eight points represent the corners of the square and the midpoint of the sides. Consider a two-dimensional problem

$$\frac{f(I)}{D} = A_1 e^{-m_1 t} + A_2 e^{-m_2 t} \quad . \tag{5}$$

For some initial guess of m_1 and m_2, the problem of estimating A_1 and A_2 can be solved by taking the partial derivatives of ε^2 with respect to A_1 and A_2, setting them equal to zero, and solving the resulting two simultaneous equations for A_1 and A_2. The coordinates of these eight points, together with the initial guess, determine the first nine pairs of values considered for m_1 and m_2. Values for A_1 and A_2 are calculated from the appropriate equations for each different pair of m_1 and m_2. The sum of squares of the residuals is determined for each of the original nine points, and the point yielding the smallest sum becomes the center of a new square. The entire procedure of setting up a new square around this point is now repeated. This procedure is repeated until the center of the square yields the smallest sum, then the size of the square is decreased and the procedure is repeated. Typically, the dimensions of the original square are determined by considering values 30 percent above and 30 percent below the initial guesses for the parameters. Each time the center of the square results in the smallest sum, the square is cut to one third of its previous size. This process is continued iteratively until a prechosen minimum square size is reached.

This method has been extended to three parameters (3). The initial guess is now located in a three-dimensional coordinate system, and it is treated as the center of a cube. One scheme uses the eight corners of the cube, the twelve midpoints of the edges, and the six centers of the sides. The authors of this method point out that this scheme of iterative guessing consumes considerable computer time when programmed for higher-order systems though theoretically this method can be extended indefinitely to include higher-order systems.

A digital computer program, MULTIFIT, was developed and implemented several years ago for fitting models comprised of a sum of exponentials (4). In the first phase of the computational procedure, the program implements an algorithm for the technique of logarithmic peeling to obtain initial estimates of the parameters. These estimates of the parameters can be improved by optionally proceeding to the second phase of the program where the technique of iterative guessing is implemented to obtain the best least-squares estimates of the parameters. This particular program assigns a weighting factor to each point computed by the formula:

$$w_j = \sigma^2_{avg} / (\sigma^2_{avg} + \sigma^2_j) \qquad [6]$$

where σ^2_{avg} is the average variance of all the data points pooled together and σ^2_j is the variance of the individual point; w_j is the weighting function used in Eq. [4]. This weighting scheme indirectly makes the weights proportional to the reciprocal of the variance of the point and at the same time avoids the problem of having an infinite weight assigned to single data points which have zero variance. In this case, the maximum weight assigned to any point is 1.0.

The iterative guessing technique is particularly cumbersome and time consuming when applied to higher-order systems. Typically, this

method requires in excess of 190 seconds of central processor time
on the CDC 6500 to obtain convergence for a third-order model. The
technique of pulse testing, a method to obtain the frequency response,
used only about 10 seconds of central processor time on the same
computer when applied to the same system. The difference in computational effort between these two methods is quite significant for
higher-order systems. For a first-order system they would rate
about even; for a second-order system the iterative guessing technique required 25 seconds whereas the pulse testing technique require 8 seconds. The computational advantage of pulse testing will
continue to increase as the order of the system increases.

2.2. Frequency Response

Frequency response analysis is a convenient means to obtain
information on the dynamics of a linear system. It is the steady-
state response of the system to sinusoidal inputs covering a wide
range of frequencies (6,8,23). The result of frequency testing is
that the amplitude of the output sinusoidal wave becomes attenuated
compared to the input wave. A measure of the degree of attenuation
is the ratio of the output amplitude to the input amplitude. This
is referred to as the *magnitude ratio*, MR. Another result of frequency testing is that the output wave will be out of phase with the
input wave, that is some *phase lag* will occur. The phase lag is
denoted by ϕ. Both the magnitude ratio and the phase lag are functions of the wave frequency, ω. This functional dependence can be
used to determine the dynamic equations representing the system.
The functional dependence is typically represented by a plot of log
MR *vs.* log ω and a semilogarithmic plot of ϕ *vs.* log ω. These plots
together are referred to as a Bode diagram. An example of the magni-
tude ratio part of the Bode diagram is given in Figure 6.

Though direct sine-wave testing is an extremely useful way of
obtaining precise dynamic data, it has the significant disadvantage
that it can be very time consuming when applied to processes that
have large time constants. Also, it must be applied only to systems
which will not be disturbed while the testing is going on. In some
cases this cannot be accomplished because of the time required to
reach a steady state. Also, the steady-state oscillation of the
system must be established at several discrete values of frequency.
Therefore, it can take several days to generate the complete frequency response curves of a slow process.

2.3. Pulse Testing

In pulse testing, a nonperiodic input of any general shape is
put into the system. This produces an output pulse. The complete
frequency response of the system can then be obtained from this information, and hence a mathematical model found. Pulse testing has
two significant advantages over direct sine-wave methods for generat-

ing frequency-response information: (i) This method disturbs the system for only a short period of time since only one input pulse of fairly short duration is called for, as opposed to sine-wave testing which requires sinusoidal inputs covering a wide range of frequencies, and (ii) this method allows the use of any general shaped input pulse, as opposed to the strict requirement of true sinusoidal inputs for direct sine-wave testing (6,8). This method has the disadvantage of increased computational effort but this is overcome by use of a digital computer.

The reason that a nonperiodic function can yield the frequency response data which would require several sinusoidal functions is that a nonperiodic pulse has a continuous frequency spectrum and therefore it excites the system at all frequencies at once, while a sine wave (of frequency ω_1) has a frequency spectrum which contains only one frequency value, ω_1. The meaning of frequency spectrum and the frequency content of a nonperiodic function is discussed in the Appendix.

2.4. Computational Methods for Pulse Testing

Consider a general pulse, $m(t)$, put into the system, and the corresponding output pulse $C(t)$, which is obtained. The output pulse may appear as soon as the input pulse is introduced, or it may exhibit a delay time, T_d, before it appears as shown in Figure 1. The delay time will not affect the duration of the output pulse, T_C, it will merely shift the output on the time axis. The transfer function, $G(s)$, for the system is then given by

$$G(s) = \frac{C(s)}{m(s)} = \frac{\int_0^\infty C(t)e^{-st}dt}{\int_0^\infty m(t)e^{-st}dt} = \frac{\int_0^{T_C} C(t)e^{-st}dt}{\int_0^{T_m} m(t)e^{-st}dt} \quad . \tag{7}$$

Considering the frequency response of the system (6), one can re-write equation [7] as

$$G(j\omega) = \frac{C(j\omega)}{m(j\omega)} = \frac{\int_0^{T_C} C(t)e^{-j\omega t}dt}{\int_0^{T_m} m(t)e^{-j\omega t}dt} \tag{8}$$

where s has been replaced by $j\omega$. Using the Euler relationship

$$e^{-j\omega t} = \cos\omega t - j\,\sin\omega t \quad ,$$

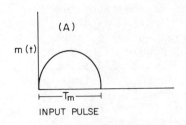

INPUT PULSE

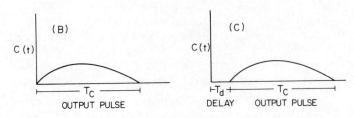

OUTPUT PULSE   DELAY   OUTPUT PULSE

FIGURE 1:   Input pulse and output response used in pulse testing.
(A) General input pulse; (B) Output response; (C) Output
response with time delay. T_m=time duration of the input
pulse, T_C=time duration of the output pulse; T_d=delay
time before output pulse appears.

Equation [8] can be rewritten as

$$G(j\omega) = \frac{\int_0^{T_C} C(t)\cos\omega t\ dt - j \int_0^{T_C} C(t)\sin\omega t\ dt}{\int_0^{T_m} m(t)\cos\omega t\ dt - j \int_0^{T_m} m(t)\sin\omega t\ dt} \qquad [9]$$

Redefining terms in Eq. [9] one obtains:

$$G(j\omega) = \frac{R_C(\omega) - jI_C(\omega)}{R_m(\omega) - jI_m(\omega)} \qquad [10]$$

where

$$R_C(\omega) = \int_0^{T_C} C(t)\cos\omega t\ dt \qquad [10a]$$

$$I_C(\omega) = \int_0^{T_C} C(t)\sin\omega t\ dt \qquad [10b]$$

$$R_m(\omega) = \int_0^{T_m} m(t)\cos\omega t \, dt \qquad\qquad [10c]$$

$$I_m(\omega) = \int_0^{T_m} m(t)\sin\omega t \, dt \quad . \qquad\qquad [10d]$$

Essentially, one needs to be able to obtain the magnitude ratio, MR, and the phase angle, ϕ, for the transfer function $G(s)$ in order to model in the frequency domain. Noting that Eq. [10] is a complex function, its amplitude is expressed as

$$|G(j\omega)| = \frac{|R_C(\omega) - jI_C(\omega)|}{|R_m(\omega) - jI_m(\omega)|} \quad ,$$

therefore,

$$MR = |G(j\omega)| = \frac{\sqrt{[R_C(\omega)]^2 + [I_C(\omega)]^2}}{\sqrt{[R_m(\omega)]^2 + [I_m(\omega)]^2}} \qquad\qquad [11]$$

For the phase angle, ϕ, Eq. [10] is first rearranged to yield

$$G(j\omega) = \frac{R_C(\omega)R_m(\omega)+I_C(\omega)I_m(\omega)+j[R_C(\omega)I_m(\omega)-R_m(\omega)I_C(\omega)]}{R_m^2(\omega)+I_m^2(\omega)}$$

Therefore,

$$\phi(\omega) = G(j\omega) = \tan^{-1}\left[\frac{R_C(\omega)I_m(\omega) - R_m(\omega)I_C(\omega)}{R_C(\omega)R_m(\omega) + I_C(\omega)I_m(\omega)}\right] \quad . \qquad\qquad [12]$$

Equations [11] and [12] must be evaluated at specific values of ω over the frequency range of interest. At each of these ω, the terms of $R_C(\omega)$, $I_C(\omega)$, $R_m(\omega)$ and $I_m(\omega)$ are found by numerical integration. Difficulty may occur with the numerical integration at high frequencies because the numerical procedures are often unable to describe the high-frequency oscillation of the sinusoidal functions without using an unreasonable number of increments. This difficulty may be overcome by first approximating C(t) and m(t) by a series of.line segments and then analytically integrating the approximation. Accuracy then depends upon the adequacy of the straight-line approximation.

An alternative to the evaluation of Eqs. [11] and [12] for frequency response data by pulse testing is to use direct numerical evaluation of Fourier transforms. That is, as shown in Appendix, the expression for the Fourier transform of a function, f(t), is

$$S(j\omega) = \int_{-\infty}^{\infty} f(t)e^{-j\omega t}dt \quad . \tag{13}$$

If the function, f(t), is zero for negative time and for times greater than some value T_1, then Eq. [13] becomes

$$S(j\omega) = \int_{0}^{T_1} f(t)e^{-j\omega t}dt \quad . \tag{14}$$

We then see that the numerator and denominator in Eq. [8] are the Fourier transfroms of function C(t) and m(t), respectively. As indicated in the Appendix, the Fourier transform is a complex function of ω which can be written as

$$S(j\omega) = |S(j\omega)| < S(j\omega) \quad . \tag{15}$$

Therefore, the amplitude ratio and phase lag of G(jω) can be expressed as

$$MR = G(j\omega) = \frac{|S(j\omega)|c}{|S(j\omega)|m} \tag{16}$$

and

$$\phi = < G(j\omega) = < S(j\omega)c - < S(j\omega)m \quad . \tag{17}$$

Computation of Fourier transforms for the input pulse, m(t) and for the output response, C(t), according to Eq. [13] using classical methods, is very time consuming and wasteful of resources on the digital computer. In recent years an algorithm for computation of Fourier transforms has been proposed which accomplishes the task much more efficiently (24-27). The basic approach used here is adapted from pulse testing techniques commonly used for chemical engineering process applications (7). This approach takes advantage of an algorithm for "Fast Fourier Transforms" of real-valued series as proposed by Bergland (24). An example which illustrates the mechanics of the "Fast Fourier Transform" is given by Hamming (28).

A digital computer program, PLTEST, was developed to compute the complete frequency response for linear systems (9). The FFT scheme used requires an input of 256 points and generates 129 frequencies at which the magnitude ratio, $|G(j\omega)|$, and phase angle, ϕ, are computed. Recall that the log-log plot of magnitude ratio *vs.* frequency and a semi-log plot of phase angle *vs.* frequency are together referred to as a Bode plot (6,8). The magnitude ratio plot in Bode plots is often normalized, that is, the magnitude ratio at each frequency is divided by the absolute value of the steady-state gain of the system. It is convenient to normalize the data because it makes most of the magnitude ratio plots start at 1 for small values of ω. By the final-value theorem,

$$\text{Steady-state gain} = \lim_{s \to 0} [G(s)] \qquad . \qquad\qquad [18]$$

This can also be given by

$$\text{Steady-state gain} = \lim_{\omega \to 0} [G(j\omega)] = G(0) \quad . \qquad\qquad [19]$$

Thus the normalized magnitude ratio is

$$\text{MR} = |G(j\omega)| / |G(0)| \qquad\qquad . \qquad\qquad [20]$$

This also provides an internal check within the program because the steady-state gain can also be computed by taking the ratio of the integral of the output response to the integral of the input pulse.

Theoretically, the most convenient input pulse would be an impulse or a Dirac delta function, $\delta(t)$. The Fourier transformation (and the Laplace transformation) of $\delta(t)$ is equal to unity at all frequencies, that is,

$$\int_{-\infty}^{\infty} \delta(t) e^{-j\omega t} \, dt = (e^{-j\omega t})_{t=0} = 1 \quad . \qquad\qquad [21]$$

Therefore, $G(j\omega)$ would simply be the Fourier transformation of the output function; no division by a small number corresponding to the input function would be required. Practically, however, there can never be an infinitely high pulse with zero width. However, one must keep in mind that the width of the pulse should be kept fairly small in order to keep its frequency content (*i.e.*, its amplitude) from becoming too small at higher frequencies. Quite often, a good rule to keep in mind is that the width of the pulse must be less than about one-half the smallest time constant of interest. If the dynamics of the process is completely unknown, it will take a few trials to establish a reasonable pulse width.

Bode plots can be used for system identification by estimating parameters in a general transfer function. One reason Bode plots are used so often is that they make it easy to combine transfer functions. A complex transfer function can be broken down into its simple elements, leads, lags, and delay-time; each of these can be separately approximated in Bode diagrams; and, finally, the total complex transfer function is obtained by adding the individual magnitude ratio and phase-angle curves (6). For every simple unit of the transfer function, one can approximate the actual frequency response by means of two asymptotes, a high-frequency asymptote and

a low-frequency asymptote. The intersection of these asymptotes
occurs at the break (or corner) frequency. On the magnitude ratio
plot the individual components of the transfer function are identified by drawing the high-frequency asymptotes with a slope of +1 for
a lead term (also referred to as a zero of the transfer function)
and a slope of -1 for a lag term (also referred to as the pole of
the transfer function). The delay time is usually easy to identify
in the time domain; often it is preferable simply to shift the origin
for the output function in the time domain to account for the delay
time. The low-frequency asymptote always has a slope of zero. It
should be noted here that, except for non-minimum phase systems, if
the magnitude ratio is specified at all frequencies the phase angle
is also completely specified (23).

A problem can occur when graphically estimating the parameters
in a transfer function from a Bode plot when two or more corner
frequencies are close together. In this case, none of the parameters
can be accurately estimated. For applications such as closed-loop
feedback control which is encountered in engineering, a high degree
of accuracy in estimating the parameters is often not essential.
Any deviation of the estimated model from the true system response
is compensated for in the feedback loop by the use of appropriate
controller action. However, for an open-loop system type of application, such as calculation of amount of drug absorbed, the parameters of the model must be estimated to a higher degree of accuracy.
This difficulty of estimating model parameters from Bode diagrams
when break frequencies are close together has been overcome by
implementing the computational scheme described below.

2.5. Refining Estimates Obtained from Bode Plots

The frequency response information obtained from the pulse
testing program is the "true" frequency response of the system as
represented by the experimental data. It is possible to fit the
experimental frequency response data to a theoretical model and obtain a best fit model for the system. A nonlinear least-squares
regression technique was used to model the system in the frequency
domain. For a majority of applications in compartment modelling,
first, second, and third-order models adequately describe the dynamics of the system. These forms of models will be considered here,
though this technique is perfectly general and can be easily extended to higher-order systems.

A first-order transfer function with unity gain, no delay time,
and no numerator dynamics is represented by

$$G(s) = \frac{1}{\tau_1 s + 1}$$
[22a]

where τ_1=the first-order time constant of the system. Similarly,
a second-order transfer function is represented by

$$G(s) = \frac{1}{(\tau_1 s+1)(\tau_2 s+1)} \qquad [22b]$$

where τ_1 and τ_2 are time constants associated with a second-order systems. Similarly, a third-order transfer function is represented by

$$G(s) = \frac{1}{(\tau_1 s+1)(\tau_2 s+1)(\tau_3 s+1)} \qquad [22c]$$

where τ_1, τ_2, τ_3 are time constants associated with a third-order system. Equations [22a,b and c] correspond to models with one, two and three exponentials in time domain, respectively. To be able to use these forms of the system model with the frequency response data, they must be expressed in terms of the amplitude ratio and phase angle. For the first-order representation,

$$MR = (\tau_1^2 \omega^2 + 1)^{-1/2} \qquad [23a]$$

and

$$\phi = \tan^{-1}(-\tau_1 \omega) \qquad . \qquad [23b]$$

For the second-order representation,

$$MR = [(\tau_1^2 \omega^2 + 1)(\tau_2^2 \omega^2 + 1)]^{-1/2} \qquad [24a]$$

and

$$\phi = \tan^{-1}\left[\frac{-\omega(\tau_1 + \tau_2)}{1 - \omega^2 \tau_1 \tau_2}\right] \qquad . \qquad [24b]$$

For the third-order representation,

$$MR = [(\tau_1^2 \omega^2 + 1)(\tau_2^2 \omega^2 + 1)(\tau_3^2 \omega^2 + 1)]^{-1/2} \qquad [25a]$$

and

$$\phi = \tan^{-1}\left[\frac{\omega(\tau_1 + \tau_2) - \omega \tau_3(1 - \omega^2 \tau_1 \tau_2)}{1 - \omega^2 \tau_1 \tau_2 - \omega^2 \tau_3(\tau_1 + \tau_2)}\right] \qquad [52b]$$

where MR represents the magnitude ratio of the system transfer function and ϕ represents the phase lag of the system transfer function. The assumed system transfer function is now fitted to the frequency response information generated from the experimental data by minimizing the sum of squares of error between them. Thus, define

$$\varepsilon_{MR}^2 = \sum_{i=1}^{N} (AR_i - MR_i)^2 \tag{26}$$

and

$$\varepsilon_{\phi}^2 = \sum_{i=1}^{N} (PH_i - \psi_i)^2 \tag{27}$$

where

ε_{MR}^2 = sum of squares error for magnitude ratio

ε_{ϕ}^2 = sum of squares error for phase angle

N = total number of points being considered

AR = amplitude ratio obtained from the pulse testing program

PH = phase angle obtained from the pulse testing program

MR = magnitude ratio obtained from Eqs. [23a], or [24a], or [25a]

ψ = phase angle obtained from Eqs. [23b], or [24b], or [25b].

As pointed out previously, for minimum phase systems, if the frequency and amplitude ratio are specified at all points, then the phase angle is also completely specified. Hence, it is sufficient to curve fit only the amplitude ratio as a function of frequency and minimize the sum of squares of error represented by Eq. [26].

The frequency response data are considered error free with respect to comparing the three transfer functions. That is, for the purposes of modelling it is assumed that any errors in the data are insignificant when compared to the error arising from the failure of a functional form to fit adequately the data. These are called errors due to lack of fit (21,29). The residual sum of squares, as given by Eq. [26] for the three models, is

$$SS_j = [\sum_{i=1}^{N} (AR_i - MR_i^2)]_j \qquad j=1,2,3 \tag{28}$$

where SS is the residual sum of squares, and the subscripts 1,2, and 3 represent the order of the transfer function representation under consideration.

Now, SS_3 will always be smaller than either SS_1 or SS_2 since the third-order representation contains more arbitrary constants in the curve fitting scheme than do the other two representations.

Since all residual sums of squares are calculated from the same number of data points, the ratios (SS_1/SS_3) and (SS_2/SS_3) will be distributed as the F distribution with $(N-1,N-3)$ and $(N-2,N-3)$ degrees of freedom, respectively. The above, of course, assumes that the errors due to lack of fit are normally distributed. Thus, the following statistical criteria can be used to select the sim-- plest adequate representation of the system (21):

Let α = the significant level of the test

and $F_{1-\alpha}$ = the 1-α point of the F distribution.

Then:

If $(SS_2/SS_3) \geq F_{1-\alpha}(N-2,N-3)$, use the third-order represent- ation.

If $(SS_2/SS_3) < F_{1-\alpha}(N-2,N-3)$, make the following test:

If $(SS_1/SS_3) \geq F_{1-\alpha}(N-1,N-3)$, use the second-order repre- sentation.

If $(SS_1/SS_3) < F_{1-\alpha}(N-1,N-3)$, use the first-order repre- sentation.

An efficient general nonlinear least-squares algorithm (30) is used in the computer program NONLINR (9) to perform the nonlinear least-squares estimation of the model parameters. This program was obtained from the software library of the computing center at Purdue. The initial estimates of the form and parameter values of the transfer function are derived from the Bode diagram; these are then used as initial guesses for the parameters in the program NONLINR.

3.   EXPERIMENTAL APPLICATION AND DISCUSSION

The techniques discussed above were applied to experimental data that were gathered in the course of a program to study the effect of chlorpromazine on pupil diameter in rabbits (9,10). The methodology to transform the pharmacologic response intensity data to biophasic drug levels is dealt with in great detail by the author in another paper (31). A brief description of the experi- mental procedure and treatment of the data is given here.

3.1. Time Course of Drug Response

The time course of intensity of pupillary diameter decrease in rabbits as a function of five bolus i.v. doses of chlorpromazine is shown in Figure 2. Pupil diameters were measured periodically

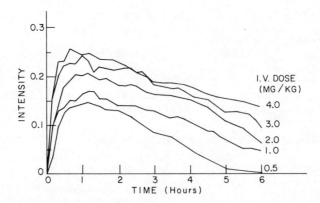

FIGURE 2:   Intensity of pupillary diameter decrease (miosis) in
rabbits measured as $(P_o - P)/P_o$, using a Whittaker Space
Sciences TV Pupillometer, as a function of five bolus
intravenous doses of chlorpromazine. Each curve represents averaged data from 3 to 4 rabbits.

---

using a Whittaker Space Sciences TV Pupillometer equipped with a
video camera and red illumination light. The video signal from the
camera was relayed to a large television screen where the pupil
diameter was easily measured with dividers and a meter stick. Each
curve represents data from 3 to 4 rabbits for each dose; the dosing
schedule was randomized according to a Mouten square experimental
design since some rabbits died during the period of experimentation.
Inspection of Figure 2 indicates that a clearly discernable graded
response profile is obtained. The times of absolute response intensity maximums are not identical, but the times all lie within a
narrow band and there is no obvious trend with dose.

It can be seen from Figure 2 that the times of maximum response
intensity do not occur at zero time. This would be expected since
the biophase containing the receptor sites are somewhat removed
from the systemic circulation compartment where the drug was administered.

3.2. Dose-Effect Curve and Relative Biophasic Drug Level

For a mean time of maximum response intensity, t_{max}, the
average values of the maximum intensity, I_{max}, and their standard
deviations are plotted against the corresponding dose. The resulting intravenous dose-effect curve is shown in Figure 3. The dose-
effect curve represents a single-valued nonlinear functional relationship between the intensity of response, $I(t)$, and the relative
biophasic drug level, $f(I)$. The observed pupillary diameter de-

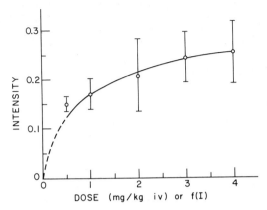

FIGURE 3:   Intravenous dose-effect curve for chlorpromazine-induced
            pupillary diameter decrease in rabbits, constructed from
            the maximum response intensities from Figure 2. Standard
            deviations are indicated for each value.

crease response intensities were transduced (or transformed) to
$f(I)$ values using the dose-effect curve. The individual $f(I)$ are
plotted as a function of time with the dose (D) as a parameter in
Figure 4. Next, the dose-normalized and average values of $f(I)$ are
plotted as a function of time as shown by the solid curve in
Figure 5. These data represent the response output of the system
to a unit impulse input and are ideally suited for mathematical
modelling to describe the dynamics of the system.

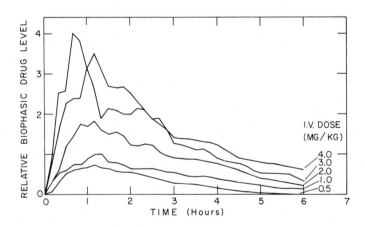

FIGURE 4:   Biophasic drug levels obtained from transforming pupil-
            lary diameter decrease response data using a dose-effect
            curve.

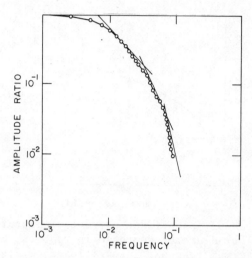

FIGURE 5:   A comparison of the models obtained from MULTIFIT and
            PULSE TESTING programs for pupillary diameter decrease,
            with the actual experimentally observed data.
            o : MULTIFIT; + : PULSE TESTING; continuous curve:
            actual response.

3.3. Model Identification

The dose normalized relative biophasic drug level data were
modelled in the time domain using the nonlinear least-squares estimation program, MULTIFIT and in the frequency domain using the
pulse testing program, PLTEST. The output from MULTIFIT directly
gives the A_i's and the m_i's of the system model. It was found that
a model consisting of a sum of two exponentials adequately described
the data. The output from the pulse testing program, PLTEST is in
the form of a Bode diagram (which is a log-log plot of amplitude
ratio *vs*. frequency) as shown in Figure 6. From the Bode diagram,
the gain of the system is found to be 188.35960 and the two poles
of the transfer function occured at frequencies of 0.0064 and 0.032
cycles per minute for a second-order model. These initial estimates
of the poles are then used as starting guesses for the nonlinear
least-squares fitting program, NONLINR which refines these estimates.
A comparison of the A_i's and m_i's obtained from the time domain
estimation and frequency domain estimation is summarized in Table 1.

It may be observed that the two models are not significantly
different. However, the MULTIFIT program used up about 25 seconds
of CPU time on the CDC 6500 whereas the combined PLTEST and NONLINR
programs executed in about 8 seconds of CPU time. Of course, this
economy becomes even more significant if computations have to be
repeated several times.

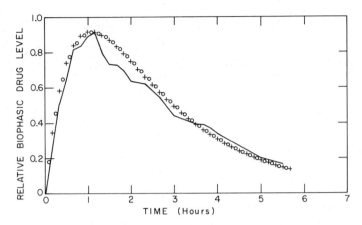

FIGURE 6:   Bode diagram constructed from f(I) dose values of
            pupillary diameter decrease. Tangent lines are drawn
            to indicate how the model parameters were determined
            from the diagram, as well as the general form of the
            model transfer function.

The savings in computation for the second-order system, as
noted above, are significant. However, the savings for a third-
order model are even greater. The time domain technique required
190 seconds while the frequency domain technique took about 10
seconds. The reduction in computation time occurs even though a
nonlinear least-squares technique is also used in the frequency
domain. This happens because the frequency domain technique requires fewer parameters to be determined by nonlinear least squares
for any order model when compared to the time domain approach.
For example, model determination for a second-order system in the
time domain requires the minimization of Eq. [4] with respect to
four parameters, A_1, A_2, m_1 and m_2. For a third-order system,

TABLE 1:   A comparison of parameters obtained from MULTIFIT and
           PLTEST for a model for chlorpromazine induced pupillary
           diameter decrease in rabbits.

Parameter	A_1	A_2	m_1	m_2
MULTIFIT	2.589	-2.589	0.008718	0.02369
PLTEST	2.358	-2.358	0.008424	0.02576

Eq. [4] must be minimized with respect to six parameters, A_1, A_2, A_3, m_1, m_2 and m_3. However, the frequency domain technique presented here requires minimization of Eq. [28] where MR_i is given by Eq. [24a] for a second-order system, or Eq. [25a] for a third-order system. This, in turn, means that only two optimal parameters are needed for a second-order system, τ_1 and τ_2, and three for the third-order system, τ_1, τ_2, and τ_3. The reason for the reduction in model parameters can be seen by considering the model for a second-order system. In the time domain it is given by

$$G(t) = A_1 e^{-m_1 t} + A_2 e^{-m_2 t} , \qquad [29]$$

where the four parameters are found by minimizing Eq. [4]. In order to model in the frequency domain, we take the Laplace transform of Eq. [29] to obtain

$$G(s) = \frac{A_1}{s+m_1} + \frac{A_2}{s+m_2} \qquad [30]$$

Equation [30] can be manipulated to produce

$$G(s) = \frac{A}{(\tau_1 s+1)(\tau_2 s+1)} \qquad [31]$$

where

$$K = A_1/m_1 + A_2/m_2$$

and

$$\tau_1 = 1/m_1 \quad , \quad \tau_2 = 1/m_2 \quad ,$$

with the constraint that $A_1+A_2=0$ when the pharmacological response does not occur at the point where the drug is introduced as in the case for the pupillary response considered here. We see that Eq. [31] requires only three parameters K, τ_1 and τ_2. However, K is readily determined by the ratio of the area under the curves of the output and input pulses. This then leaves only τ_1 and τ_2 to be determined by the minimization of Eq. [28].

The case where numerator dynamics must be considered, for example measuring drug levels in blood for a system described by a two-compartment model, will require changes in the form of Eq. [22] to [25]. Except for the modification of these equations, the modelling technique presented here is unchanged. An in-depth discussion of the various forms of transfer functions which correspond to the common compartment model configurations will be reported by the author in the near future.

APPENDIX

A discussion of the frequency content of nonperiodic function and how one is able to obtain this information from the Fourier transform is given here.

The Fourier series expansion can be used to represent a periodic function in terms of a summation of trigonometric functions. Even though there is an infinite number of frequencies represented, not all frequencies appear in the series (*i.e.*, there is an infinite, but denumerable number of frequencies). The frequency content of periodic functions will first be examined and the results will then be extended to nonperiodic functions.

A periodic function, $f(t)$ with a period T, can be represented by a Fourier series. That is,

$$f(t) = a_0 + \sum_{k=1}^{\infty} 2[a_k \cos(k\omega_1 t) - b_k \sin(k\omega_1 t)] \qquad [A1]$$

where

$$\omega_1 = 2\Pi/T = \text{the angular frequency of the first} \qquad [A2]$$
$$\text{harmonic, } i.e., \text{ k=1; it is also called}$$
$$\text{the fundamental frequency}$$

$$(\omega = 2\Pi f, \ f = 1/T, \ \omega/2\Pi = f = 1/T; \text{ therefore, } \omega = 2\Pi/T)$$

and

$$a_k = \frac{1}{T} \int_{-T/2}^{T/2} f(t)\cos(k\omega_1 t)dt \qquad k=1,2,3,\ldots \qquad [A3]$$

$$b_k = \frac{1}{T} \int_{-T/2}^{T/2} f(t)\sin(k\omega_1 t)dt \qquad k=1,2,3,\ldots \qquad [A4]$$

Equation [A1] can be written in the form of a single trigonometric function plus phase angle

$$f(t) = a_0 + \sum_{k=1}^{\infty} A_k' \cos(k\omega_1 t + \phi_k) \qquad [A5]$$

$$A_k' = 2\sqrt{a_k^2 + b_k^2} \qquad [A6]$$
$$= \text{magnitude (amplitude) of the k-th harmonic}$$

and

$$\phi_k = \tan^{-1}(b_k/a_k) \qquad\qquad\qquad\qquad [A7]$$

= the phase angle of the k-th harmonic with respect to the cosine wave.

If one uses the identities

$$\cos k\omega_1 t = \frac{1}{2}(e^{jk\omega_1 t} + e^{-jk\omega_1 t}) \qquad\qquad [A8]$$

and

$$\sin k\omega_1 t = \frac{1}{2j}(e^{jk\omega_1 t} - e^{-jk\omega_1 t}) \quad , \qquad\qquad [A9]$$

Eqs. [A1] to [A4] can be manipulated to the form

$$f(t) = \sum_{-\infty}^{\infty} F_k(j\omega_1) e^{jk\omega_1 t} \qquad\qquad\qquad [A10]$$

where it can be shown that

$$F_k(j\omega_1) = a_k(\omega_1) + jb_k(\omega_1) = \frac{A_k(\omega_1)}{2} e^{j\phi_k(\omega_1)} \qquad [A11]$$

or

$$F_k(j\omega_1) = \frac{1}{T} \int_{-T/2}^{T/2} f(t) e^{-jk\omega_1 t} dt \quad . \qquad\qquad [A12]$$

The magnitude of F_k is one-half of the magnitude of the k-th harmonic, *i.e.*,

$$|F_k(j\omega_1)| = A_k/2 \quad ,$$

and the phase angle of F_k is equal to the phase angle of the k-th harmonic with respect to the cosine wave. Therefore, if $F_k(j\omega_1)$ is known, f(t) can be defined in the frequency domain, [*i.e.*, this is analogous to using the Laplace transform with s=jω to define a time function in the frequency domain (6)].

Referring to Eq. [A12], as the period (T) of the function it represents increases, the coefficient (F_k) becomes smaller, and when T→∞, F_k→0. To represent functions with large T, a different function is defined:

$$\rho_k(j\omega_1) = T\, F_k(j\omega_1) = \int_{-T/2}^{T/2} f(t)e^{-jk\omega_1 t}\, dt \qquad . \qquad [A13]$$

Equation [A13] represents the frequency content of a periodic function, that is, for any $T<\infty$.

Now, a non-periodic function can be viewed as a periodic function with $T\to\infty$, that is, a non-periodic function is never repeated and a periodic function with $T\to\infty$ is never repeated. For this case we modify Eq. [A13] to the form

$$S(j\omega) = \lim_{\substack{T\to\infty \\ k\omega_1 \to \omega}} \rho(j\omega_1) = \lim_{\substack{T\to\infty \\ k\omega_1 \to \omega}} \int_{-T/2}^{T/2} f(t)e^{-jk\omega_1 t}\, dt$$

or

$$S(j\omega) = \int_{-\infty}^{\infty} f(t)e^{-j\omega t}\, dt \qquad\qquad [A14]$$

where

$$\omega = k\omega_1 = \text{a finite non-zero variable} \qquad .$$

Using the definition of ω_1, that is $\omega_1 = 2\Pi/T$, we see that as $T\to\infty$, $\omega_1 \to 0$ and, therefore, the spacing of the discrete frequencies, $k\omega_1$ (multiple of $2\Pi/T$) approach zero, in other words, the harmonics of the fundamental frequency ω_1 become more densely spaced until they approach a continuous spectrum.

Equation [A14] defines the Fourier transform of $f(t)$. The result, $S(j\omega)$, is complex function of ω, that is,

$$S(j\omega) = |S(j\omega)| < S(j\omega) \qquad .$$

Therefore, the frequency response characteristics of the system may be recovered from the frequency content of the input pulse, $m(t)$, and the output pulse, $C(t)$. These in turn, are obtained from the Fourier transforms of $m(t)$ and $C(t)$, that is,

$$MR = G(j\omega) = \frac{|S(j\omega)|C}{|S(j\omega)|m} \qquad\qquad [A15]$$

$$\phi = < G(j\omega) = < S(j\omega)C - < S(j\omega)m \qquad . \qquad [A16]$$

Now since $C(t)$ and $m(t)$ are functions which are zero for $t<0$ and zero for some finite time t_C or t_m, the expression for the Fourier series, Eq. [A14] is equivalent to the expression obtained from the Laplace transform with s=jω, Eq. [8].

REFERENCES

1. Am. Pharm. Assoc. Bioavailability Pilot Project (1973) "The Bioavailability of Drug Products", Am. Pharm. Assoc., Washington, D.C.
2. Ackerman, E., Strickland, E.H., Hazelrig, J.B. and Gatewood, L.C. (1967) Clin. Pharmacol. Ther. 8, 170.
3. Hazelrig, J.B., Ackerman, E. and Rosevear, J.W. (1963) "An Iterative Technique for Confirming Models to Biomedical Data", 16th Annual Conf. Eng. Med. Biol., Baltimore.
4. Schoenwald, R.D. (1971) "A Drug Absorption Analysis for the Mydriatic, Tropicamide, Using Pharmacological Response Intensities", Ph.D. Thesis, Purdue University.
5. Wagner, J.G. (1975) "Fundamentals of Clinical Pharmacokinetics", Drug Intelligence Publ., Hamilton, Ill.
6. Coughanowr, D.R. and Koppel, L.B. (1965) "Process Systems Analysis and Control", McGraw-Hill, New York.
7. Dollar, C.R. (1972) "Frequency Response Data Via Pulse Testing", Computer Programs for Chemical Engineering Education, Vol. 3, Control, CACHE.
8. Luyben, W.L. (1973) "Process Modelling, Simulation, and Control for Chemical Engineers", McGraw-Hill, New York.
9. Jhawar, A.K. (1974) "Mathematical Modelling of Physiological Data", Ph.D. Thesis, Purdue University.
10. Kuehn, P.B. (1974) "Pharmacological Response Kinetics of Chlorpromazine in Rabbits", Ph.D. Thesis, Purdue University.
11. Jacquez, J.A. (1972) "Compartmental Analysis in Biology and Medicine", American Elsevier, New York.
12. Rescigno, A. and Segre, G. (1966) "Drug and Tracer Kinetics", Blaidsell, New York.
13. Smolen, V.F. and Weigand, W.A. (1973) J. Pharmacokin. Biopharm. 1, 329.
14. Schoenwald, R.D. and Smolen, V.F. (1971) J. Pharm. Sci. 60, 1039.
15. Smolen, V.F. (1972) Can. J. Pharm. Sci. 7, 1.
16. Riggs, D.S. (1970) "Control Theory and Physiological Feedback Mechanisms", Williams and Wilkins, Baltimore.
17. Smolen, V.F., Turrie, B.D. and Weigand, W.A. (1972) J. Pharm. Sci. 61, 1941.
18. Smolen, V.F. and Schoenwald, R.D. (1971) J. Pharm Sci. 60, 96.
19. Guest, P.G. (1961) "Numerical Methods of Curve Fitting", Cambridge Univ. Press, Cambridge.
20. Westlake, W.J. (1973) J. Pharm. Sci. 62, 1579.
21. Himmelblau, D.M. (1968) "Process Analysis by Statistical Methods", Wiley, New York.
22. Nelder, J.A. and Mead, R. (1964) Comp. J. 7, 308.

23. Truxal, J.G. (1955) "Automatic Feedback Control System Syn-
        thesis", McGraw-Hill, New York.
24. Bergland, G.D. (1968) Commun. ACM, 11, 10, 703.
25. Bergland, G.D. (1969) IEEE Spectrum, 6, 41.
26. Cooley, J.W. and Tukey, J.W. (1965) Math. Comput. 19, 297.
27. Gentleman, W.M. and Sande, G. (1966) "Fast Fourier Transforms
        — for Fun and Profit", Proc. Fall Joint Computer Conf.,
        San Francisco, Calif.
28. Hamming, R.W. (1973) "Numerical Methods for Scientists and
        Engineers", McGraw-Hill, New York.
29. Davies, O.L., ed. (1956) "Design and Analysis of Industrial
        Experiments", Hafner, New York.
30. Marquardt, D.W. (1963) J. Soc. Indust. Appl. Math. 11, 2, 431.
31. Kuehn, R.B., Jhawar, A.K., Weigand, W.A. and Smolen, V.F.
        (1976) J. Pharm. Sci. 65, 1593.

DESIGN AND ANALYSIS OF KINETIC EXPERIMENTS

FOR DISCRIMINATION BETWEEN RIVAL MODELS

Bengt Mannervik

Department of Biochemistry
Arrhenius Laboratory
University of Stockholm
S-106 91 Stockholm, Sweden

ABSTRACT

Methods for mathematical modelling of experimental data are reviewed with particular emphasis on applications to kinetic data analysis. Investigations in enzyme kinetics involve mechanistic model building more frequently than empirical model building, because understanding of the underlying mechanism is usually the main objective of such studies. Methods of analysis should preferably include various graphical displays of the experimental data set as well as an efficient and reliable procedure for fitting of models to the data. Regression analysis based on the least-squares method is usually preferred, but distribution-free (nonparametric) methods could also be considered. Criteria are listed for evaluation of the goodness of fit as well as for discrimination between alternative mathematical models. These criteria imply the use of regression analysis based on the least-squares principle and statistical analysis of the results. The criteria involve consideration of convergence properties of a model, examination of parameter values, analysis of residuals, and comparison of residual sums of squares of rival models with one another as well as with the experimental variance. The design of experiments should be based on the immediate objective of the investigation, $i.e.$, model discrimination or parameter estimation. Usually a sequential procedure involving design of experiments for model discrimination and subsequent analysis of data followed by design for optimal parameter estimation and a new analysis has to be employed. Factors which should be considered in the design of experiments are the number and spacing of experimental points, and the error structure of the data. A discrimination function identifies domains in the space of reactant concentrations (or other independent variables) where maximal information for discrimi-

nation between two rival models is available. For complex models it
may be useful to restrict the analysis to a particular domain of the
factor space and examine degenerate versions of the models (*e.g.*
an asymptotic form). The error structure of the experimental data
should be evaluated by replicate measurements or by analysis of the
residuals of the fit of an adequate model. The established dependence of the experimental variance on the dependent variable should
be used for weighting in the regression analysis and for judging
the goodness of fit of a model fitted to the data. An example is
given which clearly demonstrates the importance of the design of
experiments for the feasibility of discrimination between alternative models.

1.   INTRODUCTION

Mathematical models of observed data may serve two major purposes, *i.e.*, the quantitative description of relationships between
variables and the understanding of the system investigated. The
two objectives are not mutually exclusive, but the first one is in
the forefront when the model should be used for prognosis of the
outcome of new observations (empirical model building), whereas the
second one is important when the model should be correlated with
the observable expressions of an underlying mechanism (mechanistic
model building). For prognostic uses the form of the mathematical
model is of subordinate importance as long as the model provides an
accurate description. Thus the velocity of an enzymatic reaction
under steady-state conditions, which is usually related to the independent variables by a quotient of two polynomials of reactant
concentrations, could equally accurately be described by an appropriate ensemble of exponential functions. The identification of
the "true" model may be important only if extrapolations should be
made far outside the domains of the variables that have been examined in the modelling.

For the understanding of a system, on the other hand, it is
usually imperative to try to find the "true" model, because the
nature of the mathematical expression may have implications for the
discrimination between alternative interpretations of the underlying
physical events. In principle, it is expected that the "true" mathematical model should be derivable from and correspond to a theory
of structure, function, mechanism, *etc*. The present treatment will
concentrate on the second purpose of mathematical modelling and will
use as examples some studies in the realm of enzyme kinetics. However, this restriction of the subject does not limit the generality
of the strategy described, which should be applicable when a real
variable can be expressed explicitly as a function of one or several
independent variables.

In general, the process of mathematical modelling of experimental data may be described as follows. The problem may be to

evaluate the assumed interaction between the dependent variable, y, and the independent variable, x. Hence, a functional relationship, y=f(x), is assumed. An experiment is designed to investigate the influence of different settings of the independent variable on the value of the dependent variable. Subsequent analysis may suggest one or several mathematical models, $f_j(x)$, to fit the experimental data. An interpretation of the mathematical model(s) in terms of the problem is attempted. If a single mathematical model is consistent with a satisfactory interpretation and if this model is superior to all other models which can be formulated, it remains to refine the analysis of the model in order to obtain as accurate and precise values as possible of the constants in the equation. This objective normally requires a new sequence of design, experiment, and analysis.

If, on the other hand, rival models appear to describe the experimental data approximately equally well, it will be necessary to find experimental conditions which allow discrimination between the alternative models. For this purpose a new experimental design should be introduced, followed by experiment and analysis. When a model finally has been selected it may, as above, be necessary to make a complementary experiment under conditions optimal for the determination of the constants in the equation. In practice, it is, therefore, usually necessary to employ a series of experimental designs, experiments, and analyses in order to optimize the conditions for, on the one hand, model selection and, on the other hand, parameter estimation (1). In the following treatment, some of the methods will be described which can be applied for the design and analysis of experiments with emphasis on the discrimination between alternative models.

2. METHODS

2.1. Methods for Fitting Mathematical Models to Experimental Data

The majority of kinetic studies performed in biological sciences have probably, so far, been analyzed by using graphical methods. For example, the parameters a and b of a simple exponential function,

$$y = a \cdot e^{-bx} \tag{1}$$

can be estimated from the straight line obtained by plotting log y against x. Similarly, in enzyme kinetics the simple Michaelis-Menten equation

$$v = V[A]/(K_m + [A]) \tag{2}$$

where v is velocity, [A] is substrate concentration, and V and K_m are the parameters to be estimated, is usually displayed in a linearized form from which the parameters can be estimated. These methods

may be adequate and sufficient if (a) the model is true and (b) the
experimental error is negligible. However, in practice neither of
these conditions may be taken for granted. If a model which cannot
be linearized by transformation is considered, the estimation of
the parameters may not be feasible, and if the experimental error
cannot be ignored, even the parameter estimation in a linear model
may be seriously biased. Therefore, statistical methodologies are
usually required for improved fitting of equations to data and unbiased parameter estimation. The use of statistical methods also
facilitates testing of hypotheses both concerning the choice of mathematical model and the adequacy of parameter values. However, the
realization of the power of statistics should not be taken as a pre-
text to refrain from using graphical methods of analysis. In fact,
many problems can clearly be solved in a qualitative manner by simple
plotting procedures. For example, the effect of an inhibitor on an
enzymatic reaction may be evaluated by varying the substrate concentration, and even the generalized concepts of competitive, non-
competitive, and uncompetitive effects (2) will often easily be
identified visually by any of the common plotting procedures. Figure 1 shows some typical examples of these inhibition patterns for
non-Michaelian kinetics. The corresponding rate equations may be
quite complex (2), but the graphs clearly discriminate between the
alternatives. Bardsley and coworkers (3-6) have described powerful
and elegant mathematical procedures that allow evaluation of curve
shapes in any graphical space in which one variable is plotted as
a function of another (v vs. v/[A], log v vs. log [A], etc.). Another example is the use of the concept of the tact invariant for
evaluation of cooperativity in a specified region of any type of
protein-ligand binding curve (7). Therefore, various plots of experimental data should be used to facilitate the selection of reasonable mathematical models even if statistical methods are employed
for the fitting of equations to the data.

Two types of statistical methods appear in common use in enzyme
kinetics: regression analysis based on the principle of least squares
and analysis by a distribution-free (or nonparametric) method described by Eisenthal and Cornish-Bowden (8). The latter method has
recently received considerable attention for parameter estimation
because of its simplicity in the underlying statistical assumptions
(9). A major limitation has been that the suggested plotting method
(the "direct linear plot") (8) could only be applied to Michaelian
enzyme kinetics. The significance of this restriction is stressed
by the conclusion that few, if any, enzymes really obey the Michaelis
Menten equation (10). However, this limitation of the direct plot
does not apply generally to distribution-free statistical methods.
In fact, if the "true" model is known it will suffice to make replicate measurements in as many experimental points as there are
parameters in the model. The optimal design points of the experiment can be found by use of the criterion of Box and Lucas (11).
The use of this design criterion has been described by Duggleby

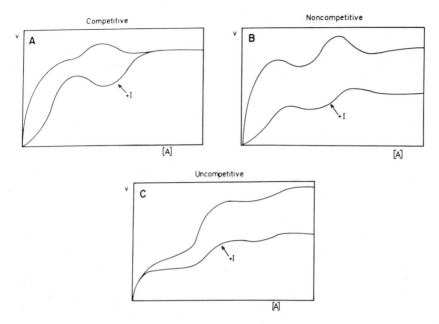

FIGURE 1: Graphs of velocity as a function of substrate concentration , [A], in the absence and presence of an inhibitor, I. Salient
features are shown for generalized (A) competitive, (B) noncompetitive, and (C) uncompetitive inhibition in cases of non-Michaelian
enzyme kinetics. The effect of the inhibitor is eliminated at high
[A] for competitive and at low [A] for uncompetitive inhibition.
An algebraic description of the different inhibition patterns has
been presented for rational rate equations (2). However, the generalized inhibition patterns are not limited to rational functions,
but could also be applied to other classes of rate laws in which
the asymptotic properties of the interaction between two ligands
can be defined.

(12-13) and Endrenyi (14). For example, if the Michaelis-Menten
equation is really the true model and the experimental error is
known to be proportional to velocity the optimal design is to make
some (>5) replicates at the highest and some at the lowest substrate
concentration practicable. With this design, plotting of the experimental data is really superfluous, because only the median (or
the mean for data having a Gaussian error distribution) of the velocities determined at the low and the median of those at the high
substrate concentration are required to calculate V and K_m in Eq.
[2]:

$$V = \frac{[A]_\ell - [A]_h}{([A]_\ell/\bar{v}_\ell) - ([A]_h/\bar{v}_h)}$$ [3]

$$K_m = \frac{\bar{v}_h - \bar{v}_\ell}{(\bar{v}_\ell / [A]_\ell) - (\bar{v}_h / [A]_h)} \qquad [4]$$

where $\bar{v}_\ell$ and $\bar{v}_h$ are the median (or mean) velocities at low substrate, $[A]_\ell$, and high substrated concentrations, $[A]_h$, respectively. In conclusion, there is little use for the direct linear plot in the analysis of experimental data, because it is not suitable for model discrimination and is in practice for large data sets substituted by a computer program for parameter estimation (*cf*. ref. 9). However, distribution-free methods and other robust estimation procedures will probably be more commonly used as soon as adequate methods for discrimination between alternative mathematical models have been developed (*cf*. refs. 15,16).

Most of the work involving fitting of models to kinetic data has been carried out by use of regression analysis based on the principle of least squares. This method was introduced for analysis of Michaelian enzyme kinetics (Eq. [2]) by Johansen and Lumry (17) and Wilkinson (18) and was thereafter made available to all interested investigators by Cleland (19) who has distributed copies of programs for this and some additional simple equations.

Of the variety of underlying statistical assumptions a few have been tested in connection with real kinetic data. One of the assumptions is a Gaussian (normal) distribution of data. Cornish-Bowden and coworkers concluded that in their experience a long-tailed (leptokurtic) distribution may be typical of enzyme kinetics (20), whereas Fromm and coworkers could not find significant evidence for deviations from a Gaussian distribution (21). In a more detailed investigation in our laboratory it was found that a Gaussian distribution was indeed applicable for a large set of experimental data (22). Thus, even if the demonstration that experimental data do not exhibit a normal distribution may be difficult to achieve, the suspicion that they follow a non-normal distribution has been a major impetus for introducing distribution-free methods of analysis (23).

Another basic assumption in the least-squares approach is that the error of the independent variable (*e.g*. substrate concentration) should be negligible, at least in comparison with the error in the dependent variable (velocity). This has been tested (22) and is probably generally true unless the velocity can be measured with very great precision. Another prerequisite is that the velocity data analyzed should have constant variance, otherwise a weighted regression analysis should be used. Although this condition has been widely recognized, attention to the actual experimental variance is often neglected judging from the literature. A procedure for determining suitable weighting factors will be discussed below (Section 2.4).

Although the regression analysis based on the least-squares
principle rests on several assumptions (additional ones being that
the errors should be uncorrelated and have a mean value of zero),
it still appears to be the most versatile method for general purposes and the following discussion will be limited to experience
collected by use of this method. For most applications nonlinear
regression is mandatory and several algorithms have been devised
for this purpose (see ref. 24 for citations of some early publications). The majority of practitioners seems to agree that most
modern programs are adequate for the problems encountered in enzyme
kinetics, and a discussion of the properties and merits of different
algorithms is outside the scope of this paper (cf. ref. 25). It is
desirable that the output includes not only the residual sum of
squares, parameters and their standard deviations, but also the
predicted velocities and residuals at the reactant concentrations
used. If so, analysis of residuals by graphical or other methods
can easily be performed. Such analyses are of utmost importance
for judgement of goodness-of-fit and discrimination between alternative models, as well as for deriving weighting factors. Two programs
for nonlinear regression analysis, BMDP3R and BMDPAR from the Biomedical Computer Program package (University of California, Los
Angeles), have been adopted in our laboratory and found well suited
to serve in the solution of most of the problems encountered.

2.2. Discrimination between Alternative Mathematical Models

When a data set has been analyzed according to an assumed mathematical model, two principal questions arise:

1.   Does the model adequately describe the data?
2.   Is there a better alternative to the model used?

The second problem often arises as a consequence of a negative
answer to the first question, and it is therefore important to define conditions which can be used to define the "goodness of fit"
of a model. First, all the qualitative and semiquantitative features
which can be evaluated by suitable plots have to be fulfilled by a
good model. For example, if a suitable graph of the experimental
points shows a maximum or a point of inflection, the mathematical
model used should allow for this feature. In this connection it is
worth repeating that different transformations for graphical display
may reveal different characteristic properties of the relationship
between dependent and independent variables and that it is worthwhile to employ a variety of plots (3).

For the remainder of the discussion let us assume that we have
chosen to employ regression analysis according to the principle of
least squares to obtain the 'best' fit of the mathematical model to
the experimental data. Some simple problems can be solved by linear
regression in which the parameters (constants) of the model enter in

a linear fashion as in the equation of a straight line. Problems
involving models in which parameters appear nonlinearly have to be
solved by iterative nonlinear regression. An example is the
Michaelis-Menten equation (Eq. [2]), in which K_m enters nonlinearly.
This equation cannot be fitted by linear regression unless the
variables are transformed. Most of the somewhat more complex models
cannot be linearized by transformation. For the sake of generality,
it will therefore be assumed that the fitting of a mathematical model
to a data set is made by nonlinear regression.

The problem of the regression analysis can be formulated as a
minimization of the regression function:

$$Y = \sum_{i=1}^{n} (v_i - \hat{v}_i)^2 = \min \qquad [5]$$

where v_i and $\hat{v}_i$ are the observed and calculated velocity (dependent
variable), respectively, of the i-th measurement (i=1,2,...,n).
For example, in fitting the Michaelis-Menten equation (Eq. [2]),

$$\hat{v}_i = \hat{V}[A]_i / (\hat{K}_m + [A]_i) \qquad , \qquad [6]$$

where the parameters $\hat{V}$ and $\hat{K}_m$ are the predicted maximal velocity
and Michaelis constant, respectively, and $[A]_i$ is the substrate
concentration of the i-th measurement. The computer program searches
for the combination of the parameters (V, K_m) which will minimize
Eq. [5]. When the calculations have been terminated, the results
can be evaluated with attention to four major points, $i.e.$, (i) convergence, (ii) parameter values, (iii) residuals, and (iv) the residual sum of squares. These points are closely interrelated, but
should be evaluated separately to gain maximal insight into the quality of the fit obtained. Both the goodness-of-fit of the model and
the discrimination between possible alternative models could be based
on such an analysis. When, in 1969, our own attempts to use regression methods in mathematical modelling of enzyme kinetic data started,
little attention had been directed towards the major problem of how
to discriminate between alternative models. It was therefore imperative to try to formulate a strategy for this purpose, and a set of
discrimination criteria based on points (i)-(iv) above were put forward (24,26). Simultaneously, similar work was carried out by
Reich $et\ al.$ (27,28), but their approach was focused on a strategy
for parameter estimation.

The suggested procedure for discrimination has been used successfully for almost a decade in our laboratory and has been incorporated $in\ extenso$ in a recent computer program by Garfinkel and coworkers (29).

Points (i)-(iv) will now be discussed in relation to the properties expected for a good model.

2.2.1. <u>Convergence</u>. Convergence is a crude measure of the success
of the minimization procedure. The usual convergence criterion is
that the minimum value of the regression function (Eq. [5]) cannot
be improved within a specified limit. Lack of convergence may occasionally result as a consequence of numerical problems. For example, in the case of linear regression of a transformed form of
the Michaelis-Menten equation, two variables were inadvertently kept
in constant ratio because only one of them was originally recognized
as an experimental variable. This condition resulted in a singular
design matrix and the linear regression of the experimental data
failed (30). This unusual case shows how clues to a better understanding of the system investigated may be obtained by the lack of
convergence (even if more decisive information may be obtained from
other evidence). However, usually a solution of the regression analysis will be provided and this solution should be examined from all
possible points of view. Plots of the predicted velocity against
reactant concentrations should closely follow the experimental data
and exhibit their characteristic curve shapes. Likewise, progress
curves showing concentrations of variables *versus* time should be
examined when data are analyzed using rate equations in their integrated forms. Many computer programs provide such plots automatically in the output. Examination of these plots will immediately
give a qualitative impression of the adequacy of the fit.

In computations involving complicated models and ill-conditioned
problems it may be very difficult to find the global minimum in the
regression analysis. These cases may require several attempts with
different sets of initial estimates of the parameters before a satisfactory fit is obtained. In addition, it may be valuable to employ
an alternative computer program using a different algorithm for the
minimization. If different programs converge to the same minimum
via different routes in the parameter space, the reliability of the
solution increases considerably. In our laboratory, a program based
on a Marquardt algorithm has been used for this purpose as a complement to the programs based on the Gauss-Newton algorithm normally
used (see above). Occasionally it may also be necessary to restrict
the regression by introducing constraints on the parameter values
to obtain an acceptable solution. In summary, if the predicted
values obtained by the regression analysis do not appear to depict
the experimental data properly, it is useless to try to evaluate
the results further. New attempts should be made to improve the
minimization or an alternative model should be considered.

2.2.2. <u>Parameter Values</u>. The fitting of a model to a set of experimental data results in estimates of the constants (parameter values)
of the equation used. An adequate model should have constants which
are meaningful. For example, an inhibition constant should have a
real, positive value. Negative parameter values may sometimes be
obtained by unconstrained regressions, even if only positive values
are acceptable (*cf*. ref. 30). This result may be prevented by in-

troduction of constraints. The cause of the appearance of unaccept-
able parameter values may be numerical problems in the calculations
or the use of an erroneous model.

By use of the variance-covariance matrix resulting from the
regression, the standard deviations of the parameters can also be
estimated. The parameter values of a good model should have low
standard deviations. In linear models the parameter values may be
compared with their corresponding standard deviations using the
t-statistic to test whether they differ significantly from zero
(24,26). In nonlinear regression the probability levels cannot be
rigorously defined (31) and the t-test is less reliable. As a rule
it can be stated that for simple models standard deviations should
be <50% of the corresponding parameter values to be acceptable. In
complex models, which to fulfil an underlying theory have many parameters, it may sometimes be necessary to accept larger standard
deviations. It should be noted that the standard deviations are
inversely proportional to the square root of the number of data in
the set ($\sqrt{n}$).

In classical modelling of data, the parameters of the model
should be independent, $i.e.$, the covariance between any two parameters in the model should be nil. This requirement cannot usually
be maintained in mathematical models of enzyme kinetics, because
the parameters are often intrinsically correlated. For example,
according to the treatment of Briggs and Haldane (32), V and K_m in
Eq. 2 are equal to $k_3[E]_o$ and $(k_2+k_3)/k_1$, respectively, and both
contain k_3 as an element. Therefore, a finite covariance is expected (unless $k_2 >> k_3$) between the estimates of V and K_m. Thus,
even if a model displaying marked covariance between parameters
cannot be rejected on this basis alone, it is nevertheless a criterion for a good model if the covariances are small. It is, therefore,
desirable to obtain in the output after regression the variance-
covariance matrix or the corresponding normalized correlation matrix
of the parameters.

2.2.3. Residuals. The difference between the observed value in the
i-th experimental point (v_i) and the corresponding value calculated
after fitting the model to the data set ($\hat{v}_i$) is the residual (q_i)
of the i-th point:

$$q_i = v_i - \hat{v}_i \qquad\qquad [7]$$

If the observed values have a Gaussian distribution then the residuals are also expected to follow this distribution after fitting
a good model to the data set. The expected mean of the residuals
is zero. Examination of the residuals after a regression analysis
is often the easiest way of detecting small systematic deviations
of the predicted values from the experimental observations. The
residuals should be plotted against each of the independent variable

as well as against the dependent variable. In each of these plots
the residuals should display a random distribution about the zero
level. Correlation between residuals may be detected by use of a
correlation coefficient (*cf*. ref. 33). Distribution-free tests involving ranking of residuals or examination of the number of sign
changes ("runs") in a given direction or simply counting of the
numbers of positive and negative residuals may be useful in evaluating randomness. It may also be useful to make two-dimensional
plots in which the residuals are mapped in a plane defined by two
experimental variables (34). For a good model such maps should not
display any domain containing an excensive number of residuals of
equal sign. Systematic deviations of the residuals from a random
distribution not only reveal the lack-of-fit of an inadequate model,
but may also, by their pattern, disclose the nature of the defect
and thereby lead to the formulation of a better model (26,28,35-37).
The analysis of residuals (of a good fit) may also provide information about the error structure of the experiments (38) - see below.

2.2.4. The Residual Sum of Squares. Upon convergence of the regression, the function Y (Eq. [5]) attains a minimum value which is
the residual sum of squares (SS). For a good fit, this value is
due to the experimental error only and should be comparable to a
corresponding estimate of the latter. To account for the number
of experimental values (n) included and the number of parameters
(p) in the model, the mean sum of squares, Q^2, is calculated:

$$Q^2 = SS/(n-p) \qquad\qquad [8]$$

For a good model, Q^2 is expected to be equal to the experimental
variance, σ^2. If an independent estimate, s^2 of the experimental
variance is available, the values of Q^2 and s^2 can be compared.
For linear models the quotient of Q^2/s^2 can be compared with the
appropriate F-statistic to answer the question of whether they
could be considered as significantly different.

When two models j and k with p_j and p_k ($p_j < p_k$) parameters,
respectively, are fitted to the same data set and yield residual
sums of squares SS_j and SS_k ($SS_j > SS_k$), the significance of the improvement of the residual sum of squares by using the model (k)
containing more parameters can be made by comparison of the quotient

$$\frac{(SS_j - SS_k) \cdot (n - p_k)}{(p_k - p_j) \cdot SS_k}$$

with the F-statistic $F(p_k - p_j, n - p_k)$ at the appropriate significance
level. If the quotient exceeds the F-value, the null hypothesis
that no significant improvement is obtained by chosing the model
containing the additional parameters can be rejected. In this case,
model k is adopted; in the opposite case, the simpler model (j)

should normally be retained. The analysis of nonlinear models by use of the F-statistic is limited by a similar uncertainty about the significance levels as that mentioned above in connection with analysis of parameter values by use of the t-test (31). The smaller the residual sum of squares, the better is the fit of the model to the set of experimental data. However, a Q^2-value lower than the estimated experimental variance is not meaningful. The p term in Eq. [8] accounts for differences in the degrees of freedom between different models, but in addition serves as a penalty against use of overdetermined models. Thus the model giving the lowest Q^2-value can usually be considered as giving the "best" fit to the data, provided that it is adequate in all other respects.

A procedure for discriminating between rival mathematical models based on the above points, can be summarized in the following criteria (24,26,30,35):

1.   Models failing to given proper convergence in the regression analysis are considered inferior.
2.   Models giving unreasonable (*e.g.*, negative) or unreliable (*e.g.*, large standard deviations) parameter values are considered inferior.
3.   Models giving residuals with a nonrandom distribution or a non-zero mean are considered inferior.
4.   The model giving the smallest mean sum of squares, Q^2, of the models satisfying the discrimination criteria (i)-(iii) is chosen as the best of the alternatives.

2.3. Design of Experiments

The design of experiments is crucial in the analysis of a system under investigation. It cannot be emphasized too strongly that no method of analysis, graphical or numerical, can extract more information than is already in the data set collected. Thus, maximal information content has to be ascertained by careful planning of the acquisition of data.

First, it should be recognized that there are different optimal designs depending on the objectives of the analysis (24). For parameter estimation it has been shown that maximal precision is obtained if the experimental points are chosen in such a manner that the determinant of the information matrix is maximized (11,39). Evaluation of this criterion requires knowledge about the error structure of the data (see below). For example, in the case of constant absolute error, the optimal design points for estimating the constants in Eq. [2] are at $[A]_1$, giving the highest velocity possible, and at $[A]_2$, giving half the velocity obtained at $[A]_1$ (in the extreme case at $[A]_2=K_m$) (12-14). In a model containing p parameters, it is, in principle, sufficient to limit the analysis to p experimental points (11,12). However, few experimenters limit their investigation to this number, and if more experimental points are included the spacing of the independent variables (concentrations, time points, *etc.*) may also influence the parameter estimation. Endrenyi and Kwong have made a detailed simulation study of the effect of using harmonic, geometric, arithmetic, and c^2-arithmetic scaling of concentrations on the estimation of the parameters of Eq. [2] and a hyperbolic binding equation (40). In case that wide ranges of the independent variables are investigated, a geometric spacing (constand ratio of successive concentrations) is usually most adequate. For practical reasons we have often employed concentrations in the ratios 1:2:5:10:20:50:100:... .

However, in many investigations the primary problem is the identification of the "true" model and not the best set of parameter values. For this purpose the optimal design will usually not be identical with that for parameter estimation. In some cases, application of the discrimination criteria (i)-(iv) listed above will make possible the selection of one out of a set of alternative models, but often a definite choice between two rival models cannot be made on the basis of a single data set. In the latter cases, new experiments have to be made in which the conditions for discrimination have been optimized. For this purpose, a discrimination function, g, has been defined which helps to find the best experimental conditions for discrimination between two alternative models, j and k (24,26,35). The rationale used in defining the discrimination function is to find a point or region in the space of independent variables in which the inferior model cannot be fitted to the data (41). Additional experiments in this region will give large nonrandom residuals ("bias error") for the inferior model that will be expressed in residual plots and a larger mean residual sum of squares (Q^2).

The theory behind the discrimination function has been described in greater detail previously (24,26,35). It suffices to formulate the discrimination function as the difference between the residuals or the difference between the predicted values of the two alternative models j and k:

$$g_i = |\hat{v}_{ij} - \hat{v}_{ik}| \qquad\qquad [9]$$

where i denotes the i-th experimental point. The square of the discrimination function is proportional to the information content for discrimination (42), and the optimal experimental conditions are those which maximize g. The maximum of g can be located by differentiation of g with respect to the independent variables (reactant concentrations, time, *etc.*). This can be done analytically or by numerical methods (*cf.* refs. 24,26,35). Alternatively, the discrimination function is evaluated directly by taking the difference between the predicted values for the two models in each experimental point (Eq. [9]). Examples are given in refs. (33) and

(36). The discrimination function is evaluated by use of the current estimates of the parameter values of models j and k. When the suitable conditions for discrimination have been defined, new experiments should be made to accomplish the task of selecting the best of the rival models. If new data can be combined with the original data set, it may suffice to make some additional experiments, but often it may be desirable to make a new complete experiment. In the latter case it has to be kept in mind that the experimental points have to include not only those which are optimal for discrimination, but also points that define the parameter values. A joint design which combines both the experimental settings optimizing parameter estimation for each of the rival models and the best conditions for discrimination may be considered. For maximal overall discrimination the sum (or the integral) of g over all experimental points should be maximized:

$$\sum_{i=1}^{n} g_i = \max \qquad\qquad [10]$$

where n is the number of experimental points $[n \geq \max(p_j, p_k)]$; p_j and p_k are the number of parameters in models j and k, respectively. The goal of maximal overall discrimination becomes prominent in the analysis of progress curves. This problem has been approached by Markus and Plesser (43). If a "p-point design" is used, $[n = \max(p_j, p_k)]$ replicate measurements have to be made in the experimental points.

A simple illustration of the discrimination function is given by Figure 2. A data set has been fitted by using Eq. [2], but an extension of the equation to account for the tendency of inhibition by excess of substrate (A) has also been considered:

$$v = \frac{V'[A]}{K_m' + [A] + K_1[A]^2} \qquad\qquad [11]$$

where K_1 is a constant. The discrimination function for these alternative models is:

$$g = \left| \frac{V[A]}{K_m + [A]} - \frac{V'[A]}{K_m' + [A] + K_1[A]^2} \right| . \qquad\qquad [12]$$

It is evident that the global maximum of g is V, which is approached when $[A] \to \infty$. Consequently, new experiments at higher substrate concentrations should be made (if practicable) to facilitate discrimination. Usually, local maxima are also obtained at intermediate substrate concentrations. These maxima result from the differences

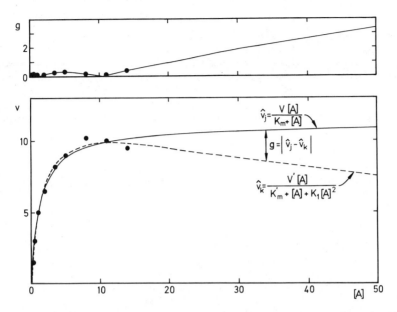

FIGURE 2: The discrimination function, g, for the Michaelis-Menten equation and the corresponding rate law describing inhibition by high substrate concentrations. The two alternative rate equations were fitted by nonlinear regression to the data points. The discrimination function is the absolute difference between the v values predicted by the two rival models according to Eq. [12] and is evaluated by the use of the current estimates of the parameters $\hat{V}$, $\hat{V}'$, $\hat{K}_m$, $\hat{K}'_m$, and $\hat{K}_1$. The experimental error (σ) is approximately ±0.3.

in V and V' as well as in K_m and K'_m obtained by fitting the models by regression methods to the data set. Such subsidiary maxima are often of the same order of magnitude as the experimental error (σ) and of little use for discrimination. More intricate examples such as the discrimination between a ping-pong and a sequential two-substrate mechanism, have been treated elsewhere (26,33,35). If a substance is considered as a possible linear competitive inhibitor, I, of an enzyme following the Michaelis-Menten equation, it should be tested at the highest concentration possible, $[I]_{max}$. The maximal difference between Eq. [2] and the equation for linear competitive inhibition is then at the substrate concentration:

$$[A] = \sqrt{K_m K'_m (1+[I]_{max}/K_i)} \tag{13}$$

where K'_m is the estimate of the Michaelis constant for the enlarged rate equation and K_i the inhibition constant (35).

For complex models containing many terms it may sometimes be advantageous to restrict the analysis to a limited domain of the space of independent variables. This approach was tested in an investigation of the kinetics of glutathione transferase A where the complete rate equation for a two-substrate random steady-state mechanism was compared with a similar equation lacking a constant term (K_1) in the denominator (44):

$$v = \frac{V_1[A][B]+V_2[A]^2[B]+V_3[A][B]^2}{K_1+K_2[A]+K_3[B]+[A][B]+K_4[A]^2+K_5[B]^2+K_6[A]^2[B]+K_7[A][B]^2} .$$

[14]

It can be shown that at low substrate concentrations, [A] and [B], this equation degenerates to:

$$v = \frac{V[A][B]}{K'+K''[A]+K'''[B]+[A][B]}$$

[15]

where

$$V=V_1/D, \quad K'=K_1/D, \quad K''=\left(K_2 - \frac{K_1 V_2}{V_1}\right)/D, \quad K'''=\left(K_3 - \frac{K_1 V_3}{V_1}\right)/D$$

and

$$D = \left[1 + \frac{V_2}{V_1}\left(\frac{K_1 K_3}{V_1} - K_3\right) + \frac{V_3}{V_1}\left(\frac{K_1 K_2}{V_1} - K_2\right)\right] .$$

The alternative equation lacking K_1 (Eq. [14]) degenerates to an equation similar to Eq. [15] lacking K'. (The explicit expression is obtainable by setting $K_1=0$.) By using these degenerate models of the original alternatives and restricting the analysis to the domain of low concentrations of [A] and [B], it could be shown more definitely that a constant term in the denominator (K_1 or K') was required, than by fitting the full rate equations to the complete data set (44).

Eq. [15] is an extension of the concept of the osculating hyperbola at the origin of a plane curve (45) to a corresponding surface in three-dimensional space. The asymptotic properties of rate equations for two-substrate reactions have previously been treated from another point of view (46).

2.4. Error Structure of Experimental Data

Both the design of experiments and the analysis of experimental data are crucially dependent on information about the experimental variance. Regarding the design of experiments, it is evident that if the experimental error is constant, observations at very low

values will contain little information because they will be swamped
by the "noise" of the error. If, on the other hand, the relative
error is constant, it is usually advisable to collect data from as
wide a range as possible. In the analysis of data it is also necessary to have information about the error structure in order to make
a proper fitting of a model to the data set and to evaluate the
goodness of fit. When the method of least squares is used, the data
fitted should have equal variance, and if the error varies with the
magnitude of the dependent variable a suitable weighting procedure
has to be adopted. The method used is to multiply each term in the
regression function (Eq. [5]) with a weighting factor, w_i, inversely
proportional to the experimental variance:

$$Y = \sum_{i=1}^{n} w_i (v_i - \hat{v}_i)^2 \qquad . \qquad [16]$$

The problem of error structure of enzyme kinetic data has been approached by several investigators, and it has been found that in
many cases the variance increases with velocity (20-22,33,38,47,48).
In a detailed study of the kinetics of glutathione transferase A
it was found that the variance of v could be approximated by

$$Var(v) = K_1 \cdot v^{\alpha} \qquad [17]$$

where K_1 and α are empirical constants which were estimated by replicate measurements at different levels of v (22). The value of
the exponent found under the conditions investigated was $\alpha=1.6$. A
similar error function was independently suggested by Siano *et al.*
(21). It was found that introduction of this error function to define weighting factors:

$$w_i \propto v_i^{-\alpha} \qquad [18]$$

was superior to the use of weighting factors independently estimated
(*cf.* ref. 49) by 4 replicate measurements in each experimental point
(33). A major drawback in using individual estimates instead of
the error function is that many replicates are required in each
experimental point (20,50). Values of $\alpha=0$ and $\alpha=2$ correspond to
the "classical" error structures of constant absolute error and
constant relative error, but experience in our laboratory shows
that any value in the range of $0 \leq \alpha \leq 2$ can be found (22,38). In fact,
some data sets yield values of $\alpha>2$. The weighting factors as defined by Eq. [18] should preferably be based on the predicted velocities ($\hat{v}_i$) and not on the experimental values (v_i), provided that
an adequate model has been used to derive the predicted values.
However, in general, no difference in the regression analysis will
be noted if instead the experimental values are used in the calculation of the weighting factors (Eq. [18]).

In the original investigations it was suggested that the error function should be derived from replicate experiments. However, an alternative method based on the analysis of the residuals of a preliminary regression was subsequently suggested (38). The rationale behind the new method is simply that if the model fitted to a data set is adequate, the residuals should just reflect the experimental error. For example, Figure 3 shows a typical example of a residual plot obtained after fitting a good model to a data set exhibiting non-constant variance. It is evident that the residuals tend to increase with increasing velocity. By ordering neighboring residuals in groups of 5 to 6 the local variance can be estimated as the mean of the squared residuals:

$$\frac{1}{m} \sum_{i=1}^{m} (v_i - \bar{v}_i)^2 \qquad\qquad [19]$$

where m is the number of residuals in the group. By treating the different estimates of the local variances as a function of velocity (or reactant concentrations if Eq. [20] is considered - see below) the constants of Eq. [17] can be estimated. When proper weighting factors have been defined by this procedure, a new weighted regression analysis can be made. This new iterative weighting method has been found to give the same results as the original procedure (38). It has the advantage of eliminating the need for extensive replication experiments. Furthermore, the error structure may change from experiment to experiment and the novel procedure eliminates the need of new replicate measurement for each experimental design, since the coefficient α can be estimated by analysis of the residuals.

It should be noted that although the error function defined in Eq. [17] is a monotonous function, some experimental conditions may give rise to more complex relations exhibiting a maximum (or minimum) of the variance at intermediate values of velocity. Therefore, it is advisable to investigate the variance at several levels of velocity (or substrate concentrations) and not limit the analysis to one estimate in the low and one in the high velocity region as practised by Siano et al. (21). In particular, it may be considered probable than an enzyme inhibited by excess of substrate should exhibit a smaller variance at a low velocity obtained at high substrate concentrations than a corresponding velocity obtained at low concentrations (where initial rates in general are more difficult to estimate).

An attempt has also been made to introduce the discrimination function, g, as a weighting factor specifically for the problem of model discrimination (as distinct from parameter estimation) (52). The rationale behind this attempt was to give emphasis to values obtained in the domains of reactant concentrations which have maximal information for discrimination. However, the application

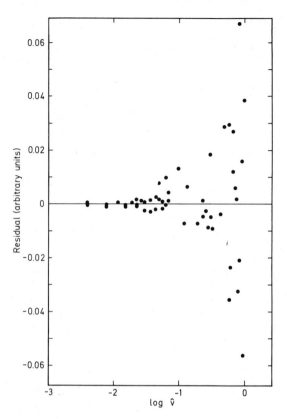

FIGURE 3: Dependence of residuals on the velocity of a reaction catalyzed by butyrylcholinesterase. Unweighted residuals $(v_i - \hat{v}_i)$ are plotted against the logarithm of predicted velocity $(\hat{v}_i)$. Residuals and predicted velocities were obtained by fitting the rate equation

$$v = \frac{V_1[A] + V_2[A]^2 + V_3[A]^3}{1 + K_1[A] + K_2[A]^2 + K_3[A]^3}$$

by weighted $(w_i = v_i^{-2})$ nonlinear regression to a set of 53 experimental data. The fit accounted for the observed rate behaviour of the enzyme and the residuals are therefore expected to express only the experimental error and not deviations caused by bias. The model was overdetermined, because V_3 and K_3 were redundant parameters. (Fitting of a corresponding 2:2 function gave a residual sum of squares which was <2% larger than that of the above 3:3 function.) The experimental conditions were the same as those described in ref. (51). The substrate (A) used was butyrylthiocholine.

of this weighting procedure is hampered by the lack of suitable statistics for evaluating the results.

It should also be considered that the velocity is a stochastic variable and not the true independent variable, which is the concentration of substrate, inhibitor, *etc.* varied in the experiment. An investigation (I. Jakobson, M. Warholm and B. Mannervik, unpublished work) of the experimental variance as a function of the concentrations of a substrate, [A], as well as an inhibitor, [I], demonstrated that, when both A and I were present, the variance could be approximated by

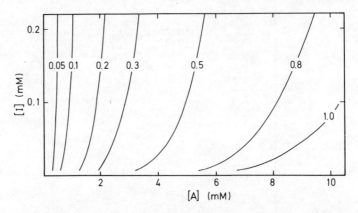

FIGURE 4: Dependence of experimental variance on the true independent variables. Variance is mapped by iso-variance contours on the factor space (*i.e.*, plane of the independent variables [A] and [I]). The local variance was estimated in different regions of the factor space by analysis of residuals in groups of six according to the method suggested in ref. (38). The residuals were obtained from a good fit of kinetic data in which the effect of the inhibitor (I) S-octylglutathione on glutathione transferase A was studied at different constant levels of the substrate (A) glutathione [*cf*. Fig. 1 in ref. (38)]. The local variance estimates were fitted by weighted nonlinear regression to (concentrations in mM units):

$$Var(v) = K_1 [A]^{\alpha} \cdot [I]^{\beta} + K_2 [A]^{\gamma}$$

but it was found that this model degenerated to one term:

$$Var(v) = 0.077 \cdot [A]^{0.94} \cdot [I]^{-0.16} .$$

The values of the exponents ($\alpha=0.94$ and $\beta=-0.16$) show that, in the domain of the factor space investigated, [A] has the major role in determining the variance. This result was also obtained by independent experiments involving replicate measurements of v as a function of [A] and [I]. Thus, Var(v) is not uniquely defined by v, but is dependent on the setting of the independent variables that generate a given v value.

$$Var(v) \propto [A]^{\alpha} \cdot [I]^{\beta} \qquad\qquad [20]$$

where α and β were empirical constants ($\alpha>0,\beta<0$). Figure 4 shows an iso-variance contour map of the error function (Eq. [20]) plotted in the concentration space (the [A]-[I] plane). These results show that the magnitude of the experimental error may be different at a given velocity level depending on the setting of the experimental variables, but it could not be demonstrated that use of weighting factors based on Eq. [20] instead of Eq. [17] gave significantly different results in the weighted regression analysis. However, with improved precision in the determination of velocity and more sophisticated model building it may in the future be necessary to describe the variance as a function of the true independent variables.

2.5. Suggested Procedure for Experimentation and Data Analysis

Scientific investigations should be carefully planned to give maximal information. Figure 5 shows an outline of different phases in the design and analysis of such studies.

First, the *problem* should be clearly *identified* and formulated. For example, the dependence of the steady-state velocity of an enzymatic reaction, v, on the concentrations of the reactants, [A],[B],..., may be sought.

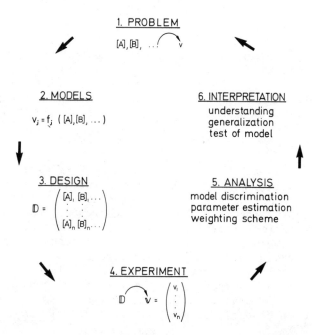

FIGURE 5: Strategy of design and analysis in mathematical modelling.

The second step is the *formulation of* suitable *models, i.e.,*
rate equations in the present example, that may describe the relationship between the dependent and independent variables. For
mechanistic model building in enzyme steady-state kinetics, the
approaches of King and Altman (53) and of Wong and Hanes (54) may
be adopted to derive a rate equation once the underlying reaction
scheme has been defined.

The third step involves the *design of the experiments* which
should be performed. For this purpose some prior knowledge of the
error structure is important. If such knowledge is lacking, a preliminary experiment should be carried out. Replicate measurements
(5 to 10) should be made at the lowest and the highest values considered for the dependent variable (v) and the variance calculated
at these levels. The hypothesis of constant variance could be
tested by comparing the ratio of the two variance estimates with
the proper F-statistic (usually a 95% confidence level is chosen).
A design matrix, $\underline{D}$, is then made up to define the experimental conditions to be utilized in solving the problem formulated. If the
mathematical model to be used is considered as definitely established,
the problem is to define the conditions that optimize the parameter
estimation according to the criterion of Box and Lucas (11) or any
other suitable criterion. If alternative models are considered, it
is usually best to cover as wide ranges as possible of the independent variables (reactant concentrations, time, *etc.*). When several
independent variables are involved in the analysis, it is usually
advantageous, both for statistical analysis and graphical display,
to select a number of levels for each variable and collect experimental data from all possible combinations (factorial design).

The fourth step is the *collection of experimental data,* which
will yield the response of the dependent variable, $(\underline{v})$, resulting
from the implementation of the design matrix. All causes of systematic experimental errors should be considered and avoided to the
extent possible. It may be useful to make the experiments defined
by the design matrix in random order to avoid the influence of
time-correlated changes in reactant solutions, in experimental conditions, or in responsiveness of the instruments used.

The fifth step is the *analysis of* the set of experimental *data*.
In general, this involves discrimination between rival models, for
which purpose the criteria mentioned in Section 2.2 may be used, if
the data have been fitted by the least-squares method. A flow chart
of the discrimination procedure is presented in ref. (30). The
model selected should also be examined with respect to goodness of
fit before being adopted as a suitable description of the phenomenon
investigated. Optimal parameter estimation should be attempted by
using refined weighting factors in the fitting of the best model
to the data set. If several data sets have been collected it may
be possible to combine all data sets for an overall analysis [see

below and refs. (55) and (56)]. The parameter values and their
corresponding standard deviations calculated in such a fit of combined data are probably more accurate estimates than those obtained
from a single data set, because the former reflect the day-to-day
variation and the latter only the variance within a single experimental series. For example, if the estimation of a K_m value for an
enzymatic reaction is attempted, it is generally found that the
estimates obtained on different days differ more than expected by
comparison with the standard deviation obtained by regression analysis of a single data set.

The sixth step is the *interpretation of the results* of the
analysis of the experimental data. Basically, an understanding of
the system investigated is attempted. The scientific process also
generally involves the generalization of the results to encompass
other conditions or other systems than those investigated. To test
the model in the system originally examined as well as to prove its
possible generalized applications new experiments are usually designed. This testing induces a new series of the six steps of experimentation and analysis (Fig. 5). However, even if such extensions are not considered, new experiments may have to be made in
order to optimize the design for model discrimination or parameter
estimation. In this connection it should be noted that only when
the best model has been chosen can the optimal design for parameter
estimation be defined.

3. EXAMPLES OF DESIGN AND ANALYSIS

3.1. Some Published Studies

Early examples of the use of methods of analysis similar to
or identical with those described above are given in refs. (17-19,
24,26-28,30,35,36,47,52,55,57-69). Reich and coworkers have made
thorough theoretical studies of the possibilities and limitations
of fitting models to data (27,28,70-73). Applications in our own
laboratory exemplify both design and analysis (24,26,30,33,35,36,
44,51,52,56,74-76). An extensive review covering many of the topics
of the present article has been published by Garfinkel *et al.* (77).

3.2. A Worked-out Example

An early application of the methods of design and analysis
outlined above was the investigation of the steady-state kinetics
of glyoxalase I (26,30,35,36). Glyoxalase I catalyzes the formation of S-D-lactoylglutathione from methylglyoxal and glutathione
(78). The problem addressed was whether the enzyme acts on methylglyoxal (M) and glutathione (G) directly or on their hemimercaptal
adduct (A) which is formed rapidly in a spontaneous reaction. A
difficulty in the analysis is that the concentrations of the three
reactants are linked by the equilibrium:

$$M + G \rightleftharpoons A \qquad\qquad\qquad\qquad [21]$$

$$K_d = \frac{([M]_o - [A]) \cdot ([G]_o - [A])}{[A]} = \frac{[M][G]}{[A]} \qquad\qquad [22]$$

where $[M]_o$ and $[G]_o$ denote total concentrations and $[M]$, $[G]$, and $[A]$ concentrations of free M, G, and A, respectively, K_d is the equilibrium (dissociation) constant (3 mM). This linkage limits the number of truly independent variables to two: the third variable is uniquely determined by the other two according to the equilibrium expression (Eq. [22]).

To cut a long story short, it can be stated that the best model so far put forward to explain the kinetics of glyoxalase I was a branching mechanism (Model III). This mechanism unites the two original alternatives of a one-substrate (A) and a two-substrate (M and G) mechanism to a hybrid model (36,79). Alternatives which were rejected included a simple Michaelis-Menten mechanism (Model I) extended by treatment of glutathione as a competitive inhibitor *versus* the hemimercaptal adduct (Model II) (36). The different reaction schemes are shown in Figure 6 and the corresponding rate equations are:

$$v = \frac{V[A]}{K_m + [A]} \qquad\qquad\qquad\qquad \text{(Model I)}$$

$$v = \frac{V[A]}{K_m(1 + [G]/K_i) + [A]} \qquad\qquad \text{(Model II)}$$

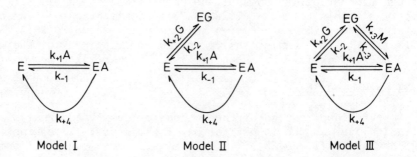

Model I   Model II   Model III

FIGURE 6: Reaction schemes of three alternative kinetic models considered for the enzyme glyoxalase I. E, G, M, and A denote enzyme, glutathione, methylglyoxal, and their hemimercaptal adduct, respectively. The curved arrow symbolizes the step that releases the product of the enzymatic reaction from the enzyme.

$$v = \frac{V_1[A]+V_2[A][M]}{K_1+K_2[A][M]+K_3[G]+K_4[M]+[A]} \qquad \text{(Model III)}$$

It should be noted that Models I and II are degenerate forms of Model III, both topologically with regard to the reaction schemes and algebraically with regard to the rate equations. In the following treatment it will be shown that the design of the experiments is of utmost importance for the demonstration of the superiority of the "best" model (III) over the rival Models I and II. The example is of a wide general interest because many metal-ion-activated enzymes act on equilibrium systems similar to that of Eq. [21] (consider, for example, the case: M=Mg^{2+}, G=guanosine triphosphate, GTP, and A=Mg-GTP). The design of kinetic experiments with such enzymes has recently been treated (80,81), but the following analysis will illustrate the types of problems which may arise owing to inadequate designs.

Kinetic experiments involving glyoxalase I from yeast were performed essentially as described previously (30). The concentrations of free reactants were calculated from Eq. [22] and the rate equations of Models I-III were fitted to the experimental data. Four different designs involving approximately the same number of experimental points (n~20) were investigated and their influence on model discrimination and parameter estimation evaluated. The different design matrices are mapped on the space of independent variables (the $[M]_0$-$[G]_0$ plane) (Figure 7).

Design 1 involves the use of a constant ratio of the total concentrations of methylglyoxal and glutathione. This kind of design has frequently been used in kinetic investigations of phosphotransferases (81). It would appear to be an appropriate design in so far as it provides for variation of [A], [G] and [M] over wide ranges. The results of fitting Models I and II to the data obtained by design 1 are shown in Table 1. It is clear that Model II degenerates to Model I, because the third parameter (K_i) is not significantly different from zero as shown by the high standard deviation. Model III has only one parameter with a standard deviation that was less than the parameter value (data not shown) and does not give a better fit than Model I (cf. Table 2). Thus, both Models II and III appear to be overdetermined and should therefore be discarded in favor of Model I. Figure 8 shows a v vs. [A] and a residual vs. predicted velocity plot. The fit of Model I to the data set is good and the residuals fall within the 95% confidence limits defined by the experimental error (which was constant) and appear to have a random distribution. The standard deviations of the parameters are acceptable and the mean sum of squares ($Q^2=0.927 \cdot 10^{-4}$) does not differ significantly from the estimated experimental variance ($\sigma^2 \approx 10^{-4}$). It would therefore appear that Model I is both adequate and sufficient as a description of the kinetics of the system. Models II and III fit the data equally well, but are inferior owing to their content of redundant parameters.

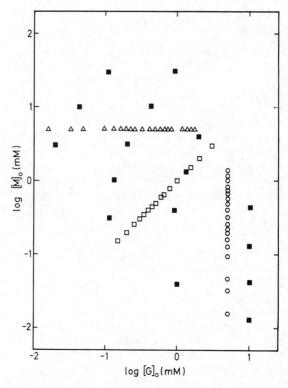

FIGURE 7: The plane of independent variables (factor space) in the analysis of the kinetics of glyoxalase I. $[G]_o$ and $[M]_o$ denote the total concentrations of glutathione and methylglyoxal, respectively. The symbols show the different design matrices mapped on the concentration space: □, design 1; o, design 2; Δ, design 3; ■, design 4.

Designs 2 and 3 represent other designs which have commonly been adopted in kinetic studies on phosphotransferases (81). In these designs a large excess of the total concentration of one of the variables is used while the concentration of the second is varied. Neither of designs 2 and 3 provided significantly better conditions for discrimination between the alternative models. As in the experiment according to design 1, those performed according to designs 2 and 3 could adequately be described by Model I. Table 2 shows the relative Q^2 values obtained by fitting Models I-III to the different data sets.

The 4th design investigated was a factorial design in which the concentration of free glutathione was varied at different constant levels of adduct. By experiments according to this design, a clear discrimination was obtained, which showed the inferiority of Models I and II in comparison with Model III. Table 2 shows that Model III have the smallest Q^2-value of the Models fitted.

TABLE 1: Results of fitting Models I and II to data set 1 (n=16)

Constant	Estimate±s.d.	
	Model I	Model II
V (ΔA/min)	0.752±0.018	0.762±0.093
K_m (mM)	0.367±0.019	0.367±0.021
K_i (mM)	nonexistent	39.2±374.0 (a)
SS ($\Delta A^2/min^2$)	0.001299	0.001298
ρ (b)	0.0552	0.0524
$n-p_j$ (c)	14	13

(a) Redundant parameter as evidenced by the high standard deviation.
(b) Serial correlation coefficient of residuals ($|\rho|<1$) (see ref. 33 for definition); a value near 1.0 shows that the residuals are highly correlated; a value near zero demonstrates lack of correlation.
(c) Degrees of freedom; n is the number of measurements and p_j is the number of parameters of the j-th model.

TABLE 2: Effect of design on discrimination between alternative models

Design (a)	Residual sum of squares (b)			Discrimination
	Model			
	I	II	III	
1	1.09	1.09	1.00	No
2	1.15	1.15	1.00	No
3	1.02	1.02	1.00	No
4	37.07	17.68	1.00	Yes

(a) The various designs are displayed in Figure 7.
(b) The residual sums of squares obtained by fitting the models to a particular data set have been normalized by division with the residual sum of squares of Model III. It should be noted that the Q^2 values are smallest for Model I in the first three experiments, because the degrees of freedom (n-p, see Eq. [8]) is largest when this model is used. Thus, according to the analysis of designs 1-3, Model I appears to be the model which should be adopted even if no clear discrimination can be based on the residual sum of squares (criterion iv). However, design 4 shows clearly the superiority of Model III (see also Figures 9 and 10).

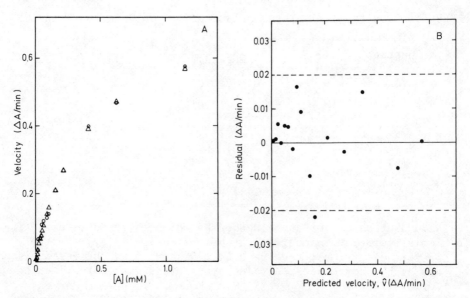

FIGURE 8:   Results of fitting Model I to the experimental velocities
obtained by use of design 1.   A: Observed (o) and predicted (Δ)
velocities plotted against the concentration of adduct of glutathione
and methylglyoxal ([A]).   B: Residuals plotted versus predicted
velocities (v̂).   The dashed lines correspond to the 95% confidence
level (± $t_{0.05}\sigma$) calculated from an estimate of the experimental
error (σ).

This Q^2-value is not significantly greater than the estimated experimental variance.   The inadequacy of Model I is shown not only
by the high Q^2-value but also by analysis of the residuals.   Figure
9 shows the nonrandom distribution of residuals in the factor space
(Figure 9A) as well as in a residual *vs*. [G] plot (Figure 9B).
Note that all positive residuals are located in one domain (Figure
9A) and that all of the residuals at the highest [G] are negative
(Figure 9B).   Furthermore, 10 of the residuals deviate more than
2σ from zero.   Figure 10, on the other hand, displays what appears
to be a random distribution of the residuals obtained after fitting
Model III to data set 4.   Of the different designs investigated
only that giving rise to data set 4 shows that Model III is superior
to Models I and II.   Simply stated, the failure of design 1 was due
to the fact that the combinations high $[G]_0$ - low $[M]_0$ and low $[G]_0$
- high $[M]_0$ were not realized and that a constant $[M]_0/[G]_0$ ratio,
in effect, reduces the independent variables from two to one.   The
weakness of designs 2 and 3 is that one of the variables was not
allowed to assume low values and that, again, essentially only one
independent variable is utilized.   Hence, it may be concluded that
the entire available concentration space should be explored by

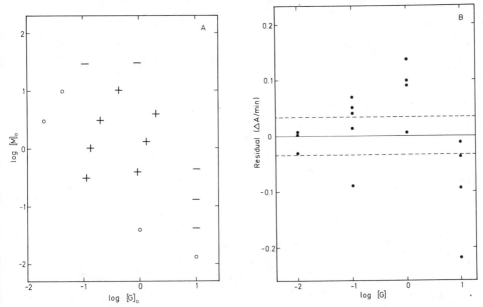

FIGURE 9: Plots of residuals obtained by fitting Model I to data
set 4. A: Residuals plotted in the plane of independent variables
(factor space). o, Residuals smaller than the experimental error
(σ); +, residuals $> \sigma$; -, residuals $< -\sigma$. B: Residuals plotted
versus [G] (mM). The dashed lines correspond to the 95% confidence
level.

independent settings of the independent variables in order to get
a complete evaluation of the dependence of velocity on reactant
concentrations.

It remained to optimize the parameter estimation. Normally
this would involve consideration of new experiments according to a
new design (24). However, in the present case advantage was taken
of the fact that the combination of the four different experiments
(designs 1-4) had covered the entire available substrate concentration space. Parameter estimation was therefore made by fitting
Model III to all four data sets simultaneously. This is permissible
if the enzyme concentration is the same in all experiments. If
different enzyme concentrations have been used, a normalizing factor
has to be introduced for each data set added to the first one, because
the numerator terms of the rate equation are proportional to the
enzyme concentration. In the present case it was shown that the
velocity of the enzymatic reaction was proportional to the enzyme
concentration, and the normalization could therefore be made by a
simple linear combination of the four data sets. The procedure,
which has previously been used successfully in our laboratory (cf.
ref. 56), is simply to multiply the rate equation considered (in

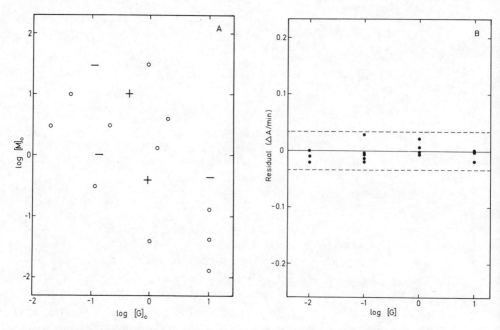

FIGURE 10:   Plots of residuals obtained by fitting Model III to
data set 4.   See legend to Figure 9.

the present case, Model III) by a factor

$$(c_1 z_1 + c_2 z_2 + c_3 z_3 + \ldots + z_m)$$

where c_j (j=1,...,m-1) are coefficients normalizing the velocities
of the j-th experiments to the enzyme concentration of the m-th
experiment.   For every data set (j) z_k-values are defined such
that:

$$z_k = 1 \quad \text{for} \quad k=j, \quad \text{and}$$

$$z_k = 0 \quad \text{for} \quad k \neq j \quad (j,k=1,\ldots,m) \quad .$$

Thus, instead of fitting the rate equation

$$v = f([A],\ldots) \tag{23}$$

to the data, the extended equation

$$v = (c_1 z_1 + c_2 z_2 + \ldots + z_m) \cdot f([A],\ldots) \tag{24}$$

is used, in which c_j are additional parameters to be estimated.
Since the new parameters enter linearly in the regression it is, in
our experience, generally not more difficult to fit the extended
equation (Eq. [24]) than to fit the original model (Eq. [23]) to

the data. An additional feature of the use of Eq. [24] is that the standard deviations of the parameters obtained are based on several independent experiments.

The analysis of data sets 1-4 by the above procedure, using Model III expanded according to Eq. [24], gave the parameter values presented in Table 3. The normalizing coefficients (c_1-c_3) were in the range of 0.8-1.6 and had standard errors which were 3-6% of these values. The corresponding asymptotic correlation matrix of the parameters in Model III is shown in Table 4. It is evident that the parameters of terms in the rate equation which contain the same reactants are highly correlated. For example, the coefficients of the [A]·[M] terms (V_2 and K_2) have the highest correlation coefficient (0.9916), and the coefficient of the [M] term (K_4) is almost as strongly correlated with V_2 and K_2, respectively. A strong correlation between these coefficients should be expected, because the "kappa" factors (53,54) which yield the terms of the rate equation always introduce the elementary rate constant for addition of a reactant in the reaction scheme in combination with the reactant concentration. Thus, V_2 and K_2 expressed in rate constants (cf. ref. 36) are identical except for a factor $k_{+4}[E]_o$ ($V_2 = k_{+4}[E]_o \cdot K_2$, where k_{+4} is the rate of release of product and $[E]_o$ is the total enzyme concentration). On the other hand, K_1 and K_3, which belong to terms that have no factor of reactant concentration in common with any other term (if we disregard the possibility of expressing [A] as [G]·[M]/K_d under equilibrium conditions - Eq. [22]), show the lowest correlations with the other parameters (Table 4).

___

TABLE 3: Parameter values obtained after fitting
Model III to the combined data sets 1-4

Constant	Estimate±s.d.
V_1 (ΔA/min)	1.296±0.174
V_2 ($\Delta A \cdot min^{-1} \cdot mM^{-1}$)	0.095±0.074
K_1 (mM)	0.355±0.057
K_2 (mM^{-1})	0.170±0.127
K_3	0.223±0.049
K_4	0.082±0.042

TABLE 4: Asymptotic correlation matrix of the parameters of
Model III obtained by fitting Model III to the
combined data sets 1-4

	V_1	V_2	K_1	K_2	K_3	K_4
V_1	1					
V_2	0.8748	1				
K_1	0.7796	0.5998	1			
K_2	0.9122	0.9916	0.6662	1		
K_3	0.8079	0.6697	0.4359	0.6846	1	
K_4	0.8514	0.9807	0.4949	0.9619	0.7016	1

The set of parameters shown in Table 3 is not completely
satisfactory, because the standard deviations of V_2 and K_2 are high
in comparison with the parameter values. It is likely that additional experiments could give more precise values, but the possibility that Model III is not the "true" model should also be considered. Recent experiments involving equilibrium binding of
ligands to the enzyme support the latter possibility (82). The
introduction of a new interpretation illustrates a typical result
of analysis, which may initiate a new cycle of design, experiment,
and analysis (Figure 5).

4. CONCLUSIONS

The strategy outlined in Figure 5 and Section 2.5 has been
applied to numerous problems in enzyme kinetics and been found to
be of great assistance in experimental work as well as in analysis.
Evidently it is not limited to this area of research, but it can
be utilized in any application in which a real variable can be
studied as a function of one or several independent variables.
Graphical analysis of experimental data should be exploited as far
as possible, because recognition of significant patterns in the
data set can often easily be made visually. In our laboratory,
nonlinear regression has been found to be an adequate fitting
procedure for various types of problems. The results of regression
analysis lend themselves to statistical evaluation by use of the
criteria discussed in Section 2.2. These criteria facilitate the
judgement of goodness-of-fit as well as the discrimination between
rival models. The discrimination is critically dependent on the
design of the experiments as shown by the example in Section 3.2.

The use of the discrimination function has been found valuable in designing experiments for discrimination between alternative models. It should be emphasized that the process of mathematical modelling of experimental data is sequential in nature as indicated in Figure 5. Initially, design and experiments as well as analysis should be focussed on selecting the "best" model. This model should explain the experimental data within the limits of the established experimental error. When an adequate model has been found, new experiments should be designed that optimize conditions for parameter estimation. These values should be consistent with the theory underlying the mathematical model. When all these conditions have been fulfilled, the modelling has succeeded.

ACKNOWLEDGEMENTS

The procedure for discrimination between alternative models was developed in collaboration with Dr. Tamas Bartfai. I am grateful for his advice throughout the years as well as his valuable comments on this paper. In addition to the co-authors of the publications from our laboratory cited here, I am indebted to Mr. Håkan Ek who carried out the experiments described in Section 3.2. I am also grateful to Ms. Véronique Yahiel and Ms. Kerstin Larson for expert assistance in the analysis of data and to Mr. Peter Westholm for drawing of the diagrams. The work cited from our laboratory has been supported by the Swedish Natural Science Research Council and the Swedish Cancer Society.

REFERENCES

1. Box, G.E.P. and Hunter, W.G. (1965) Technometrics, 7, 23-42.
2. Mannervik, B. (1978) FEBS Lett. 93, 225-227.
3. Bardsley, W.G. and Childs, R.E. (1975) Biochem. J. 149, 313-328.
4. Bardsley, W.G. (1976) Biochem. J. 153, 101-117.
5. Childs, R.E. and Bardsley, W.G. (1976), J. Theor. Biol. 63, 1-18.
6. Bardsley, W.G. and Crabbe, M.J.C. (1976) Eur. J. Biochem. 68, 611-619.
7. Bardsley, W.G. and Wyman, J. (1978) J. Theor. Biol. 72, 373-376.
8. Eisenthal, R. and Cornish-Bowden, A. (1974) Biochem. J. 139, 715-720.
9. Cornish-Bowden, A. and Eisenthal, R. (1974) Biochem. J. 139, 721-730.
10. Hill, C.M., Waight, R.D. and Bardsley, W.G. (1977) Mol. Cell. Biochem. 15, 173-178.
11. Box, G.E.P. and Lucas, H.L. (1959) Biometrika, 46, 77-90.
12. Duggleby, R.G. (1979) J. Theor. Biol. 81, 671-684.
13. Duggleby, R.G. (1980) In "Kinetic Data Analysis: Design and Analysis of Enzyme and Pharmacokinetic Experiments", (L. Endrenyi, ed.) pp. 169-179, Plenum, New York.

14. Endrenyi, L. (1980) In "Kinetic Data Analysis: Design and Analysis of Enzyme and Pharmacokinetic Experiments", (L. Endrenyi, ed.) pp. 137-167, Plenum, New York.

15. Cornish-Bowden, A. (1980) In "Kinetic Data Analysis: Design and Analysis of Enzyme and Pharmacokinetic Experiments" (L. Endrenyi, ed.) pp. 105-119, Plenum, New York.

16. Atkins, G.L. and Nimmo, I.A. (1980) In "Kinetic Data Analysis: Design and Analysis of Enzyme and Pharmacokinetic Experiments" (L. Endrenyi, ed.) pp. 121-135, Plenum, New York.

17. Johansen, G. and Lumry, R. (1961) Compt. Rend. Trav. Lab. Carlsberg, 32, 185-214.

18. Wilkinson, G.N. (1961) Biochem. J. 80, 324-332.

19. Cleland, W.W. (1963) Nature, 198, 463-465.

20. Storer, A.C., Darlison, M.G. and Cornish-Bowden, A. (1975) Biochem. J. 151, 361-367.

21. Siano, D.B., Zyskind, J.W. and Fromm, H.J. (1975) Arch. Biochem. Biophys. 170, 587-600.

22. Askelöf, P., Korsfeldt, M. and Mannervik, B. (1976) Eur. J. Biochem. 69, 61-67.

23. Nimmo, I.A. and Atkins, G.L. (1979) Trends Biochem. Sci. 4, 236-239.

24. Bartfai, T. and Mannervik, B. (1972) FEBS Lett. 26, 252-256.

25. Jennrich, R.I. and Ralston, M.L. (1979) Ann. Rev. Biophys. Bioeng. 8, 195-238.

26. Bartfai, T. and Mannervik, B. (1972) In "Analysis and Simulation of Biochemical Systems" (H.C. Hemker and B. Hess, eds.) pp. 197-209, North-Holland/American Elsevier, Amsterdam.

27. Reich, J.G. (1970) FEBS Lett. 9, 245-251.

28. Reich, J.G., Wangermann, G., Falck, M. and Rohde, K. (1972) Eur. J. Biochem. 26, 368-379.

29. Kohn, M.C., Menten, L.E. and Garfinkel, D. (1979) Comput. Biomed. Res. 12, 461-469.

30. Bartfai, T., Ekwall, K. and Mannervik, B. (1973) Biochemistry, 12, 387-391.

31. Draper, N.R. and Smith, H. (1966) "Applied Regression Analysis", pp. 263-304, Wiley, New York.

32. Briggs, G.E. and Haldane, J.B.S. (1925) Biochem. J. 19, 338-339.

33. Mannervik, B. (1975) BioSystems, 7, 101-119.

34. Ottaway, J.H. and Apps, D.K. (1972) Biochem. J. 130, 861-870.

35. Mannervik, B. and Bartfai, T. (1973) Acta Biol. Med. Ger. 31, 203-215.

36. Mannervik, B., Górna-Hall, B. and Bartfai, T. (1973) Eur. J. Biochem. 37, 270-281.

37. Ellis, K.J. and Duggleby, R.G. (1978) Biochem. J. 171, 513-517.

38. Mannervik, B., Jakobson, I. and Warholm, M. (1979) Biochim. Biophys. Acta, 567, 43-48.

39. Atkinson, A.C. and Hunter, W.G. (1968) Technometrics, 10, 271-289.

40. Endrenyi, L. and Kwong, F.H.F. (1972) In "Analysis and Simula-
     tion of Biochemical Systems" (H.C. Hemker and B. Hess, eds.)
     pp. 219-237, North-Holland/American Elsevier, Amsterdam.
41. Hunter, W.G. and Reiner, A.M. (1965) Technometrics, 7, 307-323.
42. Bard, Y. and Lapidus, L. (1968) Catal. Revs. 2, 67-112 (see
     p. 104).
43. Markus, M. and Plesser, T. (1980) In "Kinetic Data Analysis:
     Design and Analysis of Enzyme and Pharmacokinetic Experi-
     ments" (L. Endrenyi, ed.) pp. 317-339, Plenum, New York.
44. Mannervik, B. and Askelöf, P. (1975) FEBS Lett. 56, 218-221.
45. Bardsley, W.G. and Waight, R.D. (1978) J. Theor. Biol. 70,
     135-156.
46. Pettersson, G. (1972) Biochim. Biophys. Acta, 276, 1-11.
47. Haarhoff, K.N. (1969) J. Theor. Biol. 22, 117-150.
48. Morrison, J.F. (1969) In "Least Squares Methods in Data
     Analysis" (R.S. Anderssen and M.R. Osborne, eds.) pp. 63-69,
     Australian National University Computer Centre Publication
     CC2/69.
49. Ottaway, J.H. (1973) Biochem. J. 134, 729-736.
50. Jacquez, J.A., Mather, F.J. and Crawford, C.R. (1968) Biometrics,
     24, 607-626.
51. Augustinsson, K.-B., Bartfai, T. and Mannervik, B. (1974)
     Biochem. J. 141, 825-834.
52. Bartfai, T. and Mannervik, B. (1973) FEBS Lett. 32, 179-183.
53. King, E.L. and Altman, C. (1956) J. Phys. Chem. 60, 1375-1378.
54. Wong, J.T.-F. and Hanes, C.S. (1962) Can. J. Biochem. Physiol.
     40, 763-803.
55. Hurst, R., Pincock, A. and Broekhoven, L.H. (1973) Biochim.
     Biophys. Acta, 321, 1-26.
56. Uotila, L. and Mannervik, B. (1979) Biochem. J. 177 869-878.
57. Berman, M., Shahn, E. and Weiss, M.F. (1962) Biophys. J. 2,
     275-287.
58. Dammkoehler, R.A. (1966) J. Biol. Chem. 241, 1955-1957.
59. Bliss, C.I. and James, A.T. (1966) Biometrics, 22, 573-602.
60. Cleland, W.W. (1967) Adv. Enzymol. 29, 1-32.
61. Plowman, K.M. and Cleland, W.W. (1967) J. Biol. Chem. 242,
     4239-4247.
62. Hanson, K.R., Ling, R. and Havir, E. (1967) Biochem. Biophys.
     Res. Commun. 29, 194-197.
63. Lasch, J. (1969) Acta Biol. Med. Ger. 23, 747-757.
64. Mäkinen, K.K., Euranto, E.K. and Kankare, J.J. (1969) Suomen
     Kemistilehti, B42, 129-133.
65. Kowalik, J. and Morrison, J.F. (1968) Math. Biosci. 2, 57-66.
66. Arihood, S.A. and Trowbridge, C.G. (1970) Arch. Biochem.
     Biophys. 141, 131-140.
67. Pettersson, G. and Pettersson, I. (1970) Acta Chem. Scand.
     24, 1275-1286.
68. Rapoport, T.A. Höhne, W.E., Reich, J.G., Heitmann, P. and
     Rapoport, S.M. (1972) Eur. J. Biochem. 26, 237-246.

69. Kuhn, E. and Brand, K. (1973) Biochemistry, 12, 5217-5223.
70. Reich, J.G. (1974) Stud. Biophys. 42, 165-180.
71. Reich, J.G., Winkler, J. and Zinke, I. (1974) Stud. Biophys. 42, 181-193.
72. Reich, J.G., Winkler, J. and Zinke, I. (1974) Stud. Biophys. 43, 77-90.
73. Reich, J.G. and Zinke, I. (1974) Stud. Biophys. 43, 91-107.
74. Jakobson, I., Askelöf, P., Warholm, M. and Mannervik B. (1977) Eur. J. Biochem. 77, 253-262.
75. Jakobson, I., Warholm, M. and Mannervik, B. (1979) Biochem. J. 177, 861-868.
76. Jakobson, I., Warholm, M. and Mannervik, B. (1979) FEBS Lett. 102, 165-168.
77. Garfinkel, L., Kohn, M.C. and Garfinkel, D. (1977) CRC Crit. Rev. Bioeng. 2, 329-361.
78. Ekwall, K., and Mannervik, B. (1973) Biochim. Biophys. Acta, 297, 297-299.
79. Mannervik, B., Bartfai, T. and Górna-Hall, B. (1974) J. Biol. Chem. 249, 901-903.
80. Bartfai, T. (1979) Adv. Cyclic Nucleotide Res. 10, 219-242.
81. Morrison, J.F. (1979) Meth. Enzymol. 63, 257-294.
82. Marmstål, E. and Mannervik, B. (1979) FEBS Lett. 102, 162-164.

ESTIMATION OF POOLED PHARMACOKINETIC PARAMETERS DESCRIBING

POPULATIONS

L.B. Sheiner and S.L. Beal

University of California
Department of Medicine
Division of Clinical Pharmacology
San Francisco, California 94143, U.S.A.

ABSTRACT

By population pharmacokinetics we mean the typical relationships between physiology and pharmacokinetics; the interindividual
variability in these relationships, and their residual inexplicable
intraindividual variability. Knowledge of such population features
is useful both to gain insight into the drug-patient system, and
to adjust individual drug dosage. Experimental data from which
population kinetics might be estimated, often comes from only a few
atypical individuals (e.g., normal volunteers, rather than patients).
More representative data might be those coming from routine patients.
These data, however, are marked by varying quality, accuracy and
precision, as well as there being little data per patient. To
combine data from various sources, and to use routine data, one
must overcome certain data analysis problems. The standard (experimental data oriented) approach does not do so. An approach is
available that regards the population, rather than the individual,
as the primary unit of analysis. It is useful for analysis of
routine data and for combining data of varying quality. This paper
discusses the approach and contrasts it with the standard one.

1. INTRODUCTION

Many of the important questions in pharmacokinetic research
are answered by investigating population pharmacokinetics: the
pharmacokinetic characteristics of typical individuals, rather
than those of particular individuals. For example, whether a
disease state influences kinetics is determined by the knowledge
of population pharmacokinetics. To answer such questions, it may
be desirable to combine data from many sources. The different

sources may be characterized, however, by varying amounts of data
per subject, varying data quality, and by having arisen from various experimental designs (including no design at all). Standard
methods of data analysis may be inadequate to deal with such data,
or with combining it. In this paper, we describe why this is so,
and suggest an alternative approach (1,2) with which we have had
some success.

2. ESTIMATING POPULATION KINETICS

2.1. What We Wish to Know

Consider the following questions. Does drug clearance vary
with variations in renal function? Is so, what is the quantitative
relationship between the two? What is the bioavailability of a preparation? These questions concern the average relationships of
physiology to pharmacokinetics, or the average value of a pharmacokinetic characteristic in a population. Answers to these
questions are important when choosing an initial dose of drug for
a patient with renal disease, or when deciding whether a new drug
preparation should be marketed.

Estimates of certain kinds of population pharmacokinetic
parameters help answer questions. These are the quantitative
values of the proportionality between renal function (as measured,
perhaps, by creatinine clearance) and drug clearance; or the population mean bioavailability of a drug preparation. Population
pharmacokinetic parameters of this type are called fixed effect
parameters because they quantify the relationship between certain
fixed (measurable) effects such as renal function, and pharmacokinetics.

Consider some other questions. Assuming drug clearance does
vary with renal function, how well do we know an individual's
clearance if we know his renal function? What is the least bioavailability we may reasonably expect to see for a certain preparation given to a normal individual? Answers to these questions are
important when deciding how confident one is in one's initial
dosage for the renal patient (and so in choosing the frequency
of follow-up), or in considering the "worst case" behaviour of a
preparation prior to marketing. Estimates of population pharmacokinetic parameters of a different type are needed here. We need
to know the (probability) distribution of deviations of individual
values of fixed effect parameters from the population average values.
A measure of an important feature of this distribution, its spread,
is given by the standard deviation of individual parameter values
about the population mean. This is a population pharmacokinetic
parameter of a different type, an interindividual random effect
parameter; random, because individual deviations are regarded as
occurring according to the probability distribution.

Finally, we may ask, how much drug clearance varies day-to-day within an individual, or how much a steady-state drug level varies day-to-day due to all causes other than bioavailability. We might want to answer these questions in order to set reasonable sizes for dosage increments (certainly no less than twice the random day-to-day variability), or reasonable manufacturing tolerances for tablet-to-tablet variability (certainly tablet-to-tablet content may vary as much as *ca.* 25% of the inevitable day-to-day percentage variation in drug levels caused by physiological fluctuation alone). To answer these questions, yet another random effect population parameter is needed: a combined intraindividual and measurement error random effect parameter (or just "error" random effect parameter); random because day-to-day fluctuation and measurement errors are also regarded as occurring according to probability distributions. A natural choice for such a parameter is the standard deviation of the distribution of day-to-day observed drug levels about their mean value within a typical individual.

2.2. The Standard Approach

The traditional, or standard approach employs an *experimental paradigm*. The *subjects* of study are a group of normal individuals, or a group of patients, selected to represent a spectrum of severity of some condition (*e.g.*, renal insufficiency), the effects of which are to be studied. Usually only 10-30 subjects are studied. The *dosage* pattern administered to the subjects is usually of simple design, perhaps a single bolus dose, or a single infusion, often to be repeated under other conditions at another time, in a balanced design. The *sampling schedule* (over time) for the biological fluids of interest (blood, urine, perhaps others) is usually fixed, and the same for each subject. It is designed to reveal maximum information about individual kinetics, and consists of many samples (often more than 20) per subject.

Data analysis, applied to the standard study, proceeds in two stages. First, the data from each individual are analyzed to yield estimates of the individual pharmacokinetic parameters, such as drug clearance. The step is usually accomplished using (weighted or unweighted) nonlinear regression with the least-squares criterion (most pharmacokinetic models are statistically nonlinear). The individual parameter estimates, obtained in the first step, are then subject to a second step of analysis to obtain population parameter estimates. For the fixed effect parameters relating pharmacokinetics to physiology, least-squares regression is used, but here it is most often statistically linear (that is, the relation of drug clearance, for example, to creatinine clearance is usually modelled as a linear one). When estimating population mean parameters unrelated to other factors, simple averages of the individual parameter estimates are usually taken. For estimates of the interindividual random effect parameters, the standard devia-

tions of the individual parameters about the regression line (or
the mean value) are used.

When error random effect parameters are estimated, which is
rarely, the square root of the sum of the pooled squared residuals
of the initial nonlinear fits divided by the (pooled) residual
degrees of freedom is usually used.

Standard errors of the fixed effect parameter estimates when
the latter are calculated as averages of individual parameter
estimates, are usually taken to be the standard deviations of the
individual parameter estimates divided by the square root of the
number of sampled individuals. Standard errors of the interindividual random effect parameter estimates are not usually computed.
Some might be tempted to use the usual chi-square approach (3) to
compute confidence intervals, however. The same approach might be
used (with greater justification) to obtain confidence intervals
for the error random effect parameter.

Advantages of the standard approach are quite obvious. Fore-
most among these is that there is considerable accumulated experience with it, and aspects of it have been found to be quite reliable. Experimental design may sometimes pose problems, but there
is considerable experience and literature on this subject, and help
can be obtained easily. Studies can often be completed relatively
quickly, and homogeneity of data with respect to study conditions,
data quality, precision and accuracy of drug level determinations,
amount of data per individual, and so forth, is assured. Standard
computer programs are available for both stages of the data analysis.
Pharmacokineticists have considerable experience with these programs,
and know them to be reliable. Data analysis is generally rapid and
inexpensive. Statistical models, with which most kineticists are
relatively unfamiliar, are so simple that often they need not be
explicitly stated. The kineticist may thus devote most of his
attention to the kinetic models with which he is more familiar and
more immediately concerned.

The source of the data can be a major disadvantage for the
standard approach. Ethical problems are encountered in planning
a study of patients. It may be difficult to justify temporarily
witholding a drug for study purposes or causing added risk and
annoyance to an already precarious and uncomfortable patient.
Consequently, traditional studies are often undertaken in normal
individuals, or special patient groups with lesser degrees of ill-
ness. This, in turn, leads to problems with the representativeness
of the obtained information. Although a traditional study may
accurately assess the pharmacokinetics of the study group, this
group may not represent subjects of greatest concern: patients
who will be exposed to the drug in the course of their treatment.
Because the number of study subjects is usually relatively small,

population parameter estimates, even if obtained from a representative group, may deviate substantially from true population values. This is especially true of the interindividual random effect parameters.

A traditional study is quite costly. Although data analysis is not, compensation of volunteers, temporary hospitalization on a clinical research ward, assay of numerous samples, and so forth, often make the cost of a traditional study in excess of several thousands of dollars per subject.

Finally, serendipity is not possible. Careful control of diet, study conditions, etc., all undertaken in the interest of obtaining data with low variability prevents the discovery of unexpected influences on kinetics. Drug interactions or diet might be important in a clinical setting, but will never be discovered (by accident) in a carefully designed study. Theophylline is an example. The initial impression that the interindividual standard deviation for clearance was rather small (4) was contradicted by clinical experience. The original study had been done in asthmatics in remission. Only when the influence of smoking (5), heart failure (6), or severe obstruction (7) was studied, did variability due to these influences become apparent. Moreover, even after correcting for these influences, acutely ill patients seem, for unknown reasons, to vary consistently but unpredictably from expectations with an interindividual random effect standard deviation more than twice as large as that of stable asthmatics in remission, or of normals (7).

Problems with the standard approach's data analysis *per se* tend not to be major because of certain features of the experimental data. We discuss the problems here, nonetheless, in order to provide background for later discussion of the analysis of routine patient data, wherein the problems become acute, and necessitate an alternative approach to data analysis.

When ordinary least squares is used for the initial (individual-fitting) step in the data analysis, implicit assumptions (8) are mong others, that all errors intervening between the "true" level and the observed level are: (i) independent of each other (as one goes from concentration to concentration), (ii) additive, and (iii) of the same typical magnitude (*i.e.*, the error standard deviation is constant for all drug concentrations). All of these assumptions are open to question, even for the standard approach. Consder, in particular, the third assumption. Often absolute measurement error is not of constant magnitude. It begins at a "lower limit of detectability", the magnitude of the smallest typical error. It then rises, often in proportion to the drug concentration itself. Measurement errors of different magnitudes are also seen when data are of more than one type; for example,

when both urine concentrations and plasma concentrations of drug
are available. Both concentrations arise from models sharing certain parameters, and these parameters are most efficiently estimated by using all the data in a joint fit. Yet the assumption
of equal error magnitude for quantities typically differing a
hundred-fold in scale is patently wrong. A third cause of differing error magnitudes is differences among subjects: some may
have kinetics more variable across time than others.

When using usual least squares, one must use known weights to
adjust for possible inhomogeneities in error magnitude. But where
do the weights come from? Study of the measurement process by
replication, which might offer some basis for assigning relative
error magnitudes, is usually not done. Even if it were, it might
not suffice, since a considerable part of the total "error" is due
not to assay measurement, but to inadequacies of the pharmacokinetic model, or to transient variation in the subjects' kinetics.
These "errors" must be incorporated into the weights if the latter
are to be correct, yet they are not readily measurable.

Analysts often resort to weighting their data by the reciprocal
of the datum itself, or the datum squared (9). Such weighting is
meant to adjust for the increase in error magnitude associated with
higher true values. The approach may help, but if concentrations
near the minimum detectible concentration are present in the data,
such weighting will tend to markedly over-weight these points.

Problems with the assumptions may have little importance when
analyzing experimental data. This is because (i) even the largest
errors are often small compared to the total concentration range
covered by the data, so that neither non-additive error, nor
variable error magnitude nor correlated errors are important; and
(ii) the traditional weighting schemes (as above), when used for
data from various sources, often provide a reasonable balance.

The second step of the standard data analysis, the estimation
of population parameters from individual parameter estimates, also
has problems. What if one were fitting data to a pharmacokinetic
model involving four or five parameters? For subjects with fewer
concentration-time points than parameters, one might be unable to
estimate all of the individual parameters. These subjects' data
might then have to be discarded, even though they undoubtedly
contain information about the population parameters.

Even with individual parameter estimates from all subjects,
the mean and standard deviation of these estimates may be poor
estimates of the population mean and interindividual standard
deviation. This is because each individual estimate is "contaminated" to varying degrees by an irrelevant error: the error
made in estimating the parameter itself. Suppose there were scarce

data from each individual so that each individual's parameter estimates were typically far from their true values. Although individual parameter estimates might be as likely to be too high as
too low so that the estimated "population" mean might not be biased,
still, for a precise estimate of the population mean, less attention
should be paid to the poorer individual estimates. The estimate of
the "population" interindividual standard deviation will have the
same problems, but it *will* be biased upwards. It will contain
variability both from interindividual biological sources and from
estimation. Independent variabilities add, rather than cancel
each other.

It is not clear how the presence of estimation error *per se*
will influence the appropriateness of the standard errors computed
for the fixed effect parameters. We do know that if estimation
error varies with the individual parameter itself, the standard
approach to computing standard errors described previously is not
very good.

Again, the experimental design saves the standard method.
Sufficient data are gathered from everyone so individual estimates
are always possible; the amount of individual data is sufficiently
large and covers a sufficiently wide range so that parameter estimates have small error relative to interindividual variability.
Therefore, although undoubtedly somewhat biased, the estimates of
the interindividual standard deviations will not be greatly biased.
For the same reasons, the fixed effect population parameter estimates are likely to be quite good, and their standard error estimates may well be reasonable.

2.3. An Alternative Approach

Because of the limitations imposed by the data source of the
standard approach, the possibility of using data gathered directly
from patients receiving drugs of interest is attractive. In the
limit, one might attempt to extract population parameter estimates
only from data gathered by physicians in the process of rendering
care to their patients. These are "pure" routine data. One might
also consider supplementing such information with a few extra
samples, samples not necessary for patient care.

If such readings could be used, they would provide an attractive supplement to those employed in the standard approach. Ethical problems are virtually nonexistent: the data are already being
gathered for the justifiable purpose of patient care. Even when a
few additional blood or urine samples are obtained, if, as we contemplate, no alteration of dosage is imposed, it is not difficult
to justify taking them. Data are representative: they are gathered
from precisely those individuals comprising the population of
interest. The amount of data per individual will necessarily be

small, but the total number of individuals studied will be large,
so that the population *per se* should be better represented than
it is in the standard approach. Data will be inexpensive, since
most of their costs are absorbed in the justifiable expenses of
patient care; the marginal cost increment will only be for the
additional samples, and for data analysis. Finally, serendipity
is possible with these data: patients taking other drugs, varying
diets, and so forth, will all be present in the data-base, and,
with sufficient diligence, one should be able to uncover at least
clues to the effects of such factors.

The data source, and the method of gathering data (essentially
abstracting charts) is straightforward for the alternative approach;
the data analysis is not. Consider the two data analysis problems
previously discussed -- the assumption of homogeneous error magnitude, and the estimation of population parameters from individual
parameter estimates.

We may no longer be confident that errors are small, and error
magnitudes relatively, if not absolutely, homogeneous. Now data
are gathered in varying circumstances and have variable reliability:
times of sampling are subject to recording errors, as are doses,
and some patients may be more carefully studied than others. Drug
assay error may vary day-to-day as different lots of reagents,
different technicians and even different laboratories contribute to
the total data. Patients will be quite heterogeneous. Some will
have rapidly changing kinetics as disease states rapidly shift;
others may be quite stable. Therefore, attention must be paid to
varying error magnitudes because errors will often be relatively
larger than in experimental data, and certainly will be more heterogeneous in magnitude.

Not only are the above problems acute, but individual parameter estimates are now suspect as a means to population values.
There is no experimental design, so drug concentration samples
will not usually be taken at times most likely allowing accurate
parameter estimates. Varying amounts of data will be gathered
from patients, and some will surely have too little data to permit
accurate individual parameter estimates. Combining this variability
with the previously discussed variability in data quality implies
that individual parameter estimates, even where they are available,
will certainly have variable, and possibly often poor, accuracy.
Population parameters estimated from these will be significantly
"contaminated" with estimation error and the interindividual random effect parameter estimates especially may be seriously biased.
Clearly, an alternative approach to data analysis is required.

Our colleagues in other disciplines, such as econometrics,
sociology and psychology, are faced with a similar problem: the
analysis of non-experimental "naturalistic" data. We have borrowed

the techniques they and statisticians have evolved to deal with
such data. The approach involves shifting one's point of view,
and regarding the population itself as the unit of analysis. Just
as when fitting an individual's data to an individual model, the
goal is not the estimation of the "true" drug concentration for
each time point, but rather the individual's model parameters, so,
when modelling the population, the goal is the population parameter
estimates, not the individual parameters. When one makes this
shift, however, an essential assumption of the usual (weighted or
unweighted) least-squares approach is no longer tenable. When
modelling only one individual's data, each drug concentration can
be approximated as being statistically independent of any other.
Now, one must explicitly recognize that only concentrations from
different individuals are statistically independent of each other.
The concentrations from one individual all display a systematic
deviation from the expected (mean) population values due to the
individual parameter shifts (from the population mean values)
which are associated with the particular individual.

One must be able not only to account for correlations among
observations, but also for heterogeneous variability in observations beyond that previously discussed. Since the expected concentration (at a given dose and time) in an "average" patient is
now the reference point, one is forced to make provision for
variable magnitudes of the influence of "random" interindividual
deviations, as well as intraindividual and error ones.

Stated generally, the data analysis approach must allow a
general, correlated error structure, with varying error magnitude
as a function of the observable data (such as age, medical condition, time of sample after dose, etc.) and the fixed effect parameters themselves. Certain approximations may be made to simplify
matters, but no reasonable approximations will avoid the necessity
of dealing with a considerably more complex data analysis approach
than the standard one. This is the inevitable price one must pay
for devoting less (no) care to study design.

Neither ordinary, nor weighted simple least squares can be
used. An approach that can be applied, and that we do use is the
method of extended least squares as applied to a nonlinear mixed-
effect statistical model. The notion of extended least squares
is introduced in (10), and under the additional (but unnecessary)
assumption that all random effects are normally distributed, this
method, as applied to a linear mixed-effect model, is simply the
maximum likelihood method (11). Application of the method to
pharmacokinetic estimation is described in some detail in (1)
where the normality assumption is made for convenience only. Since
pharmacokinetic models are statistically nonlinear, the extended
least-squares method as applied to them [described in detail in
(2)], and as we think of it in this paper, is actually an approximate maximum likelihood method.

The method yields direct estimates of all three types of population parameters. Indeed, with this method, the random error parameter can be partitioned into several components of intra-individual variability and these components can be estimated with data resulting from the occurrence of different values of the intraindividual random effects. The asymptotic covariance matrix of these estimates may be estimated, and thus standard errors for all parameter estimates are directly provided.

We can do no more than sketch the extended least-squares method here, and indicate its relationship to ordinary or weighted least-squares. The reader is referred to the above publications on this topic for more information.

In ordinary least squares, as applied to an individual's data, one seeks the set (vector) of parameter values, $\underline{\theta}$, which minimize the sum of squared deviations, $i.e.$, the values of $\underline{\theta}$ minimizing

$$\sum_{i=1}^{n} [y_i - f(\underline{\theta},\underline{X}_i)]^2 \qquad [1]$$

where y_i is the observed response at the i-th instant, and $f(\underline{\theta},\underline{X}_i)$ is the predicted value of that response under the model, f, given the individual's pharmacokinetic parameters, $\underline{\theta}$, and the collection of independent variables evaluated at the i-th instant, $\underline{X}_i$, which includes time.

If there were differing error magnitudes, and one knew the (different) variances of the individual observations, one would estimate θ using weighted least squares; $i.e.$, one would seek $\underline{\theta}$ by minimizing

$$\sum_{i=1}^{n} \{[y_i - f(\underline{\theta},\underline{X}_i)]^2/v_i^2\} \qquad [2]$$

where v_i^2 is the variance of the i-th observation. Note that, usually, weighted least squares is appropriately used by assuming only that v_i is known up to a proportionality factor.

If, in expression [2], the v_i were unknown, they could perhaps be modelled as a function of $\underline{\theta},\underline{X}_i$, and some random effect parameters, $\underline{\xi}$. Designate the value of this function as $v_i(\underline{\theta},\underline{\xi},\underline{X}_i)$ $(=v_i)$. One would not estimate $\underline{\theta}$ and $\underline{\xi}$ by minimizing expression [2] with respect to $\underline{\theta}$ and $\underline{\xi}$ because a trivial and uninteresting solution would always result. The parameters $\underline{\xi}$ would be chosen so as to render all v_i infinite. This choice would always minimize the expression, no matter what values were assigned to $\underline{\theta}$. It is therefore clear that one needs to add some penalty to expression

[2] which increases as $\underline{\xi}$ takes values leading to larger variances, v_i^2. Such an expression arises in the maximum likelihood approach to estimation when errors are normally distributed. In this case, the maximum likelihood estimates of $\underline{\theta}$ and $\underline{\xi}$ are those minimizing

$$\sum_{i=1}^{n} \frac{[y_i - f(\underline{\theta}, X_i)]^2}{v_i(\underline{\theta}, \underline{\xi}, X_i)^2} + \ln[v_i(\underline{\theta}, \underline{\xi}, X_i)^2] \qquad [3]$$

where $\ln(z)$ is the natural logarithm of z. Note here that as the value of v_i^2 increases, so does the logarithmic term. This penalty counteracts the decrease in the left-hand, sum-of-squares term. Reasonable, bounded estimates of all parameters are obtained from this approach. In the case that $v_i(\underline{\theta}, \underline{\xi}, X_i)$ is constant for all i, minimizing expression [3] gives exactly the same estimate of $\underline{\theta}$ as results from minimizing expression [1]. Also, in the case of $v(\underline{\theta}, \underline{\xi}, X_i) = \underline{\xi}u(X_i)$ for all i and known function u, minimizing [3] gives the same estimate of θ as results from using weighted least squares with weights, $1/u(X_i)$. Even when distributions are not normal, the estimates provided by minimizing expression [3] have certain desirable statistical properties, and are reasonable choices as estimates of the underlying parameters (10).

Now let us step back one level and contemplate estimating population parameters rather than individual ones. We must now regard the symbol y_i as referring to a vector of numbers, and the symbols X_i and v_i as referring to matrices, one set of y_i, X_i and v_i for each individual studied. The length of y_i, the row dimension of X_i and both dimensions of v_i are the number of observations taken from the corresponding individual. The parameters $\underline{\theta}$ and $\underline{\xi}$ are now population parameters; $\underline{\theta}$ are fixed effect parameters, $\underline{\xi}$ contains both types of random effect parameters, *i.e.*, inter- and intra-individual random effect parameters. Some adjustments need to be made to expression [3] so that it deals with vectors rather than scalars, and the summation is across N individuals, not n observations; matrix notation must be used, but the spirit is the same as before.

One important difference, however, is present. Now the individual observations in each y_i vector will all correlate with each other through the common set of differences of the individual parameters from the population mean values. Thus, off-diagonal elements (the covariances of the observations) of the matrix v_i^2 will involve the interindividual random effect parameters, and the diagonal elements (variances of observations) will involve both the error random effect parameters, *and* the interindividual random effect ones. This is because the variances must express the typical magnitude of the deviations of the observations from their population mean expectations, not just from the individual mean expectations.

In order to make the computation of the v_i^2 matrices possible, we linearize the individual pharmacokinetic model in the individual parameters about the population parameters. The variance of a linear combination of random variables is a linear combination of the variances and covariances of these constituent variables. The v_i^2 matrix then becomes essentially a simple linear combination of the elements of ξ, the population random effect parameters.

We have implemented, tested, and are distributing a computer program, NONMEM (NONlinear Mixed Effect Model) that can be used to analyze data according to the approach outlined above. The program has certain especially desirable features. For example, it has a convenient form for input; it facilitates the modelling of the population error structure ($i.e.$, v_i); it provides for constraints on the parameter estimates; it provides an estimate of the asymptotic covariance matrix of all the estimators; it performs derivative-free estimation; it produces flexible and informative tables and graphs; and others. A tape of the program and detailed didactic documentation may be obtained by writing to the authors.

The main advantage of the alternative approach has already been described: it uses an alternative source of data which has the advantage of being representative of the population of interest. There are also other advantages. When analyzing even experimental data, the ability to state a general parametric model for the error structure and have the analysis estimate its parameters frees the analyst from the task of specifying known weights for the data analysis. The provision of standard errors for all parameters, including the random effect ones, means that one can avoid using the methods of the standard approach, methods that are theoretically incorrect as applied to population parameters estimated from individual nonlinear parameter estimates. The ability to use quite fragmentary data from any individual is also a major advantage.

There are some important disadvantages. There is little accumulated experience with the approach in biomedical applications. Analysts unfamiliar with it, because it is so general and flexible, are likely to specify statistical models incorrectly and obtain incorrect and misleading results. The models that must be specified, are more complex than in the standard approach. There the model-building takes place in stages. First one thinks of the individual pharmacokinetic model, and only later of the population model for pharmacokinetic parameters. In the alternative approach, the entire model must be specified at once, along with overt specification of a (possibly complex) statistical error model. Although this may be seen as an advantage in that all assumptions are made explicit, it is a difficult task, and lack of scrupulous attention to it can lead to unfortunate errors. The linearization approximation mentioned above could theoretically lead to biased estimates of population parameters; its effects are not well known at present. Finally, because many parameters and all data must be dealt with simultaneously, the data analysis requires a large general-purpose computer, and is costly. This cost may be more than offset, however, by the cost savings realized in not paying for the data.

CONCLUSIONS

Population pharmacokinetic parameters are of three types. Fixed effect parameters describe the relationship of physiology to kinetics. Interindividual random effect parameters measure the magnitude of the random individual variability in the relations of the first type. Intraindividual and measurement error random effect parameters measure the magnitude of the random residual variation in observations of drug concentrations about their expected value, the latter derived from knowledge of individual kinetic parameters.

A number of interesting and important questions can be answered by knowledge of population pharmacokinetic parameters, and their investigation is therefore a matter of some interest. The standard approach to obtaining information on them involves an experimental paradigm. In this approach, although reasonably good estimates of the pharmacokinetic parameters of some (or even all) individuals can be obtained, the population parameter estimates resulting from these may not be representative of the population of interest, or if they are representative, they may be very imprecise due to the relatively small number of individuals studied.

To analyze routine data generated in the course of patient care, or to combine data from various sources, one is faced with certain data analysis problems. These are also present when analyzing experimental data, but are probably of less import in this case. When the standard techniques are applied to routine data, or data from various sources, however, the problems become acute. An alternative data analysis approach that regards the population as the unit of analysis is available. It is implemented in an exportable computer program. This approach to analysis allows meaningful assessment of routine data, but is costly and requires greater sophistication on the part of its users. A reasonable compromise is to use the standard approach wherever it seems applicable, and to use the alternative approach when the standard one will not suffice, as in the analysis of routine data. In this manner, the approaches need not be seen as competitive, but rather as complementary. Together, they may allow fuller use of the varied types of data available to pharmacokinetic investigators.

REFERENCES

1. Sheiner, L.B., Rosenberg, B. and Marathe, V.V. (1977) J. Pharmacokin. Biopharm. 5, 445-479.
2. Sheiner, L.B., Beal, S., Rosenberg, B. and Marathe, V.V. (1979) Clin. Pharmacol. Therap. 26, 294-305.
3. Snedecor, G.W. and Cochran, W.G. (1967) "Statistical Methods", 6th Ed., p. 74, Iowa State Univ., Ames, Iowa.
4. Mitenko, P.A. and Ogilvie, R.I. (1973) N. Engl. J. Med. 289, 600-603.
5. Hunt, S.N., Jusko, W.J. and Yurchak, A.M. (1976) Clin. Pharmacol. Therap. 19, 546-551.
6. Piafsky, K.M., Sitar, D.S., Rangno, R.E. and Ogilvie, R.I. (1977) Clin. Pharmacol. Therap. 21, 310-316.
7. Powell, J.R., Vozeh, S., Hopewell, P., Costello, J., Sheiner, L.B. and Riegelman, S. (1978) Am. Rev. Resp. Dis. 118, 229-238.
8. Watts, D.G. (1980) In "Kinetic Data Analysis: Design and Analysis of Enzyme and Pharmacokinetic Experiments". pp. 1-24,   Plenum, New York.
9. Boxenbaum, H.C., Riegelman, S. and Elashoff, R.M. (1974) J. Pharmacokin. Biopharmaceut. 2, 123-148.
10. Beal, S.L. (1974) "Adaptive M-Estimation with Independent Nonidentically Distributed Data", Unpublished Ph.D. dissertation, Los Angeles, Calif.
11. Jennrich, R.T. and Sampson, P.F. (1976) Technometrics, 18, 11-17.

ANALYSIS OF SETS OF ESTIMATES FROM PHARMACOKINETIC STUDIES

B.E. Rodda

Merck Sharp & Dohme Research Laboratories
Rahway, New Jersey   07065
U.S.A.

ABSTRACT

The objective of most pharmacokinetic studies is to obtain
estimates of the parameters which define the assumed pharmacokinetic
model. Little attention is usually paid to the interrelationships
of these estimates and the impact that such patterns may have on
the interpretation of the data. Because each parameter is often
considered as an independent entity, hyperspecification of the model
is not often considered, despite its prevalence. This manuscript
addresses some considerations which must be made in these cases
and some statistical procedures which, although simple, have not
been extensively employed in the practical application of pharmacokinetic analyses.

1.   INTRODUCTION

The estimation techniques used in pharmacokinetic modelling
are usually multivariate in nature. Parameters of mathematical
models are not estimated individually, and most estimation procedures produce sets of interrelated estimates. The relationships
among the estimates may or may not represent true patterns, but in
any case can provide an index of the dimensionality of the model.
In many cases the multivariate nature of these sets of estimates
is ignored, and each estimate is evaluated as though it represented
the single, most important parameter of the model.

Most pharmacokinetic models possess three types of parameters
which need to be estimated. First, there are the rapid rate constants such as absorption and distribution. To estimate a rapidly
changing quantity with accuracy requires many data in the region

285

of maximum gradient, data which are not easily obtained in pharmacokinetic studies. Secondly, slow rate constants must be estimated;
these are much easier to estimate because there is less uncertainty
with less change. And thirdly, there are the "invisible" parameters
— those parameters which are estimated by default. Examples are
parameters associated with peripheral, non-observed compartments
in multicompartment models. Sets of these parameters will often
be related, and these relationships can be employed to augment the
single parameter analyses.

2. ESTIMATION PROCEDURES

In the conventional pharmacokinetic study, a model is specified
which may include two, three or sometimes as many as nine or ten
parameters which describe the body's effect on the fate of a pharmaceutical entity.

In these studies, pharmacokinetic parameters are not estimated
individually but in sets. Their interdependence may best be appreciated by a brief presentation of conventional nonlinear least-
squares estimation, the procedure most often employed in pharmacokinetic studies. The properties associated with the technique and
the analysis of sets of estimates are general, regardless of the
number of parameters in the model.

The additive error form of a pharmacokinetic model can be
represented as

$$Y_{ij} = C(t_i, \underline{\theta}_j) + \varepsilon_{ij}$$

where Y_{ij} is the plasma concentration observed for subject j at
time t_i; $C(t_i, \underline{\theta}_j)$ is the concentration predicted by the model for
subject j at time t_i; and ε_{ij} is the deviation between the observation and the model. The errors ε_{ij} are usually assumed to have an
expectation of zero and to be uncorrelated for all i and j, and
some variance structure is often assumed which may subsequently
result in a weighted or unweighted estimation procedure. The para-
meter vector, $\underline{\theta}_j = (\theta_1, \theta_2, \ldots, \theta_p)'_j$ is often assumed to be identical
for each subject, in which case the subscript is unnecessary and
intersubject variability is included in ε_{ij}. On the other hand,
these parameters may be affected by treatment or another systematic
effect, and different vectors may exist for each of a variety of
conditions. In either case, the estimation procedure remains the
same; because of this, the subscript j will be ignored in subsequent
notation.

The procedure most often used in nonlinear least-squares pro-
grams consists of expanding $C(t_i, \underline{\theta})$ in a Taylor series to the first-
order terms about the current set of estimates, say $\hat{\underline{\theta}}$. Thus,

$$C(t_i,\underline{\theta}) = C(t_i,\underline{\hat{\theta}}) + \sum_{k=1}^{P} \frac{\partial C(t_i,\underline{\theta})}{\partial \theta_k} (\theta_k - \hat{\theta}_k) + \varepsilon_i$$

or

$$C(t_i,\underline{\theta}) - C(t_i,\underline{\hat{\theta}}) = \sum_{k=1}^{P} \frac{\partial C(t_i,\underline{\theta})}{\partial \theta_k} (\theta_k - \hat{\theta}_k) + \varepsilon_i \quad .$$

This is now in the form of the general linear model

$$\underline{Y} = \underline{X}\,\underline{B} + \underline{\varepsilon}$$

where

$$\underline{Y} = \begin{pmatrix} C(t_1,\underline{\theta}) - C(t_1,\underline{\hat{\theta}}) \\ C(t_2,\underline{\theta}) - C(t_2,\underline{\hat{\theta}}) \\ \vdots \\ C(t_m,\underline{\theta}) - C(t_m,\underline{\hat{\theta}}) \end{pmatrix}_{m \times 1}$$

$$\underline{X} = \begin{pmatrix} \dfrac{\partial C(t_1,\underline{\theta})}{\partial \theta_1} & \dfrac{\partial C(t_1,\underline{\theta})}{\partial \theta_2} & \cdots & \dfrac{\partial C(t_1,\underline{\theta})}{\partial \theta_p} \\[2ex] \dfrac{\partial C(t_2,\underline{\theta})}{\partial \theta_1} & \dfrac{\partial C(t_2,\underline{\theta})}{\partial \theta_2} & \cdots & \dfrac{\partial C(t_2,\underline{\theta})}{\partial \theta_p} \\[2ex] \vdots & \vdots & & \vdots \\[2ex] \dfrac{\partial C(t_m,\underline{\theta})}{\partial \theta_1} & \dfrac{\partial C(t_m,\underline{\theta})}{\partial \theta_2} & \cdots & \dfrac{\partial C(t_m,\underline{\theta})}{\partial \theta_p} \end{pmatrix}_{m \times p}$$

and

$$B = \begin{pmatrix} \theta_1 - \hat{\theta}_1 \\ \theta_2 - \hat{\theta}_2 \\ \vdots \\ \theta_p - \hat{\theta}_p \end{pmatrix}_{p \times 1}$$

Solutions for the components of B are derived using linear regression techniques, and the procedure is continued until a reasonable estimate of θ is obtained. Assuming that the procedure converges to an absolute minimum squared error (not a necessary result), the asymptotic properties of least squares provide comfort that estimates of the model parameters should be nearly unbiased. [It is well known that nonlinear least squares estimates are biased, *e.g.* (1).] However, as stated previously, model parameters are not estimated independently, but as an interrelated set. Because of this factor, individual estimates can be seriously misleading, and estimates of variation and covariation can be irresolvably wrong.

Consider the variance estimation problem first. Because of the linearization used to obtain estimates of the model parameters, the variance-covariance matrix of the estimates is

$$V(\underline{B}) = (\underline{X}'\underline{X})^{-1}\sigma^2$$

$$= \underline{V}^*\sigma^2 \, .$$

Thus $V(\hat{\theta}_i) = v_{ii}^* \hat{\sigma}^2$ and $\text{Cov}(\hat{\theta}_i, \hat{\theta}_j) = v_{ij}^* \hat{\sigma}^2$. On the surface, the procedure appears straightforward, but estimating variances in this fashion can lead to incorrect conclusions. This can be appreciated by a simple example using the one-compartment, mammillary pharmacokinetic model with first-order absorption. This model is one of the simplest pharmacokinetic models, but is representative.

The general additive error form of this one-compartment open model can be expressed as

$$Y_{ij} = \frac{A\alpha}{\alpha-\beta}\left(e^{-\beta t_j} - e^{-\alpha t_j}\right) + \varepsilon_{ij} \quad ,$$

where A represents the fraction of the administered dose which appears in the blood (or plasma or serum) divided by the apparent volume of distribution; α and β represent first-order appearance and disappearance rate constants, respectively; Y_{ij} is the observed concentration for subject i at time t_j, and ε_{ij} is the deviation between the observed concentration and that predicted by the model. In this notation, $\underline{\theta}$ represents the common ordered parameter vector $(A,\alpha,\beta)'$.

The components of the general linear model, $\underline{Y}=\underline{X}\underline{B}+\underline{\varepsilon}$, in this case are

$$\underline{X} = \begin{pmatrix} \frac{\alpha}{\alpha-\beta}\left(e^{-t_1\beta}-e^{-t_1\alpha}\right), & A\left[\frac{\{t_1\alpha(\alpha-\beta)+\beta\}e^{-t_1\alpha}-\beta e^{-t_1\beta}}{(\alpha-\beta)^2}\right], & A\left[\frac{\{t_1\alpha(\beta-\alpha)+\alpha\}e^{-t_1\beta}-\alpha e^{-t_1\alpha}}{(\alpha-\beta)^2}\right] \\[2ex] \vdots & \vdots & \vdots \\[2ex] \frac{\alpha}{\alpha-\beta}\left(e^{-t_m\beta}-e^{-t_m\alpha}\right), & A\left[\frac{\{t_m\alpha(\alpha-\beta)+\beta\}e^{-t_m\alpha}-\beta e^{-t_m\beta}}{(\alpha-\beta)^2}\right], & A\left[\frac{\{t_m\alpha(\beta-\alpha)+\alpha\}e^{-t_m\beta}-\alpha e^{-t_m\alpha}}{(\alpha-\beta)^2}\right] \end{pmatrix}_{m\times 3}$$

and

$$\underline{B} = \begin{pmatrix} A - \hat{A} \\ \alpha - \hat{\alpha} \\ \beta - \hat{\beta} \end{pmatrix}_{3\times 1}$$

For a given case, the estimation of the covariance matrix is straightforward. One substitutes the final estimates of A, α, and β and the time points at which samples were taken into this representation of $\underline{X}$, multiplies by the transpose, and inverts. The residual variability is easily calculated from the model deviation.

The procedure and associated conclusions may best be appreciated by a numerical example. For the one-compartment open model assume that the following set of parameters represents the unknown "truth":

$$A = 50 \ \mu g/ml$$
$$\alpha = 0.05 \ min^{-1}$$
$$\beta = 0.005 \ min^{-1} \quad .$$

These data were used previously (2) and are characteristic of the pharmacokinetics of many drugs. The pharmacokinetic model is then

$$Y_t = 55.56 \ (e^{-0.005t} - e^{-0.05t}),$$

which is presented graphically in Figure 1.[*] Given specific sampling times the estimated covariance matrix can be determined. The following sampling times were selected for this example: 5, 10, 15, 30, 45, 60, 90, 135, 180, 240, 360, 480, 600, and 720 minutes.

[*]Although time is often considered as a discrete variable in pharmacokinetic studies, it is conceptually continuous. To simplify notation, subscripts for time will not be used henceforth.

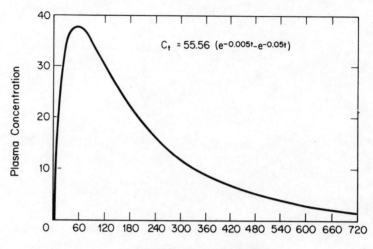

FIGURE 1: Time course of plasma concentration for one-compartment open model.

With this information, we can derive an approximate covariance matrix for A, α, β. In the specific example,

$$V(\underline{B}) = (\underline{X'X})^{-1}\sigma^2$$

$$= \begin{bmatrix} 2.892 & -3.284\times10^{-3} & 3.565\times10^{-4} \\ -3.284\times10^{-3} & 7.525\times10^{-6} & -4.218\times10^{-7} \\ 3.565\times10^{-4} & -4.218\times10^{-7} & 5.227\times10^{-8} \end{bmatrix}\sigma^2 \quad .$$

The correlation coefficients for this model are

$$\rho_{A\alpha} = -0.704$$
$$\rho_{A\beta} = 0.917$$
$$\rho_{\alpha\beta} = 0.673 \quad .$$

Thus, for this example, the estimation procedure used in most pharmacokinetic studies would produce these estimates of the correlations among the parameters of interest — regardless of the data. The fact that substantial functionally generated correlation exists must be considered when interpreting results of pharmacokinetic studies.

It is clear from the above that the correlations between model parameters are biased to a substantial degree, and no straightforward resolution of the problem exists.

An additional question of even greater relevance is, "Are the estimates of individual parameter variabilities accurate?" To address this question we examined the effects of "assay" variability and "physiologic" variability.

Assay variation is a characteristic of the plasma (or serum, *etc*.) sample and not of the functional model. It is often of a proportional nature, the error being a fraction of the assay results.

To characterize the marginal distributions of the estimates in the presence of such variation we assumed that each concentration, Y_t, in the observation vector was a normally distributed random variable with $E(Y_t)=C_t$ and $\sigma(Y_t)=0.1C_t$ in one case, and $\sigma(Y_t)=0.25C_t$ in the second, the errors being uncorrelated for $t_i \neq t_j$ in each curve. [$E(Y_t)$ denotes the expectation of Y_t and $\sigma(Y_t)$ denotes the standard deviation of Y_t.] For each case, 500 curves were generated, and the parameters were estimated by nonlinear least squares. The observed means and variances of the three estimates are presented in Table 1 for the cases when $\sigma(Y_t)=0.1C_t$ and

TABLE 1: Observed means and variances of least-squares estimates

Condition	Statistic	A	α	β
10% assay and no physiologic variation	Mean Variance	50.59 15.09	5.00×10^{-2} 5.32×10^{-5}	5.09×10^{-3} 2.18×10^{-7}
25% assay and no physiologic variation	Mean Variance	53.03 117.02	5.12×10^{-2} 3.44×10^{-4}	5.47×10^{-3} 2.04×10^{-6}
10% assay and 10% physiologic variation	Mean Variance	50.15 39.55	5.04×10^{-2} 6.98×10^{-5}	5.01×10^{-3} 5.01×10^{-7}
10% assay and 25% physiologic variation	Mean Variance	50.73 172.86	4.96×10^{-2} 2.17×10^{-4}	5.09×10^{-3} 1.86×10^{-6}

$\sigma(Y_t)=0.25C_t$. The results in this table indicate that the least-squares estimates of A, α, and β are, for practical purposes, relatively unbiased.

The average sample variances derived within each run with a 10% coefficient of variation were 9.51, 4.67×10^{-5} and 3.32×10^{-7} for A, α and β, respectively. These differ from the corresponding variance estimates in Table 1. When 25% assay variation was used, the within curve variance estimates where 51.63, 50.85×10^{-5}, and 50.09×10^{-7}. These within curve estimates also differed from the sample variances of the estimates. In neither case did we observe the proportionally expected from the $(\underline{X}'\underline{X})^{-1}$ matrix, nor a similar increase in variability for all parameter estimates from the 10% to 25% assay case. This suggests that the linearization used in estimating the model parameters may be inadequate for estimating their variances and may produce spurious confidence intervals.

Although knowledge of the distributional properties of the parameter estimates for a single population provides a baseline against which comparisons can be made, a more important consideration is the behaviour of these estimates in the more realistic situation where physiologic as well as assay variation is present. To evaluate this effect, we generated 500 curves in a similar manner to that previously described, but additionally assumed that A, α, and β each followed independent normal distributions with means of 50, 0.05, and 0.005, and standard deviations of 5, 0.005, and 0.0005, respectively, in one case and 12.5, 0.0125, and 0.00125 in another. Ten percent assay variation was incorporated in each curve. This is a reasonable, if somewhat conservative figure based on experience, and we felt that it should provide basic information about the effect of such variation on the estimates. Estimates of model parameters, as would be expected from the preceding paragraph, were practically unbiased. The variation due to the physiologic component was additive to the assay components to produce the observed variances in Table 1.

The foregoing make the point that interrelationships among the elements in a set of estimates should not be gleaned from an unconsidered reading of a computer printout. However, the associations of the population parameters are maintained quite well in sets of sample estimates. That is, for a given model, estimate the entire set of parameters for each curve and consider each set of estimates as a sample vector from a multivariate population. When analyzed with this approach, the patterns in the population can be estimated and can provide some valuable insight to the underlying pharmacologic or pharmacokinetic processes.

3. ANALYTIC PROCEDURES

One of the vices practiced by many kinetic scientists is an aversion to parsimony in their mathematical models. Indeed, one feels rather hyposcientific if the proposed model does not define the underlying process as completely as possible. However, it is not uncommon to read where a scientist has fitted a complex, six- or seven-parameter model to eight plasma samples (exsanguination is usually frowned on). This is much like fitting a straight-line to two data points — it looks like it fits well, but we have no measure of adequacy or variability.

Often in complex models, patterns among the p parameters indicate that the dimensionality of the parameter space can be adequately represented by some smaller number say (p-k). In this case, a new, simpler model should be considered. Estimates can be derived with greater confidence, and the fit should be nearly as good. In fact, residual variability is frequently smaller in simple models than in more complex models. The pooled variance estimate for deviation from the proposed model is defined as the sum of squared deviations divided by (n-p). If n is not several times larger than p, a reduction in the number of parameters in the model often has a much greater effect on the residual variance than the associated increase in the residual sum of squares.

If the parameters are uncorrelated, we can visualize the parameter space with p perpendicular axes, the axis lengths being proportional to the population variances. In most cases, however, the model parameters are correlated to some degree. If two parameters are highly correlated, then knowledge of one allows accurate prediction of the other. In the geometric framework outlined above, the axes of highly correlated parameters would be nearly coterminous, and thus one parameter would be redundant.

To estimate the dimensionality of the model parameter space, principal component analysis can be used. Principal component analysis effects an orthogonal rotation of the parameter space to a new set of parameters which are linear combinations of those in the model. These derived linear combinations are mutually orthogonal, and their axis lengths are proportional to their variability. Because the total variability of the system is unchanged, the new axes represent the same variability as the original parameter space, but allow orthogonal, additive, ordered components.

Ordinarily 80 or 90 percent of the total variability can be identified with a small number of these orthogonal axes. Using this technique to identify a parsimoniously necessary number of parameters can yield better estimates, more confidence in their accuracy, and accurate fitting and subsequent prediction.

For example, suppose we are fitting a six-parameter, two-compartment open model to some plasma concentration — time curves. This model is depicted graphically in Figure 2.

All the parameters associated with the peripheral compartment (V_2, k_{12}, k_{21}) are estimated by default. We can observe only the central compartment and, therefore, direct estimates are possible only for this compartment. Estimates of parameters for the peripheral compartments are often based on deviations of one or two observations during the post-peak period from a one-compartment hypothesis. Clearly little faith can be assigned to estimates derived in this fashion. In these circumstances, a principal component approach may provide an objective, scientific rationale for reducing the complexity of the pharmacokinetic model.

Although a more complex model may be intuitively satisfying, estimates of pharmacophysiologic parameters without knowledge of their variability may be nearly useless. An investigator who feels a need to fit a model perfectly without the value of knowing the variability of his estimates should consider an (n-1)-parameter polynomial model — it will fit perfectly, and the parameters will be hardly less interpretable from a scientific standpoint than an over-specified pharmacokinetic model.

Assuming that the investigator has selected an appropriate model, and has derived a valid set of estimates, interest is then directed toward what to do with them. This interest is usually in one or more of three areas: (i) describing the interrelationships of the model parameters and their variability, (ii) discriminating between two populations on the basis of the model parameters, and

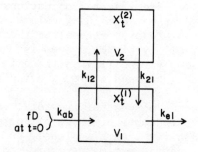

FIGURE 2: Kinetic scheme of two-compartment open model.

(iii) performing statistical tests to determine whether two populations are identical or differ in some identifiable way.

As discussed previously, simply estimating a set of parameters
provides little information to the investigator if he has no accurate measure of the variability of the estimates. In the linear
case, the set of estimates which minimizes the residual sum of
squares is unique, and as a rule, the estimates are orthogonal to
each other. However, in the nonlinear case, many sets of estimates
may be associated with an equally small residual sum of squares.
In fact, if the sum of squares surface is relatively flat in the
area of the derived estimate set, it may make no difference (within
some constraints) which set of estimates is used. In this case,
the fitting procedure will provide a very imprecise set of pharmacokinetic estimates.

Here the concept of estimate variability is two-fold. First,
the univariate variability of the estimate itself must be considered.
If the point representing our set of estimates lies in a very flat
region, then the variability associated with any of the parameters
will be large. The second topic to consider is the general variability of the point itself. If the variability about the point
representing this set of estimates is quite large, there is no con-
vincing reason for selecting one set of estimates over another within the same region.

Because of the functionally generated bias in the calculation
of the sample covariance matrix, using that matrix to characterize
the nature of the space may be misleading. In all aspects of estimation in pharmacokinetic models, the covariance matrix should not
be gleaned from the nonlinear estimation program. A set of estimates should be derived for each curve, and the set considered as
a sample vector from a multivariate population. These vectors
should then be used to calculate the sample covariance matrix.

Given sets of estimates and their corresponding covariance
matrices, the kineticist can derive univariate or multivariate confidence regions by using classical statistical techniques. In most
cases the set of estimates is assumed to be a sample from a multivariate normal population. Computation of confidence regions is
straightforward and will not be addressed here. The reader is
referred to any book on multivariate analysis [*e.g.*, Anderson (3)
or Morrison (4)].

There are many reasons for distinguishing between two populations on the basis of their pharmacokinetics; one, and probably
the most common reason, is simply to identify whether or not a population receiving one formulation differs from a population receiving an alternative formulation. A similar question would address
whether the pharmacokinetics of a particular drug for one type of

patient is the same as for a different type. In these cases univariate or multivariate statistical techniques can be used to test
the null hypothesis of no difference.

Another and sometimes more interesting reason for discriminating between two populations on the basis of their kinetics can be
appreciated by the following example: Suppose that we have observed
50 patients, 25 of whom we classify as good responders to therapy,
and 25 who are classified as poor responders. For each patient we
have estimated the set of pharmacokinetic parameters. Our objective
is to characterize the differences between the two groups in a way
which will permit us to classify confidently new patients as either
good or poor responders on the basis of their pharmacokinetic estimates. If we assume that the covariance matrices for the two groups
are similar, it is a straightforward mathematical exercise to derive
the linear discriminant function as

$$Z = (\hat{\underline{\theta}}_g - \hat{\underline{\theta}}_p)' \underline{V}^{-1} \underline{\theta}$$

where $\hat{\underline{\theta}}_g$ and $\hat{\underline{\theta}}_p$ are the average estimate vectors for the good and
poor responders, respectively; $\underline{V}^{-1}$ is the inverse of the sample
covariance matrix of the estimates and θ is the parameter vector.
We then classify those patients with values of Z above a certain
number as probable good responders, and those patients with a value
of less than or equal to this cut-off as probable poor responders.
The determination of the cut-off point is a function of the probabilities of both false positive and false negative classification.
The coefficients of the parameters in the equation give a representation of the relative importance of that parameter in the distinction process.

Additional insight can be gleaned by comparing the covariance
patterns between two groups. If covariance matrices are computed
for the good responders and the poor responders, there may be elements of distinction. To make these matrices more comparable, they
can be standardized to represent correlation matrices. In these
matrices the correlations between any pair of pharmacokinetic estimates can be compared between the two groups of convetional statistical means. In addition, subsets of these correlations can also be
compared using canonical procedures.

4. CONCLUSIONS

Parameters of pharmacokinetic models can be estimated in a
number of ways. Each methodology is associated with certain characteristics, and these characteristics should be considered in the
interpretation. Rarely are sets of estimates from nonlinear kinetic
models unique, and in some instances overspecification of the model
can produce estimates which, although appearing physiologically

reasonable on the surface, can be completely meaningless. Each
model should be approached uniquely, with no standard model used
according to a blind routine. In each case, the minimum model consistent with the data and the kinetic theory should be employed.
The estimates derived from such a model should be considered as a
set of estimates representing a sample from a multivariate population, and subsequent analyses should address specific questions and
utilize the interrelationships among the parameters. Examining the
patterns which exist among the parameters can provide far greater
understanding than independent evaluation of each component estimate.

REFERENCES

1. Box, M.J. (1971) J. Roy. Stat. Soc. B33, 171-201.
2. Rodda, B.E., Sampson, C.B. and Smith, D.W. (1975) Appl.
        Statist. 24, 309-318.
3. Anderson, T.W. (1958) "Introduction to Multivariate Statistical
        Analysis", Wiley, New York.
4. Morrison, D.F. (1967) "Multivariate Statistical Methods",
        McGraw-Hill, New York.
5. Bard, Y. (1974) "Nonlinear Parameter Estimation", Academic
        Press, New York.

SOME CURRENT PROBLEMS IN THE INTERPRETATION OF ENZYME KINETIC DATA

D. Garfinkel

University of Pennsylvania
Moore School of Electrical Engineering
Department of Computer & Information Science
Philadelphia, PA. 19104

ABSTRACT

In this chapter we discuss problems which come to notice
when data obtained in experiments with isolated enzymes are combined into complex metabolic models. The need for using experimental conditions reflecting those *in vivo* when studying isolated
enzymes is discussed. By now there is an extensive literature
concerned with analyzing enzyme kinetic data. The consensus is
that the traditional linearized methods yield inaccurate results,
and that some form of nonlinear regression should be used instead.
However, nonlinear methods, which require computing, are now
readily available only to experts. This situation is likely to
be changed considerably due to rapid decreases in computer costs
and the development of "friendly" software which can readily be
used by workers who are not computer experts. Together, these
should make proper computation much easier and cheaper, compared
to experiment, than it is at present.

1. INTRODUCTION

In this chapter we discuss problems which become particularly
evident when we combine published information on individual enzymes
into a model representing a significant portion of the metabolism
of a tissue or organ. Although a complete model is more than the
sum of its parts, it is possible to decompose the model and study
the behaviour of one enzyme at a time. We generally find problems
in two areas: experimental design and data analysis. Experiments
are often not performed under physiological conditions resembling
those within the cell, *e.g.*, the solvent used may not resemble
the cellular fluid content. Kinetic constants derived from

experimental measurements by traditional means are rarely checked
for consistency. The rigorous methods of data analysis so ably
described in other papers in this volume are mostly limited to use
by experts. Experimenters who have little access to or knowledge
of computers cannot readily use these methods until an appreciable
programming effort has made them available.

The cost of computing is now falling rapidly. This permits
the development of "friendly" software, which is easy to use for
persons who are not computer experts, at the price of being inef-
ficient computationally (as well as requiring considerable programming effort). Some of the author's contributions in this direction
are described below; it seems likely, owing to availability of funds
for this purpose, that additional contributions can be made in the
near future. The time may soon come when it is both faster and
more convenient to use computers to analyze rigorously experimental
data than to analyze them by the traditional noncomputer methods.

2.   ON THE NATURE OF THE OPERATIONS PERFORMED IN ENZYME KINETIC CALCULATIONS

The discussion which follows is directed at the traditional
way in which most quantitative information regarding enzyme kinetics is processed. There are notable exceptions to this tradition,
including participants in this symposium.

Although many enzymologists are not aware of it, enzyme
kinetics is inherently a form of quantitative model building. In
the course of analyzing their data, they fit experimental measurements to (linearized) curves, make logical decisions about the
nature of kinetic mechanisms, and then evaluate the associated
numerical parameters. Frequently the aim of the work is biological
and qualitative, to determine the enzymatic mechanism, and the
numerical part of the work is a means to this end. However, if
the numerical analysis is wrong, the qualitative conclusions can
also be wrong. The primarily mathematical part of this process has
been subjected to rigorous analysis. This has been performed, and
ably outlined, by the authors of other chapters in this symposium
volume. If the experimentally derived numbers were checked for consistency and reasonableness, we would have fewer problems due to such
causes as differences in the behaviour of different enzyme samples.
Doing this does not require an imitation of the Indian number
theorist, Ramanujan, who was described as feeling that every
integer was like a personal friend. However, sufficient attention
should be given to the numbers to verify that they are reasonable,
and that the conclusions drawn really do follow from them.

Information intended to be biologically valuable should make
sense on a physiological level. Ultimately the information
obtained in studies with enzymes defines a physiologically

meaningful system, and yet all too often it is difficult or impossible to find out from the relevant literature how an enzyme
was found to behave under (mammalian) physiological conditions
-- 37°C, near-neutral pH, ionic strength about 0.1, Mg^{++} less than
about 1 mM, *etc*. It is probably not feasible to replicate the
small-molecule composition of the enzyme's natural location, but
those small molecules whose concentrations must be experimentally
varied should take physiologically realistic values in some of the
experiments which are done. This discussion is not concerned with
such problems as whether enzyme behaviour is the same *in vivo* as
in vitro and how enzyme behaviour is modified by immobilization
or binding to membranes. These effects cannot be evaluated until
the properties of the soluble enzyme have been defined.

Sometimes, less frequently than the above, one meets with
insufficient attention to the chemistry involved in an enzymatic
reaction. Such reactions obey the general laws of chemistry as
they apply to the molecules which actually participate in the
reaction. Sometimes the real participants in the reaction or some
of their properties are ignored. As an example, the creatine
phosphokinase reaction is sometimes written as

$$\text{creatine phosphate} + ADP = \text{creatine} + ATP.$$

This is incorrect. The reaction actually catalyzed is

$$\text{creatine phosphate} + MgADP^- + H^+ = \text{creatine} + MgATP^{--}.$$

The reaction rate is pH-sensitive because the proton is
required to "activate" creatine phosphate for the transfer of its
phosphate to ADP. The rate is Mg^{2+}-sensitive because Mg^{2+}-chela-
tion "activates" ATP for transfer of its terminal phosphate to
creatine.

It is a common finding that metal ions in enzyme kinetics are
not handled well. It also happens, all too frequently, that the
actual molecular species involved in a reaction is not determined
by some common experimental designs. Mannervik [1] has considered
such questions in his work. Nothing in this paragraph defies
calculation, although it may be inconvenient. Storer and Cornish-
Bowden have written a computer program to calculate metal-ion
equilibria, for example [2].

Another problem here is excessive fragmentation of attention.
Properties of enzymes are studied one at a time, with little or
no attempt to interrelate these properties. In extreme cases,
when the effect of a modifier is considered, only one substrate
among several substrates is varied; and the experimenters may not
even specify the concentration of the other presumably saturating
substrate(s).

If the enzyme in question is allosteric, there is likely to be some kind of synergism or antagonism between its various effectors, which may not receive as much attention as it should. In order to understand systems of enzymes, which make up living material, we need to know the properties of the individual enzymes involved, not just the behaviour of their individual sub-functions. An enzyme is a sufficiently small and simple entity to be a tractable subject of investigation. While it may be convenient to study one of its properties at a time, the story does not end there.

To some extent many of the above are really manifestations of a single problem: a desire for simple answers to complex problems. This is a common problem throughout the sciences. A considerable portion of the underlying process is explained by DeMillo, Lipton and Perlis (3), who are concerned with the social process whereby mathematical theorems are proved and validated. One would think that the proof of a mathematical theorem is the "hardest" information in this world. Nevertheless, DeMillo *et al*. point out that the process of incorporating and accepting theorems into the fabric of mathematics is very much of a social process. They quote an estimate by the mathematician S. Ulam that nowadays perhaps 200,000 theorems are "proved" by mathematicians each year. Some of these are rejected and found to be faulty, some are used, and the majority are simply ignored. DeMillo *et al*. suggest that one very important criterion for the acceptance of a mathematical theorem is its simplicity. In particular, the desire for simplicity is built into the human mind. Cognitive psychologists have shown that the number of factors people can follow simultaneously is not large: 7 ± 2 simple factors and even fewer complicated factors (4). Unfortunately, biological systems and even enzymes have not been designed according to this limitation.

3.   INTERNAL CONSISTENCY OF ENZYME KINETIC DATA AND THE MEANS
OF HANDLING IT

Generally, the available methods for checking the internal consistency of parameters in enzyme kinetics are ignored. An extreme example follows. In a recent model building effort, we constructed a model of the pathway that degrades fatty acids to acetyl CoA and feeds that into the Krebs cycle. We found it necessary to construct individual models for a total of 15 enzymes using literature information. For each of these enzymes we gathered data, built some kind of model, and then checked its internal consistency by using Haldane relationships. If the information about an enzyme is internally consistent, the relevant set of these equations should be satisfied. This actually happened in none out of the 15 cases. Although some of these enzymes have not really been well studied, this consistency check is rarely reported in the literature, even for well-studied enzymes.

This situation may be generalized. One can define other consistency checks which would be desirable to apply to enzyme kinetic data. The frequency with which they are performed is best described by a line from Gilbert and Sullivan: "What, never? Well, hardly ever".

1. Although enzymes are not indefinitely stable in solution, even when cold, experimenters do not always assay activity periodically and correct their measurements for the observed decline in activity. This may be unnecessary after it is shown that the enzyme in question is sufficiently stable (under the conditions involved), but it should be done more often than it is.

2. One batch of enzyme may not quite be identical with another prepared according to the same protocol. Checking that the specific activity is the same from batch to batch is desirable. An extreme instance where this can cause difficulty is when an enzyme that is allosterically affected by ammonium ion is prepared by ammonium sulfate precipitation followed by dialysis to remove the ammonium sulfate. In this situation, any variation in the completeness of the dialysis can cause difficulty, and cross-checking is particularly important.

3. It is usually assumed that the distribution of errors in enzyme velocity measurements is normal as this allows one to use much simpler statistics. Here a rigorous experimental determination, which would involve perhaps 100 replications of the same measurement, would be unreasonable. Nevertheless, an experimenter needs some knowledge of the experimental error in order to choose proper weighting factors and methods of analyzing data. Mannervik and Cornish-Bowden have studied this problem (5-7).

4. After initial velocity measurements are made, the commonest processing of the results is to plot them in a double-reciprocal (Lineweaver-Burk) plot. Then one draws a straight line through the points plotted. It is almost automatically assumed that such lines are straight, unless they very obviously are not. It is rare to see an explicit check for their straightness, although statistical tests for linearity do exist. It is historically interesting to note that what must have been the first application of a double-reciprocal plot (8) duly included a verification using chi-square tests that the lines obtained were indeed straight.

5. Perhaps 95% of the measured data of enzyme kinetics is evaluated using the double-reciprocal plot of Lineweaver and Burk (9). It has been much criticized (10,11,12), to the point where it is widely regarded as the worst of the plots used in enzyme kinetics (13). Dowd and Riggs (12) even suggested that it be abandoned, which has certainly not happended. As Cornish-Bowden has pointed out, Lineweaver and Burk (9) subjected their method to a thorough

analysis, including proper weighting of the experimental points, and were aware of its shortcomings. Is it too much to ask that workers who insist on using their method at least read their paper first?

6. When data are processed by plotting their results with the double-reciprocal (Lineweaver-Burk) plot, it is assumed that the enzyme in question follows the Michaelis-Menten (14) model of enzyme kinetics.

This assumption tends to focus attention on the saturating regions of concentration space where such behaviour is likely to be observed. Regions where concentrations are low are likely to be neglected, even though they contain important information.

Hill, Waight and Bardsley (15), on surveying the recent enzyme literature, found that the vast majority of experimenters working with enzymes could detect deviations from Michaelis-Menten behaviour in their enzymes under suitable experimental conditions. They concluded that a true Michaelis-Menten enzyme is rare. Knowing that deviations from the Michaelis-Menten kinetics are to be expected, we find that the means of processing the data is precisely that which is *least* likely to find such deviations: the Lineweaver-Burk plot, as determined by Dowd and Riggs (12).

7. On completion of the analysis, resulting in the determination of a mechanism, the parameters obtained should fit into the rate law defining the mechanism and yield a reasonable numerical fit. A quantitative measure of how well a model reproduces observed behaviour is more valuable than a subjective, possibly biased, opinion of an enzyme's mechanism. All too often the rate law is derived only under extreme simplifying conditions, and not properly checked numerically. Computer programs to derive rate laws are available (16).

Computers are inherently capable of performing some of these consistency checks, and of reminding humans to do others.

4. NEW METHODS OF PROCESSING ENZYME KINETIC INFORMATION, AND WAYS IN WHICH COMPUTERS CAN HELP

To some extent the methods of processing enzyme kinetic data described above may be described as "following the path of least resistance" in a search for simple answers. Although computers cannot simplify the inherent complexities of enzymes, they can very definitely decrease the resistance involved in processing data correctly. The correct usage of an appropriate computer program may be the easiest and cheapest as well as the best way to handle such data.

The Michaelis-Menten model or any other more complicated model of an enzyme is a nonlinear model. With present computer techniques and economics it can be handled as a nonlinear problem. We made the point in a recent review in *Critical Reviews in Bioengineering* (17) that this could not only be treated as a nonlinear problem but that it could be handled as a nonlinear problem about as cheaply as treating it as a linear problem with all the difficulties that come with linearization. We have recently completed and distributed a program called PENNZYME (for PENNsylvania enZYME program) (18) which does this nonlinear regression so that it is not dependent on good initial guesses for starting values. The regression starts with the simplex method of Nelder and Mead (19), which can use a very poor initial estimate, even an arbitrary initial estimate if nothing better is available. Once the general region where a minimum is likely to be has been found, we use the much better conjugate gradient method of Fletcher and Powell (20) to find the exact minimum. Economically this method appears to be competitive with linear regression. It should also be noted that this program is a "friendly" program: it assumes a minimal knowledge of computing or mathematics by the user. Basically you give it a candidate rate law to be fitted by a set of data and tell it very simply what you want it to do. It will even prompt you to extract the necessary information. Friendliness is a property of computer programs that is costly both in computing efficiency and in the program preparation and documentation. However, it is probably necessary if programs are to be widely used by present-day workers who have not been trained in computing (but presumably do know their enzyme kinetics). Present trends in computer economics are favourable for such use.

Despite the negative things that have been said here about the Lineweaver-Burk plot, it may be useful for visual examination to check for trends. Certain types of inhibition are best shown with such plots. However, regression analysis should be used to obtain accurate values of the constants. According to one editor of the *Journal of Biological Chemistry* there used to be more regression analysis several years ago than there is now. Someone at the 11th International Congress of Biochemistry, held in 1979, estimated that in the last six months of the *Journal of Biological Chemistry* no more than 20% of the enzyme kinetic papers had proper regression analysis, and the rest were simply "eyeballed" (fitted visually). There is some feeling that visual examination yields much the same results as linear regression; this is supported by the investigation of Dowd and Riggs (12) for unweighted points. It may be necessary to define further conditions where this seems to apply to weighted data.

Another thing computers are likely to do for enzyme kinetics is to improve experimental design. Our experience (18) as well as that of Mannervik (21) indicates that multivariate designs are

more effective than univariate ones in discriminating between
possible mechanisms. Also proper design will permit the recovery
of considerably more information from a given experiment (22).
Proper experimental design is much more important than post-experi-
mental analysis because such analysis cannot recover information
that is not there owing to faulty design. Why is this so important? In part, we get a lot of inconclusive experiments which
probably could have reached conclusive results with better design.
Sometimes we can not perform many experiments and must extract
maximum information from experiments with small samples or unstable
preparations. A current example is work at the author's institution
with pancreatic islet cells. Here it is possible to isolate only
a few milligrams of islet cells which is then the starting point
for enzyme isolations.

We also have the problem of properly designing the assays
by which we measure the velocities of the enzymes. It has been
pointed out by Carl Frieden, who is an authority on phosphofructo-
kinase, that this enzyme is so sensitive to everything that he could
get almost any kind of kinetics he wants by manipulating assay
conditions appropriately. Another extreme example is offered by
the work of Stadtman and associates on glutamine synthetase (23).
The abstract of their symposium paper given at this Biochemical
Congress states: "By means of allosteric and substrate site interactions, cascade enzymes can collectively sense simultaneous
changes in the concentrations of at least 40 different metabolites."
This is an extreme case, however. Accordingly more attention to
proper assay conditions seems to be needed. Here the work has
been done primarily by clinical chemists and clinical pathologists;
it is reported in such journals as *Clinical Chemistry*.

Computers can, of course, aid in deriving information from
progress curves as reported by Markus and Plesser in this volume
(24). It does take appreciably more computing to work with progress curves. Computers will also enable us to attack problems
arising from X-ray crystallography. An outstanding example was
shown by Madsen in the *Canadian Journal of Biochemistry* that was
distributed to registrants at the 11th International Congress of
Biochemistry (25). This shows the changes in conformation of
phosphorylase as it is activated and inactivated, and it must have
taken an enormous amount of computing. Also, a better job of
keeping track of the properties of enzymes could be done. We are
now carrying out a prototype effort in designing appropriate
data base management systems for enzyme kinetics.

One problem which was mentioned and which will require
computer assistance, although it is not primarily a computer
problem, is finding a workable means for determining the error
structure of a body of experimental data. As was pointed out by
the statisticians among the preceding speakers, the error structure

may depend on what the mechanism of the enzyme is, and require
reformulation and retesting as one tries various hypothesis regarding the mechanism.

This leads to the last, most advanced, and furthest-removed
application of computers: application of artificial intelligence
techniques to formulate systematically hypotheses and determine
whether a given mechanism fits a given body of data, with a
reasonable error structure. Although artificial intelligence is
mostly used to manipulate symbols, it could also be used to guide
calculations such as regressions, and could remove much of the
tedium that would be associated with checking through all relevant
possibilities and performing all necessary consistency checks on
the data. A computer can be made to go through a list of possibilities and determine which are feasible and which are not. In
particular, artificial intelligence has recently begun to be used
in model building, and it is to be hoped that techniques will
develop in the direction required for this application.

We appear to be on the verge of a very sharp decrease in
computer costs in the next few years, perhaps even one or two
orders of magnitude, according to the computing literature. This
would effectively make a reasonably large and powerful computer
less expensive than a spectrophotometer. This development would
let you hang a computer on the spectrophotometer or dedicate one
to reducing that spectrophotometer's data and determining after
each experimental point what the next experiment ought to be. It
could do in "real time" computations that now take days and include
even applications of such techniques as artificial intelligence.
Furthermore, the programs to do this can be made "friendly" as
indicated above because we can then pay the price of computing
inefficiency. Hopefully by proper application of all this computer
power and techniques, we can do a better job of studying our
enzymes, at greater speed and with less expense and less trouble,
and come out with a better indication of what it is that they
actually do.

The application of computers to enzyme kinetic studies is
well summed up by Watt (26), even though he was writing about
ecological modelling: "Thus we have a cyclical sequence of
operations, in which each loop consists of performing experiments
to test the computer model, refining the computer model, and obtaining new output to suggest further experimental procedures.
This cyclical mode of operation exploits the feedback principle to
enormously speed up the rate at which we can obtain insight into
a complicated system."

ACKNOWLEDGEMENT

Preparation of this chapter was supported by NIH Grant HL 15622.

REFERENCES

1. Mannervik, B. (1980) In "Kinetic Data Analysis: Design and Analysis of Enzyme and Pharmacokinetic Experiments" (L. Endrenyi, ed.) pp. 235-270, Plenum, New York.
2. Storer, A.C. and Cornish-Bowden, A. (1976) Biochem. J. 159, 1-5.
3. DeMillo, R.A., Lipton, R.J. and Perlis, A.J. (1979) Commun. Assn. Comput. Machin. 22, 271-280.
4. Miller, G.A. (1956) Psych. Rev. 63, 81-97.
5. Askelöf, P., Korsfeldt, M. and Mannervik, B. (1976) Eur. J. Biochem. 69, 61-76.
6. Storer, A.C., Darlenson, M.G. and Cornish-Bowden, A. (1975) Biochem. J. 151, 361-367.
7. Cornish-Bowden, A., Porter, W.R. and Trager, W.F. (1978) J. Theor. Biol. 74, 163-175.
8. Lineweaver, H., Burk, D. and Deming, W.E. (1934) J. Am. Chem. Soc. 56, 225-230.
9. Lineweaver, H. and Burk, D. (1934) J. Am. Chem. Soc. 56, 658-666.
10. Wilkinson, G.N. (1961) Biochem. J. 80, 324-332.
11. Endrenyi, L. and Kwong, F.H.F. (1972) In "Analysis and Simulation of Biochemical Systems" (H.C. Hemker and B. Hess, eds.) pp. 219-237, North-Holland, Amsterdam.
12. Dowd, J.G. and Riggs, D.S. (1965) J. Biol. Chem. 240, 863-869.
13. Ainsworth, S. (1977) "Steady-State Enzyme Kinetics", p. 177, University Park Press, Baltimore.
14. Michaelis, L. and Menten, M.L. (1913) Biochem. Z. 49,333-369.
15. Hill, C.M., Waight, R.D. and Bardsley, W.G. (1977) Mol. Cell. Biochem. 15, 173-178.
16. Cornish-Bowden, A. (1977) Biochem. J. 165, 55-59.
17. Garfinkel, L., Kohn, M.C. and Garfinkel, D. (1977) CRC Crit. Rev. Bioeng. 2, 329-361.
18. Kohn, M.C., Menten, L.E. and Garfinkel, D. (1979) Comput. Biomed. Res. 12, 461-469.
19. Nelder, J.A. and Mead, R.A. (1964) Comput. J. 7, 308-313.
20. Fletcher, R. and Powell, M.J.D. (1963) Comput. J. 6, 163-168.
21. Mannervik, B. (1979) Abstrs. XIth Intl.Congr. Biochem., Toronto, p. 313.
22. Endrenyi, L. (1980) In "Kinetic Data Analysis: Design and Analysis of Enzyme and Pharmacokinetic Experiments" (L. Endrenyi, ed.) pp. 137-167, Plenum, New York.
23. Stadtman, E.R., Chock, P.B. and Rhee, S.G. (1979) Abstrs. XIth Intl. Congr. Biochem., Toronto, 1979.
24. Markus, M. and Plesser, Th. (1980) In "Kinetic Data Analysis: Design and Analysis of Enzyme and Pharmacokinetic Experiments" (L. Endrenyi, ed.) pp. 317-339, Plenum, New York.
25. Fletterick, R.J., Sprang, S. and Madsen, N.B. (1979) Can. J. Biochem. 57, 789-797.
26. Watt, K.E.F. (1968) "Ecology and Resource Management", p. 426, McGraw-Hill, New York.

ERRORS ASSOCIATED WITH THE DETERMINATION OF THE INITIAL

VELOCITIES OF ENZYME-CATALYZED REACTIONS

I.A. Nimmo and G.L. Atkins

University of Edinburgh Medical School
Department of Biochemistry
Teviot Place
Edinburgh EH8 8AG, Scotland

ABSTRACT

Possible ways are discussed of reducing systematic errors in initial velocities (v), principally those caused by drift and non-linear progress curves. Information on the nature of the random experimental error in v is also summarized. The error tends to increase with the magnitude of v, but generalizations about its distribution are less easily made.

1. INTRODUCTION

The steady-state kinetics of an enzymic reaction are usually established by measuring the initial velocity of the reaction (v) at different pre-determined concentrations of the substrates, inhibitors and so on. The independent variables (concentrations of substrates etc.) are assumed to be known exactly, whereas the dependent variable (v) is acknowledged to contain random error. This error must be assessed, especially if distribution-dependent statistical methods such as least-squares or maximum likelihood are to be used to fit the algebraic models being tested to the data. Least squares, for example, depends on the individual errors following a Gaussian distribution, and in both methods the points should be weighted according to their relative precisions. If they are not, the final fit may well be biased (1a). Systematic erros in v, in contrast to the random ones, cannot be tolerated, as their presence could mislead one into choosing the wrong model.

This article reviews some of the ways of reducing the likely systematic errors in initial velocity data, and then summarizes what is known about the nature of the random errors in them.

2. SYSTEMATIC ERRORS

2.1. Effect of Enzyme Concentration

The initial velocity of the reaction should be directly proportional to the concentration of the enzyme, and this ought to be demonstrated before any further kinetic work is undertaken (2a). Usually, departures from direct proportionality are caused by artefacts in the assay system, but they may also reflect the properties of the enzyme itself (2a).

2.2. Drift

The initial velocities will "drift" if the properties of one of the reagents (including the enzyme) change slowly with time, or if the instruments being used are not entirely stable. The systematic error thereby incorporated into v can be recognized and reduced by running the assays in duplicate, in palindromic array (*e.g.*, 1,2,3,4,4,3,2,1) (3). In principle, the average of each pair of duplicates is measured at the same point in time, so that the effects of the drift will be reduced or even eliminated. On the other hand, in any subsequent statistical analysis the two duplicates must not be treated as independent estimates of the same quantity.

2.3. Nonlinear Progress Curves

Initial velocities are often estimated from progress curves, which may be either continuous traces (*e.g.* of the absorbance at 340 nm, for an NAD^+-linked dehydrogenase) or a series of (usually 4 or more) discrete points. The simplest way of finding v is to draw the tangent at the origin and measure its slope. One might expect this to be entirely reliable if the initial part of the trace were linear, but less so if it were curved, "noisy" or missing. However, it seems that one's eye is not as objective as one might suppose, because tangents drawn in this way tend to underestimate v, even when the traces are only gently curved (4-6). The systematic error involved can easily reach 5-10%, and could therefore invalidate the more refined of kinetic experiments.

Provided that the curvature of the trace is not caused by the enzyme's being hysteretic (2b), a possible solution is to fit a polynomial to the data; *e.g.*:

$$p = a_o + a_1t + a_2t^2 + a_3t^3 + \ldots$$

where p is the concentration of the product at time t. Differentiating the polynomial and putting t = 0 shows that $v = a_1$. The

method is not as straightforward as it might appear, because there
are a number of imponderables. For example, the polynomial can be
expressed either as a series in t, as above (7), or as a series
in p (5). Further, it is not obvious what order of polynomial one
should choose; if the order is too low the overall fit may be poor,
and if it is too high the slope of the polynomial may fluctuate
rapidly near t = 0. In general a cubic or a quartic should be
satisfactory (5,6) so long as it is not extrapolated much beyond
the data (unfortunately in most assays some extrapolation is inevitable).

    Alternatively, the integrated Michaelis-Menten equation can
be fitted to the data and v calculated from the fit. Cornish-
Bowden (8) has described how to do this for a single-substrate
reaction by a distribution-free procedure, and has demonstrated
that the estimate of v is not affected much by the progressive
denaturation of the enzyme or the accumulation of an inhibitory
product. A potential snag is that the integrated approach requires
two more pieces of information than the polynomial one; these are
the equilibrium constant of the reaction and the origin of the
progress curve (*e.g.* the absorbance of the reaction mixture at
zero time), neither of which may be readily available. Nevertheless, we believe the integrated approach to be preferable to the
polynomial alternative, at least for single-substrate reactions
(6).

2.4. Single-point Assays

    Constraints of expense or time can force one to measure the
concentration of product formed (Δp) after only a single period
of incubation (Δt).   One automatically sets v to $\Delta p/\Delta t$, and its
substrate concentration (s) to that present at the start of the
reaction (s_0). The obvious difficulties are that since substrate
is consumed during the reaction, s_0 must be an overestimate of s,
and that since the velocity of a reaction usually declines with
time, $\Delta p/\Delta t$ will probably be an underestimate of v. A better
practice is to keep $\Delta p/\Delta t$ as the estimate of v, but to relate
it to the mean concentration of substrate present during the assay
(= $s_0 - \Delta p/2$) (9). Even this, however, does not remove the bias
introduced by the progressive inactivation of the enzyme or the
formation of a product that is a powerful competitive inhibitor
of the reaction. In these instances one could probably reduce
the bias by stopping the reaction at two times instead of one and
using the integrated Michaelis-Menten equation to calculate v, but
at the cost of considerable loss of precision (10).

2.5. Coupled Assays

    These are often used when no change in absorbance occurs
during the reaction of interest (the primary reaction). For

instance, hexokinase can be coupled with either glucose-6-phosphate
dehydrogenase or with pyruvate kinase plus lactate dehydrogenase.
The progress curves of these assays are characterized by a transient phase, followed by a period during which the rate of reaction
is steady, and then one in which it declines. The initial velocity
of the primary reaction is assumed to be equal to the steady rate,
which implies that the coupling enzyme must have been added at a
concentration high enough to preclude its being rate-limiting. This
is an impossible ideal, so that, in practice, observed rates which
are 99% of the primary ones are considered acceptable; they should
be checked by increasing the concentration of the coupling enzyme.
Equally, the presence of too much coupling enzyme is undesirable,
being wasteful and, should the enzyme not be entirely pure or
specific, potentially dangerous to boot: for example, glucose-6-
phosphate dehydrogenase can act on glucose as well as glucose-6-
phosphate (11). Fortunately, rules have been derived to help one
choose the optimum concentration of coupling enzymes (11,12).

     Two other pitfalls associated with coupled assays deserve
mention. The first is that a steady rate of reaction does not
necessarily imply that the coupling enzyme is not rate-limiting,
unless it lasts at least ten times as long as the transient phase
(11). The second pitfall is that substrates or inhibitors of the
primary reaction may reduce the activity of the coupling enzyme,
the inhibition of glucose-6-phosphate dehydrogenase by Mg.ATP
(one of the substrates for hexokinase) being a good illustration
of this (13). However, since the transient time is not affected
by the activity of the primary enzyme, an increase in it indicates
that the activity of the coupling enzyme has declined (13).

2.6. Errors in Substrate Concentration

     These could arise when the stock solution is made up, or
when it is diluted and dispensed. The implications of the former
are obvious and need no further comment. The latter, on the other
hand, seems to be analogous to the "Berkson-type" error in the
independent variable of a linear regression (14), and therefore
may not lead to an overall bias in the estimates of the kinetic
constants. (A "Berkson-type" error occurs when the independent
variable is set to a series of fixed values and, owing to technical errors, the true values of the variable differ from the set ones
by a constant magnitude; it does not lead to the slope of the
regression being underestimated. If, however, the values of
the independent variable are unknown and have to be measured by
a method subject to error, the slope of the regression will be
underestimated (14).)

3.   RANDOM ERRORS

Systematic errors in v must be eliminated, but random errors cannot be. Rather, they have to be quantified if distribution-dependent mathematical methods are to be used. Two points need to be established: (i) the form of the distribution of the error (Gaussian, log-Gaussian, *etc.*); and (ii) the relationship between v and its variance.

3.1. Distribution of Experimental Error

This is not easy to determine, principally because up to 100 replicate assays are required (1b), and factors such as expense, time and the instability of the reagents or the operator's patience prevent their being run. Nevertheless, a few enzymes have been studied, including α-chymotrypsin (15), glucokinase (15), hexokinase (16), glutathione S-transferase A (17) and acetylcholinesterase (18). Perhaps inevitably, no common trends emerged: in two instances the errors seemed to be Gaussian in nature (16,17), whereas in the other two (15,18) a number of non-Gaussian distributions was observed. Since in these experiments it was difficult to exclude the presence of small systematic errors such as the progressive inactivation of the enzyme, we conclude that the errors in assays run by a careful operator are likely to be approximately Gaussian, although they may also incorporate the occasional outlier.

3.2. Relative Magnitude of Experimental Error

Several attempts have been made to find out how the size of the experimental error in v varies with the magnitude of v. Usually the variance of v (Var(v)) is taken as the index of size, because when least squares is used to fit kinetic models, each value of v should be weighted according to the reciprocal of its variance.

The first generalization is that the local variance should be calculated from at least 5 or 6 replicates; unfortunately duplicates or triplicates do not give an answer that is accurate enough to be of much value (15). This has prompted the search for formulae expressing the variance of v in terms of v. Not surprisingly, the details again depend on the assay in question, but in general it seems that Var(v) tends to increase with v. The relationship is often of the form $Var(v) \propto v^{\alpha}$, the value of α ranging from 1.0 through 2.0 (constant coefficient of variation) to greater than 2.0 (15-18). Certainly there is no evidence to support the common assumptions about the sizes of errors, namely that either their absolute or their relative magnitude is constant (α = 0 and α = 2, respectively). Consequently the errors must be determined experimentally, either by measuring replicates as indicated above, or, rather neatly, from the residuals about a curve fitted using first estimates for the weighting factors (19).

4. CONCLUSIONS

Estimating initial velocities accurately is not particularly easy, especially when the progress curve is not straight or its early part is missing. If ones tries to draw a tangent at the origin of the curve one will usually underestimate v because one tends to draw a chord rather than a tangent (20). Numerical methods, such as fitting a polynomial or the integrated Michaelis-Menten equation to the curve, can help in some instances but are not a universal remedy, because the equations themselves are only approximations. Indeed, one wonders to what extent the presence of even small systematic errors in v is responsible for the claim that only a handful of enzymes actually obeys the Michaelis-Menten equation (21). Small systematic errors in the concentration of substrate, on the other hand, would seem to be of less concern.

The random errors in v do not conform to any general pattern. However, it is likely that any departures from a Gaussian distribution will not be great, so that least-squares methods of curve-fitting are theoretically acceptable. On the other hand, such methods require v to be weighted according to the reciprocal of its variance. Consequently, this must be calculated, for example: (i) from 5-6 replicates of each v; or (ii) from a pre-determined relationship between v and its variance (in which case Var(v) should be found from the predicted value of v and not the observed one); or (iii) from the residuals about the fitted curve (19). The alternative is to use some distribution-free method of curve-fitting, such as the direct linear plot (22,23), because these make fewer assumptions about the nature and magnitude of the random error (24). They are not the complete answer, however, because they tend to be less powerful than least squares and can only be applied to a relatively small number of problems (25).

We thank David Apps for his helpful comments.

REFERENCES

1. Colquhoun, D. (1971) "Lectures on Biostatistics", (a)
      pp. 214-272, (b) p. 71, Clarendon, Oxford.
2. Dixon, M. and Webb, E.C. (1979) "Enzymes", 3rd ed.   (a)
      pp. 48-55, (b) pp. 454-460, Longman, London.
3. Fisher, R.B. and Gilbert, J.C. (1970) J. Physiol. (London)
      210, 277-286.
4. Walter, C. and Barrett, M.J. (1970) Enzymologia, 38, 147-160.
5. Philo, R.D. and Selwyn, M.J. (1973) Biochem. J. 135, 525-530.
6. Nimmo, I.A. and Atkins, G.L. (1978) Biochem. Soc. Trans. 6,
      548-550.
7. Elmore, D.T., Kingston, A.E. and Shields, D.B. (1963) J. Chem.
      Soc. (London), 2070-2078.
8. Cornish-Bowden, A. (1975) Biochem. J. 149, 305-312.
9. Lee, H.-J. and Wilson, I.B. (1971) Biochim. Biophys. Acta,
      242, 519-522.
10. Atkins, G.L. and Nimmo, I.A. (1978) Biochem. Soc. Trans. 6,
      545-548.
11. Storer, A.C. and Cornish-Bowden, A. (1974) Biochem. J. 141,
      205-209.
12. Cleland, W.W. (1979) Anal. Biochem. 99, 142-145.
13. Easterby, J.S. (1973) Biochim. Biophys. Acta, 293, 552-558.
14. Snedecor, G.W. and Cochran, W.G. (1967) "Statistical Methods",
      6th ed., pp. 164-166, Iowa State Univ., Ames, Iowa.
15. Storer, A.C., Darlison, M.G. and Cornish-Bowden, A. (1975)
      Biochem. J. 151, 361-367.
16. Siano, D.B., Zyskind, J.W. and Fromm, H.J. (1975) Arch.
      Biochem. Biophys. 170, 587-600.
17. Askelöf, P., Korsfeldt, M. and Mannervik, B. (1976) Eur. J.
      Biochem. 69, 61-67.
18. Nimmo, I.A. and Mabood, S.F. (1979) Anal. Biochem. 94, 265-
      269.
19. Mannervik, B., Jakobson, I. and Warholm, M. (1979) Biochim.
      Biophys. Acta, 567, 43-48.
20. Cornish-Bowden, A. (1979) "Fundamentals of Enzyme Kinetics",
      pp. 40-42, Butterworths, London.
21. Hill, C.M., Waight, R.D. and Bardsley, W.G. (1977) Mol. Cell.
      Biochem. 15, 173-178.
22. Cornish-Bowden, A. and Eisenthal, R. (1978) Biochim. Biophys.
      Acta, 523, 268-272.
23. Cornish-Bowden, A. (1980) In "Kinetic Data Analysis: Design
      and Analysis of Enzyme and Pharmacokinetic Experiments"
      (L. Endrenyi, ed.) pp. 105 - 119, Plenum, New York.
24. Nimmo, I.A. and Atkins, G.L. (1979) Trends Biochem. Sci. 4,
      236-239.
25. Atkins, G.L. and Nimmo, I.A. (1980) In "Kinetic Data Analysis:
      Design and Analysis of Enzyme and Pharmacokinetic Experi-
      ments" (L. Endrenyi, ed.) pp. 121 - 135, Plenum, New York.

PROGRESS CURVES IN ENZYME KINETICS:

DESIGN AND ANALYSIS OF EXPERIMENTS

Mario Markus and Theodor Plesser

Max-Planck-Institut fuer Ernaehrungsphysiologie
4600 Dortmund 1, Rheinlanddamm 201
Federal Republic of Germany

ABSTRACT

Three shortcomings in progress curve analysis are discussed, and solutions or at least optimal strategies to overcome these problems are presented. (i) Systematic deviations in the progress curve data due to errors in the initial concentrations are taken into account in a linear approach by a proper weighting matrix in the parameter optimization. A transformation matrix, derived from the weighting matrix, leads to uncorrelated errors when applied to the data. The statistical tools developed for independent measurements can thus be applied to the transformed data. (ii) Model development by progress curve analysis is very cumbersome, since a progress curve does not reflect clearly the properties of the kinetics by visual inspection. Furthermore, the integration of the rate law leads to long computation times. These problems can be alleviated by determining the rates from the derivatives of a functional approximation of the progress curves. After developing a model using the rates, parameter refinement can be performed by fitting the original progress curve data. This procedure has the further advantage that optimization using the rate data is much less sensitive to the initial parameter estimates. (iii) The lack of inference from the visual inspection of progress curves also affects their experimental design. Methods mentioned in the literature that are applicable to the design of initial rate experiments, are extended for progress curves. The best results are obtained with a discrimination function that includes the statistical expectations of the minimum sum of squares.

1. INTRODUCTION

The earliest investigations on enzymes at the end of the last century were based on progress curves, $i.e.$, on the analysis of the enzymic reactions as they progress in time. Following the contribution of Michaelis and Menten (1), biochemists preferred to restrict the analysis to the rates obtained from the initial linear part of the progress curves because of the difficulties encountered in the evaluation of the full time course of the reaction. But new developments, especially in computer technology and numerical computation, have allowed a renaissance of the progress curve method in the last years.

The most gratifying advantage of progress curves, as compared to initial rates, is the large amount of experimentation time that one may save. This is illustrated by our investigations, showing that seven progress curves -aimed to determine six model parameters- led to a higher parameter accuracy than 128 initial rate measurements for pyruvate kinase from yeast (2). Several authors (3,4,5) describe how the maximum rate V and the Michaelis constant K_m can be obtained from only one progress curve experiment. This experimental economy is also illustrated by the investigations with integral reactors, as discussed by Quiroga et al. (6). Another advantage of progress curve analysis is its applicability to the study of slow molecular processes implying explicit time dependences of the rates.

General reports on the application of progress curves have been published by Garfinkel et al. (7), Orsi and Tipton (5) and Boiteux et al. (8).

When analyzing progress curves, one encounters the following difficulties, which in most cases can be avoided in initial rate measurements:

1.   The rate law must be integrated for fitting to the measured progress curves. Efficient subroutines are nowadays available to solve this problem numerically (9).

2.   Effects like reverse reactions, formation or consumption of solutes affecting the reaction rate, loss of enzyme activity, and changes of ionic strength and pH may have to be taken into account in the calculations.

3.   Progress curves present difficulties in qualitative diagnostics; for example, it is not easy to see the extent of sigmoidicity or to compare K_m's when comparing progress curves by visual inspection. This situation can be improved by determining the reaction rates from the slopes of the progress curves. Balcolm and Fitch (10) obtained the rates from the chords corresponding to adjacent time

intervals. But it has been shown that this method leads to wrong
weighting of the residuals (11). A more reliable method consists
in evaluating the time derivatives of a functional approximation
of the progress curves; Elmore *et al*. (12) and Sorenson and Schack
(13) used orthogonal polynomials; we have obtained satisfactory
results using a spline function approximation. The advantages and
disadvantages of the analysis of the rates obtained in this way are
discussed in Section 3, 'The Integral and Differential Method'.

Due to the difficulties in visual diagnostics of progress
curves, it is also important to have efficient algorithmic procedures for the design of experiments. In Section 4 we describe two
types of methods: (a) methods aimed to increase optimally the
accuracy of unknown parameters, and (b) methods to discriminate
optimally between rival models. These methods can be used with two
purposes: (i) to determine initial conditions for progress curve
experiments, or (ii) to choose data points for evaluation on the
progress curves.

4.   The errors of the data points of a progress curve consist of
the noise at every data point and a systematic deviation due to
errors in the initial concentrations. The errors in the initial
concentrations propagate in time, so that the usual statistical
methods of data analysis for independent data points cannot be
applied. A weighting matrix can be introduced that implies a
linear transformation of the data leading to independent data
errors, as described in the next section.

2.   WEIGHTING OF DATA FROM PROGRESS CURVES

2.1. General Formalism

Data from initial rate experiments are usually fitted by minimizing the objective function

$$\phi = \sum_{i=1}^{N} \omega_i (y_i - \eta_i)^2 \quad , \tag{1}$$

where N is the number of measurements, y_i and η_i (i=1,...,N) are
the experimental and theoretical values, and the ω_i are the
weighting factors given by

$$\omega_i = 1/\sigma_i^2 \quad , \tag{2}$$

where the σ_i^2 are the variances of the observations.

The minimization of ϕ, as given by Eq. [1], leads to the best
unbiased quasi-linear parameter estimates, provided some conditions
are fulfilled (see textbooks on statistics). One of these conditions

is that the data errors must be independent. But, in general, the
errors of the data obtained from progress curves cannot be considered as independent because the errors in the initial concentrations of the solutes propagate along the progress curve and cause
systematic deviations.

Newman *et al.* (14) found that the influence of systematic
errors in progress curves on the parameter estimates can be substantial and suggested to treat the initial concentrations as additional unknown parameters. This procedure is only reasonable if
the number of initial solute concentrations that are subject to
error and the number of evaluated progress curves is small. In
order to avoid the introduction of an unknown parameter for each
uncertain solute concentration in each progress curve, we prefer
to approach this problem by a proper modification of the objective
function ϕ.

First, we consider the general case of Λ progress curves.
There are $m(\lambda)$ initial solute concentrations $a_j^{(\lambda)}$ $(j=1,\ldots,m(\lambda))$
having random errors $\Delta a_j^{(\lambda)}$. These errors are independent of each
other. In addition to the systematic deviation resulting from the
propagation of the errors $\Delta a_j^{(\lambda)}$ in time, we assume a time-indepen-
dent random noise $\rho^{(\lambda)}$, having a standard deviation $\sigma^{(\lambda)}$ $(\lambda=1,\ldots,\Lambda)$.
On each curve, $n(\lambda)$ points $(\lambda=1,\ldots,\Lambda)$ are evaluated $(n(\lambda)>m(\lambda))$.

The objective function to be minimized is the quadratic form
of residuals given in matrix notation by

$$\phi = \sum_{\lambda=1}^{\Lambda} [\underline{y}^{(\lambda)} - \underline{\eta}^{(\lambda)}]' \underline{W}^{(\lambda)} [\underline{y}^{(\lambda)} - \underline{\eta}^{(\lambda)}] \qquad [3]$$

$(15,16)$, where $\underline{y}^{(\lambda)}$ and $\underline{\eta}^{(\lambda)}$ are vectors defined for each progress
curve by

$$\underline{y}^{(\lambda)} = (y_1^{(\lambda)},\ldots,y_{n(\lambda)}^{(\lambda)})' \qquad , \qquad [4]$$

$$\underline{\eta}^{(\lambda)} = (\eta_1^{(\lambda)},\ldots,\eta_{n(\lambda)}^{(\lambda)})' \qquad . \qquad [5]$$

A prime (') indicates the transposed vector.

Each index from 1 to $n(\lambda)$ corresponds to one data point taken
for evaluation from the progress curve.

The weighting matrices $\underline{W}^{(\lambda)}$ are given by $\underline{W}^{(\lambda)} = [\underline{V}^{(\lambda)}]^{-1}$. The
$\underline{V}^{(\lambda)}$ are the $n(\lambda) \times n(\lambda)$ matrices of observation covariances, which
take into account the interdependence of the errors of the observa-

tions in each progress curve; their elements are defined by[*]

$$V_{ij} = E\{[y_i-E(y_i)][y_j-E(y_j)]\} \qquad [6]$$

$$(i,j=1,\ldots,n) \quad .$$

E is the symbol of mathematical expectation. $\underline{V}$ is a diagonal matrix if the covariances of observations are zero. In this case, Eq. [3] reduces to Eq. [1].

In order to obtain the best unbiased quasi-linear estimates of the unknown model parameters p_j $(j=1,\ldots r)$, we assume that the Jacobian matrix $\underline{X}$ given by

$$X_{ij} = \partial\eta_i/\partial p_j \qquad [7]$$

$$(i=1,\ldots,n;j=1,\ldots,r)$$

has rank r.

If we denote by $\underline{T}$ a matrix such that $\underline{W}=\underline{T}'\underline{T}$, then we can write Eq. [3] as

$$\phi = \sum_{\lambda=1}^{\Lambda} [\underline{T}^{(\lambda)}\underline{y}^{(\lambda)}-\underline{T}^{(\lambda)}\underline{n}^{(\lambda)}]'[\underline{T}^{(\lambda)}\underline{y}^{(\lambda)}-\underline{T}^{(\lambda)}\underline{n}^{(\lambda)}] \quad . \qquad [8]$$

ϕ thus becomes a sum of squares of the components of the vectors $\underline{T}\ \underline{y} - \underline{T}\ \underline{n}$.

In a linear approximation we can write

$$y_k = \rho + E(y_k) + \sum_{\nu=1}^{m} g_{k\nu}\ \Delta a_\nu \qquad [9]$$

$$(k=1,\ldots,n)$$

where the $g_{k\nu}$ are the coefficients of the linear term of the Taylor expansion. Inserting $y_k-E(y_k)$ as obtained by Eq. [9] with k=i and k=j, into Eq. [6] leads to

$$\underline{V} = \underline{I}\sigma^2 + \underline{H}\ \underline{H}' \qquad , \qquad [10]$$

where $\underline{I}$ is the unit matrix. The n×m matrices $\underline{H}$ are defined by

---

[*]In what follows, the index (λ) is left out if the formulas apply to each of the Λ progress curves.

$$H_{ij} = g_{ij}\sigma_{a_j}$$ [11]

$$(i=1,\ldots,n;\ j=1,\ldots,m)\ .$$

We assume that H has rank m. The σ_{a_j} are the standard deviations of the errors Δa_j. The matrix $\underline{V}$, as given by Eq. [10], can be inverted if $\sigma \neq 0$ using Rao's formula (15), leading to the weighting matrix

$$\underline{W} = [\underline{I} - \underline{H}(\underline{H}'\underline{H} + \underline{I}\sigma^2)^{-1}\underline{H}']/\sigma^2\ .$$ [12]

The $g_{k\nu}$, needed for the determination of the matrices $\underline{H}$, can be estimated by measuring the variations δy_k due to variations δa_ν and calculating

$$g_{k\nu} \simeq \delta y_k/\delta a_\nu$$ [13]

$$(k=1,\ldots,n;\ \nu=1,\ldots,m)\ .$$

But difficulties arise in the choice of the δa_ν: if the δa_ν are smaller than, or of the order of the errors Δa_ν then the estimates $g_{k\nu}$ will have a high error. If the δa_ν are larger than the Δa_ν then the linear approximation might no longer be valid. We therefore make the approximation

$$g_{k\nu} \simeq \partial \eta_k/\partial a_\nu\ .$$ [14]

When using this approximation, the matrices $\underline{H}$, and therefore the weighting matrices, are dependent on the model and its parameters.

If the covariance matrix, as defined by Eq. [6], is calculated for the components of the transformed vectors $\underline{T}\ \underline{y}$, one obtains the diagonal matrix $\underline{I}\sigma^2$, *i.e.*, the errors of the transformed measurements can be considered as independent. This opens the path for the application of statistical tools, such as tests of goodness of fit, relying on the assumption of uncorrelated data errors (7,17, 18,19).

If the random errors ρ can be neglected, Eq. [10] reduces to

$$\underline{V} = \underline{H}\ \underline{H}'\ .$$ [15]

In this case, the n×n matrix $\underline{V}$ has the same rank as the matrix $\underline{H}$. This rank is equal to m and is smaller than n; therefore, $\underline{V}$ is a singular matrix and cannot be inverted. We have shown in co-operation with S. Schach (unpublished results) that the parameters can then be estimated in the following two way:

1.   An arbitrary but small random error is assumed, so that $\underline{V}$ can be written as in Eq. [10] and inverted to yield the weighting matrix. It can be shown that best linear unbiased parameter estimates are obtained in the limit of vanishing random error.

2.   The following weighting matrix is used:

$$\underline{W} = \underline{I} - \underline{H}(\underline{H}'\underline{H})^{-1}\underline{H}' \qquad . \qquad\qquad [16]$$

To obtain this equation, we first add a matrix $\underline{I}\varepsilon^2$ to both sides of Eq. [15]; we then invert both sides of the resulting equation. In this way, we obtain an equation whose right-hand side is identical with that of Eq. [12], with ε instead of σ; ε can then be left out from the denominator because it affects ϕ only by a constant factor and has thus no influence on the minimum of ϕ; therefore the limit $\varepsilon \to 0$ results in Eq. [16]. We have shown (unpublished results) that weighting with the matrix given by Eq. [16] leads to unique estimates of the parameters if the $n \times (m+r)$ matrix $\underline{J}$ defined by

$$J_{ij} = X_{ij} \qquad \text{for } j=1,\dots,r \atop i=1,\dots,n \qquad , \qquad\qquad [17]$$

$$J_{ij} = g_{ik} \qquad \text{for } j=r+1,\dots,r+m; \quad k=j-r \atop i=1,\dots,n \qquad\qquad [18]$$

has rank equal to $m+r$ for each progress curve.

Situations where the case just discussed applies, $i.e.$, where the random errors of progress curves are negligible as compared to the systematic error due to the errors in the initial solute concentrations, have in fact be seen in titration experiments with pyruvate kinases from $E.$ $coli$ and yeast (20,21). In these investigations, we found that the systematic error is mainly due to the error in the concentration of the solute that is added to the solution in order to start the reaction, and therefore $m=1$. The weighting matrix given by Eq. [16] reduces in this case to the form

$$\underline{W} = \underline{I} - \underline{h}\,\underline{h}'/\, |\underline{h}|^2 \quad , \qquad\qquad [19]$$

where $\underline{h}$ is a vector with components $h_i = g_{i1}$, $(i=1,\dots,n)$.

2.2. Analytical Example: Michaelis-Menten Kinetics

We want to illustrate the exposed ideas with a simple example. The enzyme is assumed to obey Michaelis-Menten kinetics and the progress curve is assumed to be measured by recording the concentration of a product as a function of time. Furthermore, we assume that the only error is due to the error in the initial substrate concentration. The model to be fitted to the experiments is the integrated form of the Michaelis-Menten equation (5):

$$Vt = \eta + K_{0.5} \ln[S_0/(S_0 - \eta)] \quad , \tag{20}$$

where η is the concentration of the product and S_0 is the initial substrate concentration. The weighting matrix can be calculated using Eq. [19]. The components of the vector $\underline{h}$ are estimated by $h_i = d\eta_i/dS_0$ where the η_i ($i=1,\ldots,n$) are the product concentrations at the points evaluated on the progress curve. Differentiating both sides of Eq. [20] with respect to S_0, we obtain

$$h_i = K_{0.5}\eta_i/[S_0(K_{0.5} + S_0 - \eta_i)] \quad . \tag{21}$$

If, for example, three points are evaluated on the progress curve, Eq. [19] yields the following weighting matrix:

$$\underline{W} = \frac{1}{h_1^2 + h_2^2 + h_3^2} \begin{pmatrix} h_2^2 + h_3^2 & -h_1 h_2 & -h_1 h_3 \\ -h_1 h_2 & h_1^2 + h_3^2 & -h_2 h_3 \\ -h_1 h_3 & -h_2 h_3 & h_1^2 + h_2^2 \end{pmatrix} \quad , \tag{22}$$

where h_1, h_2 and h_3 are given by Eq. [21]. $\underline{W}$ can be decomposed in a matrix product $\underline{T}'\underline{T}$ where

$$\underline{T} = \frac{1}{D_2} \begin{pmatrix} (h_2^2 + h_3^2)/D_1 & -h_1 h_2/D_1 & -h_1 h_3/D_1 \\ 0 & h_3 & -h_2 \\ 0 & 0 & 0 \end{pmatrix} \tag{23}$$

$$D_1 = (h_1^2 + h_2^2 + h_3^2)^{1/2} \quad \text{and} \quad D_2 = (h_2^2 + h_3^2)^{1/2} \quad .$$

In order to apply the least-squares fit, one has to transform the experimental and theoretical values with the matrix $\underline{T}$. From the experimental values y_1, y_2 and y_3, one obtains the new values

$$\bar{y}_1 = \frac{h_2^2 + h_3^2}{D_1 D_2} y_1 - \frac{h_1 h_2}{D_1 D_2} y_2 - \frac{h_1 h_3}{D_1 D_2} y_3 \tag{24}$$

$$\bar{y}_2 = (h_3/D_2)y_2 - (h_2/D_2)y_3 \tag{25}$$

$$\bar{y}_3 = 0 \quad . \tag{26}$$

The theoretical values η_i are transformed in the same manner.

The values $\bar{y}_3$ and $\bar{\eta}_3$ make no contribution to the sum of squares. We are thus left with only two transformed data points $\bar{y}_1$ and $\bar{y}_2$ for the determination of the two parameters $K_{0.5}$ and V. The loss of one data point is related to the fact that we have eliminated one unknown, namely the error ΔS_0 in the initial concentration. To understand this elimination of ΔS_0 we make the following Taylor expansion for the y_i:

$$y_i = \eta_i^{true} + h_i \Delta S_0 \quad . \tag{27}$$

η_i^{true} is the theoretical product concentration obtained from Eq. [20] with the true values of the parameters. Inserting Eq. [27] in Eq.'s [24] to [26] one obtains

$$\bar{y}_1 = \frac{h_2^2 + h_3^2}{D_1 D_2} \eta_1^{true} + \frac{h_2^2 + h_3^2}{D_1 D_2} h_1 \Delta S_0 - \frac{h_1 h_2}{D_1 D_2} \eta_2^{true} - \frac{h_1 h_2^2}{D_1 D_2} \Delta S_0$$
$$- \frac{h_1 h_3}{D_1 D_2} \eta_3^{true} - \frac{h_1 h_3^2}{D_1 D_2} \Delta S_0 \tag{28}$$

$$\bar{y}_2 = \frac{h_3}{D_3} \eta_2^{true} + \frac{h_3 h_2}{D_2} \Delta S_0 - \frac{h_2}{D_2} \eta_3^{true} - \frac{h_2 h_3}{D_2} \Delta S_0 \tag{29}$$

$$\bar{y}_3 = 0 \quad . \tag{30}$$

Now we clearly see that all the terms containing ΔS_0 cancel out. This result means that the error plays no role in the transformed experiments $\bar{y}_i$ in this first-order approximation. The sum of squares of the transformed residuals, as given by Eq. [8], results in

$$\phi = \left(\frac{h_2^2 + h_3^2}{D_1 D_2} (\eta_1^{true} - \eta_1) - \frac{h_1 h_2}{D_1 D_2} (\eta_2^{true} - \eta_2) - \frac{h_1 h_3}{D_1 D_2} (\eta_3^{true} - \eta_3) \right)^2$$
$$+ \left(\frac{h_3}{D_2} (\eta_2^{true} - \eta_2) - \frac{h_2}{D_2} (\eta_3^{true} - \eta_3) \right)^2 \quad . \tag{31}$$

The best fit will consequently lead to $\phi = 0$ with $\eta_i = \eta_i^{true}$ ($i = 1, 2, 3$), i.e., one obtains the true values of the parameters. In practical cases, however, such ideal results will not be obtained for several reasons: (i) There are always other experimental errors besides those arising from the errors in the initial concentrations of solutes; (ii) The true model is never exactly known; (iii) There are deviations from the linear approximations given by Eq.'s [9] and [27]; (iv) Numerical inaccuracies arise in the integration and optimization processes.

2.3. Numerical Examples: Hill Equation and a Random Bireactant System

We have made a numerical analysis of further examples given in Table 1. The rate laws referred to in this table are the Hill equation:

$$\tilde{v} = \tilde{V}/[1+(\tilde{K}_{0.5}/\tilde{S})^{n_H} \quad , \tag{32}$$

and a rate law obtained by assuming random binding of two substrates A and B (22):

$$\tilde{v} = \frac{\tilde{V}(\tilde{A}/\tilde{K}_A')(\tilde{B}/\tilde{K}_B)}{1+(\tilde{A}/\tilde{K}_A)+(\tilde{B}/\tilde{K}_B)+(\tilde{A}/\tilde{K}_A')(\tilde{B}/\tilde{K}_B)} \quad . \tag{33}$$

In these examples, we used simulated data assuming that the measured quantity is the volume V_T of added titrant (one mole of titrant for each mole of consumed substrate) and that the sole source of error is the error in the initial concentration of one substrate (S, when using Eq. [32], and A, when using Eq. [33]). The dilution due to the titrant was taken into account in the calculation. $K_{0.5}$ is the half-saturation constant. S, A and B are the substrate concentrations. K_A and K_A' are the dissociation constants of A

TABLE 1: Cases Analysed Using Simulated Data

Case No.	Rate Law	Initial concentration and error	True parameter values	Number of analyzed points
1	Eq. [32]	$\tilde{S}_0$=5.0, 4%	n_H=2.5, $\tilde{K}_{0.5}$=1	4
2	Eq. [32]	$\tilde{S}_0$=5.0, 4%	n_H=1 , $\tilde{K}_{0.5}$=1	4
3	Eq. [32]	$\tilde{S}_0$=0.5, 4%	n_H=1 , $\tilde{K}_{0.5}$=1	4
4	Eq. [32]	$\tilde{S}_0$=0.5, 2%	n_H=1 , $\tilde{K}_{0.5}$=1	4
5	Eq. [32]	$\tilde{S}_0$=0.5, 2%	n_H=1 , $\tilde{K}_{0.5}$=1	12
6	Eq. [33]	$\tilde{A}_0$=9.0, 1% $\tilde{B}_0$=12.0	$\tilde{K}_A$=$\tilde{K}_A'$=1	4

before and after binding of B, respectively; K_B is the dissociation constant of B before binding of A. The tilde $\sim$ in equations [32] and [33] indicates that the quantities were normalized to make them dimensionless.

The cases 1 and 2 in Table 1 differ in n_H. The cases 2 and 3 differ in the initial substrate concentration. The cases 3 and 4 differ in the error of the initial substrate concentration. The cases 4 and 5 differ in the number of analyzed data points

Points for evaluation were chosen in such a way that the reaction velocities would be equidistant.

The fitting of the progress curves was performed using the subroutine VA05A from the Harwell Subroutine Library (23). The unknown parameters were $\tilde{K}_{0.5}$ and n_H in cases 1 to 5 and $\tilde{K}_A$ and $\tilde{K}_A'$ in case 6 ($\tilde{V}$ and $\tilde{K}_B$ were assumed to be known and set equal to 1). The unknown parameters, with the exception of n_H, were transformed into their logarithms before optimization. The following arguments were used in the subroutine VA05A: $H=10^{-3}$ (parameter increment for the calculation of derivatives with respect to parameters), DMAX=2 (estimates of the distance between the initial estimates of the parameters and the minimum), ACC=$10^{-6} \cdot \bar{q}^2 \cdot n$ (ACC is the accuracy of the sum of squares; $\bar{q}$ is the average fitted quantity; n is the number of data points). The components of the vector $\underline{h}$, needed for the determination of the weighting matrix, were estimated by numerical differentiation of the theoretical values with respect to $\tilde{S}_0$ or $\tilde{A}_0$. The numerical integrations of the progress curves were performed using a subroutine from Gear (24).

After fitting, the bias was calculated as the Euclidean distance between the values of the parameters used to simulate the data (true values) and the values obtained from the optimization. In Table 2, we compare the biases obtained without and with considering the covariances of observations. When the observation covariances were not considered, we set $\underline{V}=\underline{I}$.

We see that a drastic reduction of parameter bias can be obtained by considering the weighting matrix, *i.e.*, by transforming the theoretical and experimental values with the matrix $\underline{T}$. This transformation should lead to zero bias in the ideal case discussed in the calculations above. The deviations from zero in Table 2 (especially for case 1) are caused by deviations from the linear approximations and by inaccuracies in the numerical integrations and optimizations.

TABLE 2:   Biases After Parameter Optimizations With Progress Curves

| | Bias | |
Case No.	without consideration of data covariances	with
1	0.32	0.14
2	0.23	0.013
3	0.34	0.0057
4	0.20	0.0029
5	0.13	0.0027
6	0.90	0.021

The values given are the Euclidean distances between true and optimal parameter values, after taking the logarithms of all parameters, except n_H. The cases correspond to those of Table 1.

---

3.   THE INTEGRAL AND THE DIFFERENTIAL METHOD

In this section, we compare two evaluation methods: fitting the integrated rate law directly to the measured progress curves ("integral" method), and fitting the rate law to the rates obtained from the tangents of the progress curves ("differential" method). The differential method has the advantage over the integal method that it is easier to understand plots of rates *vs*. concentrations than plots of concentrations or volumes *vs*. time. It also has the advantage that the optimization process is much faster because integrations do not have to be performed. Furthermore, we have found in all the investigated cases that the convergent set of the global minimum of the sum of squares is larger when using the differential than when using the integral method. The convergent set of a minimum (25) is defined by the set of all the initial parameter estimates which lead to that minimum when the optimization is performed. The notion of the convergent set is important in nonlinear optimization problems because more than one minimum is possible.

When using the differential method, we obtained the rates from the derivatives of a spline function approximation of the progress curves using the subroutine TS01A from the Harwell Subroutine Library (23); the fits of these rates were performed with the same subroutine and its arguments as the fits of the progress curves (see above in Section 2.3).

In Table 3 we show the computed relative areas of the convergent sets obtained with the differential and the integral method. In addition to the 6 cases described in Table 1, we considered a

TABLE 3:   Constraints for the Initial Parameter Estimates and Areas
of the Convergent Sets Relative to the Areas Defined by
these Constraints

Case No.	Origin of data	Constraints for the initial parameter estimates	Relative areas of the convergent sets	
			Integral method	Differential method
1	Simulations	$0<n_H<4.0$; $10^{-3}<\tilde{K}_{0.5}<10^3$	79%	99%
2	Simulations	$0<n_H<2.5$; $10^{-3}<\tilde{K}_{0.5}<10^3$	71%	92%
3	Simulations	$0<n_H<2.5$; $10^{-3}<\tilde{K}_{0.5}<10^3$	45%	84%
4	Simulations	$0<n_H<2.5$; $10^{-3}<\tilde{K}_{0.5}<10^3$	58%	89%
5	Simulations	$0<n_H<2.5$; $10^{-3}<\tilde{K}_{0.5}<10^3$	70%	98%
6	Simulations	$10^{-4}<\tilde{K}_A<10^4$; $2\times10^{-4}<\tilde{K}_A'<10^3$	47%	82%
7	Experiments*	$10^{-4}<\tilde{K}_{ADP}<10^3$; $10^{-4}<\tilde{K}_{PEP}^R<10^3$	60%	91%

The areas were computed after taking the logarithm of all parameters, except n_H. Observation covariances were not considered.
The first 6 cases correspond to those of Table 1.

*Pyruvate kinase from *Escherichia coli.*

seventh case in Table 3, dealing with experimental data. In this seventh case, we fitted 71 points from 14 progress curves obtained by the pH-stat method with pyruvate kinase from *Escherichia coli* (26). In this case, the convergent sets of the absolute minimum were determined by keeping constant all model parameters during the optimizations, with the exception of $\tilde{K}_{ADP}$ and $\tilde{K}^R_{PEP}$, which are the

normalized dissociation constants of ADP with the enzyme and of PEP with the conformational state that forms the active complex. These dissociation constants were normalized with respect to 1 mM.

Table 3 clearly demonstrates that when using the differential method, the initial estimates can be farther away from the minimum than when using the integral method.

Since the rates obtained from the tangents of the progress curves are affected by systematic errors due to the errors in the initial solute concentrations, a non-diagonal weighting matrix should be used in the optimization. All the general formalism of Section 2 can be applied to the differential method by setting the values y_i (i=1,...,n) equal to the rates obtained from the progress curves and the η_i equal to the theoretical values obtained from the rate law.

Table 4 shows, analogously to Table 2, the biases obtained with the differential method without and with considering the co-variances of observations.

The biases given in Table 4 are, in general, higher than the corresponding biases from the integral method in Table 2 because of the error arising from the numerical determination of the rates from the progress curves.

TABLE 4:   Biases After Parameter Optimization With Rates
           Obtained From Progress Curves

| | Bias | |
| Case No. | without | with |
	consideration of data covariances	
1	0.37	0.18
2	0.25	0.042
3	0.37	0.060
4	0.17	0.036
5	0.15	0.026
6	1.0	0.061

The values given are the Euclidean distances between true and opti-mal parameter values, after taking the logarithms of all parameters, except n_H. The cases correspond to those of Table 1.

4. DESIGN OF EXPERIMENTS

4.1. Design Strategy

Once a set of measurements has been performed, one can save time and material by performing a design indicating optimal initial conditions for a new progress curve measurement and optimal points to be evaluated on this progress curve. Such a design can be aimed either to discriminate between rival models (discrimination design) or to reduce the errors and correlation coefficients of the parameter estimates (estimation design). In general, these two types of design will alternate with each other and with experiments (27, 28).

Once a model has been chosen after a discrimination design, an estimation design is performed in order to increase the parameter accuracy for the chosen model; but it may happen that experiments designed to obtain more accurate parameters cannot be fitted within experimental error by the chosen model. In that case, a new model has to be derived, and if there is more than one possibility, a discrimination design has to be performed. It may also happen that experiments aimed to discriminate between rival models lead to a satisfactory decrease of the errors and the correlation coefficients of the parameters of the chosen model, so that a specific estimation design is not necessary.

A design of initial conditions for a progress curve measurement consists, in general, of the following steps: (i) A criterion is established by defining some function (design function) that should attain an extremum at the conditions of the new measurement; (ii) Simulated progress curves (test functions), with initial conditions varying in predefined set, are used to evaluate the design function until the extremum of the design function is found; (iii) The new measurement is performed at the initial conditions where the design function has its extremum.

After the new progress curve measurement has been performed, data points have to be selected on this progress curve for evaluation. This can be done sequentially by looking for the time after the reaction start where the design function reaches its extremum. In some cases, a selection of data points has to be performed before measuring the progress curves, $e.g.$, to determine the time at which samples should be taken, or to fix the titration steps in a titration equipment.

4.2. Estimation Design

Different methods for optimal estimation design have been proposed (29,30). We have used the method of D-optimality for the design of progress curve experiments (2). This method consists in the maximization of the determinant D_Γ of Fisher's information matrix Γ. For progress curves, we have

$$\Gamma = \sum_{\lambda=1}^{\Lambda} (\underline{X}^{(\lambda)})' \underline{W}^{(\lambda)} \underline{X}^{(\lambda)} \qquad . \qquad [34]$$

The weighting matrix $\underline{W}^{(\lambda)}$ and the Jacobian matrix $\underline{X}^{(\lambda)}$ have been defined above.

The determinant D_Γ is inversely proportional to the volume of the hyperellipsoid of the parameter errors (29) and its maximization implies a kind of average reduction of parameter errors and correlation coefficients; D-optimality is therefore called a "volume criterion". If one is interested in increasing the accuracy of only k parameters (k<r), then one should maximize the determinant of the information matrix corresponding to these parameters.

Another method of estimation design, aimed to increase the accuracy of each parameter separately, is sensitivity analysis, whose applicability to biochemical kinetics has been studied by Kanyár (31). This method evaluates the conditions where the absolute values of the derivatives of the theoretical values with respect to the parameters reach their maxima.

If one is interested in obtaining smaller correlations among parameters, one should keep in mind that these can be irreducible due to the model form (32). When high correlations are caused by a poor experimental design, they may be reduced by performing measurements at the minimum of the condition number (defined as the ratio of the square roots of the highest and lowest eigenvalues of the information matrix (33)) or the root mean square of the pairwise correlations (32). These last two criteria are called "shape criteria" because the design influences the lengths of the axes of the error ellipsoid, relative to each other.

4.3. Discrimination Design

For the discrimination design, a simple method has been successfully applied by Mannervik (34) for initial rate experiments. This method maximizes a discrimination function given by

$$g = |\eta_1 - \eta_2| \qquad\qquad [35]$$

where η_1 and η_2 are the theoretical values obtained with the rival models 1 and 2, respectively. For the design of progress curves, we should use a generalized discrimination function by considering several points of the predicted progress curve. This generalization could be done by defining, for example, a discrimination function

$$g_{ABS} = \sum_{i=1}^{n(\hat{\lambda})} |n_{1,i}^{(\hat{\lambda})} - n_{2,i}^{(\hat{\lambda})}| \qquad\qquad [36]$$

or

$$g_{SQ} = \sum_{i=1}^{n(\hat{\lambda})} (n_{1,i}^{(\hat{\lambda})} - n_{2,i}^{(\hat{\lambda})})^2 \quad . \qquad\qquad [37]$$

$\hat{\lambda}$ is the index of the progress curve that is used as test function. g_{ABS} and g_{SQ} have the same maximum if $n(\hat{\lambda})=1$, as in initial rate analysis.

Another form of discrimination design maximizes the discrimination function

$$g_{EXP} = W_1 E_1 (\phi_{min,2}^{+} - \phi_{min,1}^{+}) + W_2 E_2 (\phi_{min,1}^{+} - \phi_{min,2}^{+}) \qquad [38]$$

(29). $\phi_{min,1}^{+}$ and $\phi_{min,2}^{+}$ are the minima of the sum of squares of residuals obtained after fitting with model 1 and model 2, respectively, including the measurements and the test function in Eq. [3]. E_1 and E_2 are the mathematical expectations assuming model 1 and model 2 to be true, respectively; these expectations are calculated by taking into account the uncertainty associated with the prediction of the function $\phi_{min,i}^{+}$, (i=1,2) due to errors and covariances of the measurements and of the model parameters. Fedorov (35) derives analytical expressions for the expectations when $n(\hat{\lambda})=1$. We generalized Fedorov's expressions for the design of an arbitrary number of points on the progress curves, *i.e.*, for $n(\hat{\lambda}) \geq 1$ (unpublished results).

W_1 and W_2 are weighting factors. If the loss occurring when the true model is rejected is equal to the loss occurring when the false model is accepted, then we can set

$$W_1 = e^{-\phi_{min,1}^{0}/2} \quad , \qquad W_2 = e^{-\phi_{min,2}^{0}/2} \quad , \qquad [39]$$

(29). $\phi_{min,1}^{0}$ and $\phi_{min,2}^{0}$ are defined analogously to $\phi_{min,1}^{+}$ and $\phi_{min,2}^{+}$, but without including the test function in Eq. [3].

In order to test the discrimination functions g_{ABS}, g_{SQ} and g_{EXP} (*cf.* Eq.'s [36], [37] and [38]) we used simulated progress curves and the following rival models:

Model 1   (ordered equilibrium) with the rate law

$$v = \frac{\tilde{S}_1\tilde{S}_2}{\tilde{\theta}'_0\tilde{S}_1\tilde{S}_2 + \tilde{\theta}'_2\tilde{S}_1 + \tilde{\theta}'_{12}} \quad . \tag{40}$$

Model 2   (ping-pong) with the rate law

$$v = \frac{\tilde{S}_1\tilde{S}_2}{\tilde{\theta}''_0\tilde{S}_1\tilde{S}_2 + \tilde{\theta}''_2\tilde{S}_1 + \tilde{\theta}''_1\tilde{S}_2} \quad . \tag{41}$$

All quantities with the sign ~ are considered to be normalized in such a way that they are dimensionless. S_1 and S_2 are substrate concentrations. $\tilde{\theta}'_0$, $\tilde{\theta}'_2$, $\tilde{\theta}'_{12}$, $\tilde{\theta}''_0$, $\tilde{\theta}''_2$ and $\tilde{\theta}''_1$ are unknown parameters.

We assumed that the measured quantity was the concentration of a product as a function of time. 4 points, equidistant in time, were taken on each progress curve for evaluating the test functions and for optimizing the parameters.

Simulations were performed assuming model 2 to be true, setting $\tilde{\theta}''_0 = 1$, $\tilde{\theta}''_2 = 5$ and $\tilde{\theta}''_1 = 5$. Random errors with constant standard deviation σ were added on each simulated point. The present investigation was made with $\sigma = 0.2$, 0.5, 1.0 and 1.5. Systematic errors resulting from errors in the initial solute concentrations were not considered. Owing to this error structure, we could optimize by minimizing ϕ as given by Eq. [1] with $\omega_i = 1$.

In order to mimic the actual laboratory work, where replicates are performed to obtain estimates of the standard deviations of the errors, we simulated three replicate progress curve experiments for each set of initial substrate concentrations.

For comparing the results of the fits with the estimated experimental error using the F-test, we determined

$$F = (\phi^+_{min} - zs^2)/[(f-z)s^2] \tag{42}$$

(36,37). s^2 is an estimate of the variance of the random errors of the simulated measurements. s^2 was determined using the replicates (z degrees of freedom) for each σ. ϕ^+_{min} is the minimum of the sum of squares of the fits of the progress curves (f degrees of freedom). When F, as given by Eq. [42], is below the corresponding value of a tabulated F-value, then the fit is considered as being within experimental error.

We assumed that two preliminary progress curves (three replicates of each) had been measured before the discrimination design, one curve with initial concentrations $\tilde{S}_{1,0}=50$ and $\tilde{S}_{2,0}=10$ and the other curve with $\tilde{S}_{1,0}=50$ and $\tilde{S}_{2,0}=30$. The two curves corresponding to these initial conditions are depicted by arrow lines in the upper left corner of the substrate concentration plane in Figure 1. These curves could be fitted within experimental error with both models; this means that the preliminary curves were simulated in a parameter region where no discrimination between the rival models is possible.

In our discrimination design, we looked for a simulated third progress curve (test function) - having its starting point on the plane shown in Figure 1 - such that the discrimination function

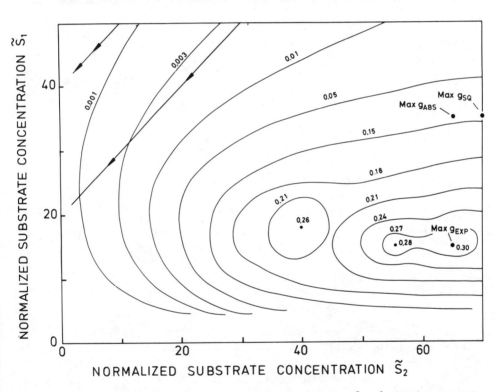

FIGURE 1: Discrimination design in the plane of substrate concentrations. The curves connect initial concentrations leading to equal values of the discrimination function g_{EXP}. The maxima of the discrimination functions g_{EXP}, g_{ABS} and g_{SQ} are given. The two arrow lines indiate the course of the preliminary experiments.

reached its maximum value. The curved lines in Figure 1 were obtained by connecting the starting points of the test functions leading to the same value of g_{EXP}. The maximum of g_{EXP} is shown at the lower right of Figure 1 (g_{EXP}=0.30). The maxima of the discrimination functions g_{ABS} and g_{SQ} are also shown in the figure.

The two preliminary curves plus the designed curve (three replicates of each curve) were fitted using model 1 and using model 2. These fits were performed for the different σ and the different discrimination functions. The F-value, as given by Eq. [42], was determined after each fit. When using model 2, we found that the fit was always within experimental error, which was to be expected because model 2 was assumed to be true in the simulations. When using model 1, we obtained the F-values shown in Figure 2, where we plotted F^{-1} as a function of σ for the different

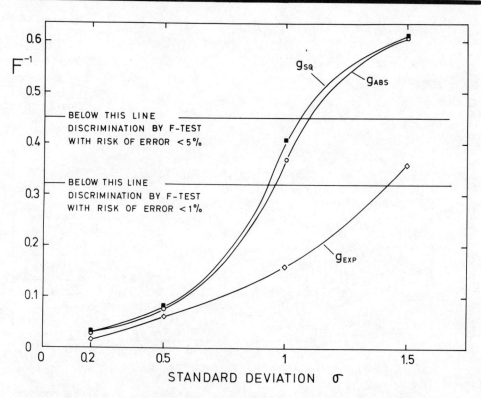

FIGURE 2: Reciprocal of F-values obtained from Eq. [42] with the wrong model after designing with three different discrimination functions (*cf*. Figure 1) in dependence of the standard deviation of the errors of the simulated measurements.

discrimination functions. The two horizontal lines in Figure 2
indicate the values obtained from an F-test table at two different
confidence levels. Figure 2 shows that in some cases (for example,
$\sigma=1.5$, risk of error: 5%) the fit of the wrong model may be within
experimental error if the experiment is planned using g_{ABS} or g_{SQ},
while a discrimination is possible when using g_{EXP}. In other words,
the more complicated discrimination function g_{EXP} may allow discrimination, where the other functions fail to do so.

5.   CONCLUSIONS

The analysis of properly designed progress curves makes this
type of experiments a valuable tool in enzyme kinetics. Three
shortcomings of this technique are eliminated in this work:

1.   Systematic deviations in the progress curve data due to random
errors in the initial concentrations are taken into account in a
linear approximation by a weighting matrix in the parameter optimization. This matrix is calculated from the covariances of the
data. From this matrix a transformation matrix is derived which
transforms the data with systematic errors and uncorrelated errors
into a new data set with uncorrelated errors. The methods applicable to uncorrelated data, e.g., tests of goodness of fit, can
therefore be adopted for the transformed progress curve data. It
is demonstrated by examples that an analysis without the weighting
matrix results in a substantial bias of the estimated parameters.

2.   Model development by progress curve analysis can be very cumbersome and time consuming, since visual inspection of a progress
curve does not reflect clearly the properties of the kinetics.
Also, the integration of the rate law requires long computing times
when the model parameters are optimized. From our experience in
progress curve analysis, we propose the following evaluation procedure.

First, the rates obtained from the slopes of the progress
curves are fitted. This optimization is faster (no integrations
of the rate law are necessary) and the convergent set is bigger
(i.e., the initial estimates can be farther away from the minimum).
Moreover, visual diagnostics of the rates is easier.

Finally, a refinement of the parameters is performed by fitting
the original progress curve data.

3.   The difficulties arising in the interpretation of progress
curve data by visual inspection affect also the experimental design,
so that elaborate design methods are required.

The different types of design given in the literature can be modified for applying them to progress curves. This applies to the estimation design (for estimating parameters) and to the discrimination design (for discriminating among rival models). The properly modified design methods allow a selection of optimal initial substrate concentrations as well as optimal points on the progress curves for measurement and for evaluation.

The best results in the discrimination design are obtained by comparing the statistical expectations of the minimum sum of squares evaluated from the rival models; these sums of squares are computed by considering the preliminary measurements plus a progress curve that serves as test function; the expectation values are computed by considering the parameter and data errors affecting the test function.

REFERENCES

1. Michaelis, L. and Menten, M.L. (1913) Biochem. Z. 49, 333-369.
2. Markus, M. and Plesser, Th. (1976) Biochem. Soc. Trans. 4, 361-364.
3. Halwachs, W. (1978) Biotechnol. Bioeng. 20, 281-285.
4. Yun, S.L. and Suelter, C.H. (1977) Biochim. Biophys. Acta, 480, 1-13.
5. Orsi, B.A. and Tipton, K.F. (1979) In "Methods in Enzymology" (S.P. Colowick and N.O. Kaplan, eds.) Vol. 63, pp. 159-183, Academic Press, New York.
6. Quiroga, O.D., Gottifredi, J.C., Mercado Fuentes, L. and de Castillo, M.E.C. (1977) Lat. am. j. chem. eng. appl. chem. 7, 89-101.
7. Garfinkel, L., Kohn, M.C. and Garfinkel, D. (1977) CRC Crit. Rev. Bioeng. 2, 329-361.
8. Boiteux, A., Markus, M., Plesser, Th. and Hess, B. (1980) In "Kinetic Data Analysis: Design and Analysis of Enzyme and Pharmacokinetic Experiments" (L. Endrenyi, ed.) pp. 341-352 , Plenum, New York.
9. Garfinkel, D., Marbach, C.B. and Shapiro, N.Z. (1977) Ann. Rev. Biophys. Bioeng. 6, 525-542.
10. Balcolm, J.K. and Fitch, W.M. (1970) J. Biol. Chem. 245, 1637-1647.
11. Cornish-Bowden, A.J. (1972) Biochem. J. 130, 637-639.
12. Elmore, D.T., Kingston, A.E. and Shields, D.B. (1963) J. Chem. Soc. London, 2070-2078.
13. Sørenson, T.S. and Schack, P. (1972) In "Analysis and Simulation of Biochemical Systems" (H.C. Hemker and B. Hess, eds.) pp. 169-195, North-Holland, Amsterdam.
14. Newman, P.F., Atkins, G.L. and Nimmo, I.A. (1974) Biochem. J. 143, 779-781.

15. Rao, C.R. (1973) "Linear Statistical Inference and its Applications", Wiley, New York.
16. Patengill, M.D. and Sands, D.E. (1979) J. Chem. Educ. $\underline{56}$, 244-247.
17. Bartfai, T. and Mannervik, B. (1972) FEBS Lett. $\underline{26}$, 252-256.
18. Reich, J.G., Winkler, J. and Zinke, I. (1974) Studia biophys. $\underline{43}$, 77-90.
19. Atkins, G.L. (1976) Biochem. Soc. Trans. $\underline{4}$, 357-361.
20. Boiteux, A., Hess, B., Malcovati, M., Markus, M. and Plesser, Th. (1976) X. Congress of Biochemistry, Hamburg, Abstract 07-4-101.
21. Markus, M., Plesser, Th., Boiteux, A. and Hess, B. (1978) 12th FEBS Meeting, Dresden, Abstract 2753.
22. Segel, I.H. (1975) "Enzyme Kinetics", p. 275, Wiley, New York.
23. Hopper, M.J. (1979) "Harwell Subroutine Library. A Catalogue of Subroutines". Computer Science and Systems Division, AERE Harwell, Oxfordshire, England.
24. Gear, G.W. (1971) "Numerical Initial Value Problems in Ordinary Differential Equations", Prentice-Hall, Englewood Cliffs, N.J.
25. Dixon, L.C.W., Gomulka, J. and Szegoe, G.P. (1975) In "Towards Global Optimization" (L.C.W. Dixon and G.P. Szegoe, eds.) pp. 29-54.
26. Markus, M., Plesser, Th., Boiteux, A., Hess, B. and Malcovati, M. (1960) Biochem. J. $\underline{189}$, 421-433.
27. Reich, J G. (1970) FEBS Lett. $\underline{9}$, 245-251.
28. Froment, G.F. (1975) AIChE Journal, $\underline{21}$, 1041-1056.
29. Fedorov, V.V. and Pazman, A. (1968) Fortschr. Physik, $\underline{24}$, 325-355.
30. Endrenyi, L. (1980) In "Kinetic Data Analysis: Design and Analysis of Enzyme and Pharmacokinetic Experiments", (L. Endrenyi, ed.), pp. 137 - 167, Plenum, New York.
31. Kanyár, B. (1978) Acta Biochim. Biophys. Acad. Sci. Hung. $\underline{13}$, 153-160.
32. Pritchard, D.J. and Bacon, D.W. (1978) Chem. Eng. Sci. $\underline{33}$, 1539-1543.
33. Hurst, R., Pincock, A. and Broekhoven, L.H. (1973) Biochim. Biophys. Acta, $\underline{321}$, 1-26.
34. Mannervik, B. (1975) Biosystems, $\underline{7}$, 101-119.
35. Fedorov, V.V. (1972) "Theory of Optimal Experiments", Academic Press, New York.
36. Retzlaff, G., Rust, G. and Waibel, J. (1975) "Statistische Versuchsplanung", pp. 116-118, Verlag Chemie, Weinheim.
37. Draper, N.R. and Smith, H. (1967) "Applied Regression Analysis", Wiley, New York.

APPLICATIONS OF PROGRESS CURVES IN ENZYME KINETICS

Arnold Boiteux, Mario Markus, Theodor Plesser, and
Benno Hess

Max-Planck-Institut fuer Ernaehrungsphysiologie
46 Dortmund, Rheinlanddamm 201
Federal Republic of Germany

ABSTRACT

Measuring progress curves and analyzing the data with elaborate
numerical tools is a powerful technique in enzyme kinetics. The
experience with this technique gained from the investigation of seven
non-cooperative as well as cooperative enzymes is presented.

1. INTRODUCTION

The increasing number of papers reporting analyses of enzyme
kinetics based on progress curve measurements indicates that this
method will become a more and more important supplement of the classical initial rate technique. Advantages and disadvantages of progress curve measurements are discussed by Markus and Plesser (1) and
in (2-5). In this paper we present a summary of our experience in
deriving rate laws or discarding mechanisms by progress curve analysis. Our experience is based on the investigation of the following
enzymes:

Glucose-6-phosphate 1-epimerase (EC 5.1.3.15) (6); glucosephosphate
isomerase (EC 5.3.1.9) (7); bacteriorhodopsin from the purple membrane of *Halobacterium halobium* (8); adenylate kinase from pig
muscle (EC 2.7.4.3); pyruvate kinase type I from *Escherichia coli*
(EC 2.7.1.40) (15); pyruvate kinase from brewer's yeast (EC 2.7.1.40)
(9).

The aim of our work is the quantitative analysis of the regulatory mechanisms in metabolic networks. Experiences, as for example
in (10), with the glycolytic pathway show that model predictions of
regulatory phenomena can only be verified in biochemical experiments

341

if the kinetics of the involved enzymes are accurately formulated
by a rate law. The rate law has to describe the activity in an experimentally meaningful concentration region. For regulatory key
enzymes the number of relevant ligands is in most cases above four.
The difficulty of deriving such a rate law with high accuracy at all
points of the concentration region is substantially increased by
the fact that regulatory key enzymes may adopt several conformational
states. For this reason, the number of initial rate measurements
would be enormous. Because of the expected reduction in experimentation time we performed progress curve measurements. In addition,
progress curves are sensitive to time-dependent effects (*e.g.*,
hysteresis), which may play an important role in the regulation of
metabolic pathways.

2.   METHODS

2.1. Recording Techniques

Measuring a progress curve means that, as a function of time, a
signal is recorded which is related to the progress of the reaction
under investigation. A simple and generally applicable technique
consists in withdrawing samples from the reaction mixture at appropriate time intervals and subsequently determining the substrate
and/or product concentrations (of the reaction) in those samples.
A certain disadvantage is the rather time-consuming procedure and
the chance for man-made errors due to separate sample analysis.
In some cases, however, the sampling technique definitely is the
method of choice and can lead to significant results; for an example
see (7).

More elegant, naturally, is a continuous recording technique,
which makes use of intrinsic signals of the reaction system itself.
If substrates or products show specific absorption, emission or
polarization of light, the progress of the reaction can conveniently
be recorded by an optical signal; for examples see (6,8,11). In
principle, secondary optical signals can be obtained from many biochemical reactions by enzymatic coupling to the UV-absorbing nico-
tinamide adenine dinucleotide system. However, the coupling technique should be handled with care to avoid serious problems. The
auxiliary chemicals must not influence the kinetics of the enzyme
under investigation and the coupled enzyme system must be able to
follow the reaction without time lag over the concentration range
covered by the progress curve.

Suitable signals for continuous recording can also be obtained
from a multiplicity of physico-chemical measurements; for instance,
calorimetric, coulometric, polarographic or potentiometric techniques can be used, depending on the problem to be solved. Potentiometric measurements with ion-sensitive electrodes and especially
the glass electrode are very convenient for the recording of progress

curves of enzymic reactions, because most of those reactions are
accompanied with proton shifts.

To show the broad applicability of the simple glass electrode
for the continuous recording of progress curves, we shall discuss
two types of experiments with this indicator device in greater
detail:

1.   In the pH-stat technique, the pH is kept constant in a well-
stirred thermostated reaction cell by automatic titration of the
reaction mixture according to the progress of the reaction. In this
case, the volume of titrant consumed is the recorded signal. One
example for the time course of such a signal, from our work on pyruvate kinase from E.coli, is given in Figure 1a. The applicability
of the pH electrode is obvious for the pyruvate kinase reaction,
since in this reaction one proton is consumed per turnover. However,
in the adenylate kinase reaction AMP + ATP = 2 ADP, the free proton
concentration changes because the three adenine nucleotides and
their magnesium complexes have different dissociation constants for
protons. Despite the fact that the pH signal for the adenylate
kinase reaction is about one order of magnitude lower than for the
pyruvate kinase reaction, it can be recorded very well (Figure 1b).

2.   The second method using pH electrodes is the direct recording
of the pH signal as a function of time. Figure 1c shows a record
of the pH for the adenylate kinase reaction. This technique is
recommended when the enzyme is not sensitive to pH changes in the
pH range of the experiments. The reproducibility of these experiments is shown in Figure 2. We measured the rate of pH change,
$\Delta pH/\Delta t$, as a function of total magnesium concentration for constant
nucleotide concentrations: AMP=10 mM; ADP=0.05 mM; ATP=10 mM. From
four independent measurements with different enzyme activities, we
found the abscissa intercept ($\Delta pH/\Delta t=0$) at 5.88±0.04 mM total magnesium concentration. This intercept is independent of the kinetics
of the enzyme but very sensitive to the dissociation constants of
the phosphate groups of the adenine nucleotides for magnesium,
potassium and protons. Variation of the nucleotide concentrations
leads to a series of data, from which a reliable set of dissociation
constants can be calculated by nonlinear data fitting. The accuracy
of the dissociation constants of the complexes is crucial for the
analysis of progress curves measured by the pH electrode as an indicator device.

2.2. Instrumentation and Computer Programs

Most of the equipment for the recording of progress curves
using pH electrodes as an indicator device is commercially available
(Figure 3). On-line connection to a powerful computer is not necessary for collecting the data; but it is very useful to calculate and
print out the reaction rate and meaningful solute concentrations
synchronously to the measurement.

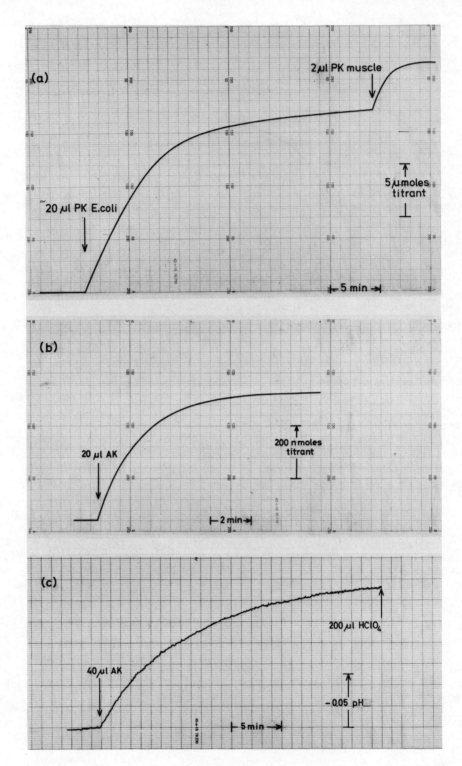

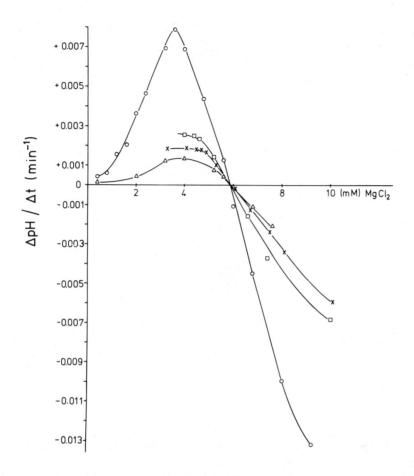

FIGURE 2: Initial rates of proton shift due to the turnover of the adenine nucleotides after addition of adenylate kinase. The nucleotide concentrations are identical for all measurements, AMP=10 mM, ADP=0.05 mM, ATP=10 mM. The symbols o, □, × and Δ represent experiments with four different enzyme preparations. The intersection ΔpH/Δt=0 is at 5.88±0.04 mM total magnesium, independent of the different enzyme activities.

FIGURE 1: Records of progress curves. (a) Pyruvate kinase from E.coli; signal: volume of titrant. (b) Adenylate kinase from pig muscle; signal: volume of titrant. (c) Adenylate kinase from pig muscle; signal: pH.

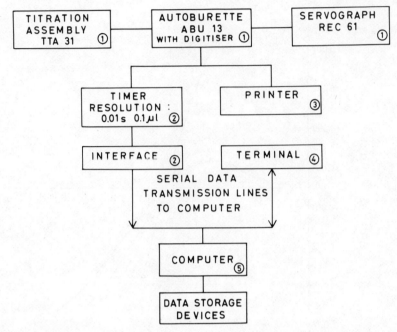

① RADIOMETER, COPENHAGEN   ② IN HOUSE DEVELOPED

③ NEWPORT LABORATORIES   ④ DIABLO 1620

⑤ PERKIN-ELMER INTERDATA 7/32

FIGURE 3:   Equipment for progress curve experiments with a pH
            electrode as indicator device.

Our computer programs for data collection and design and analysis of progress curves are written in FORTRAN. Double precision
(64 bit) programming is required throughout. The basic routines
for matrix inversion, nonlinear optimization, integration of differential equations and other numerical problems were taken from
the Harwell Subroutine Library (12). These routines and some features useful in every day work, were put together in our package
FIDELIO (FItting and DEsign of Laboratory Input Output).

2.3. Experimental Design

Though a single progress curve contains a large amount of information, this information is not so obvious and not so easily
available as in initial rate experiments. In the latter case, the
characteristic parameters such as maximum rate, concentration at

half-maximum rate and an estimate of the sigmoidicity can easily be
obtained from a plot of the data. For progress curve experiments,
one has to design carefully the initial conditions, which guarantee
that the progress curve covers the interesting part of the concentration region. This task can be done comfortably by computer simulation of the experiments using special design criteria (1). In
addition, a number of experimental criteria have to be met. For
example, in titration experiments at constant pH the effectiveness
of stirring, the titrant concentration and the maximum reaction
rate must be well balanced. For optimal resolution, the titration
steps have to be much smaller than the corresponding concentration
for half-maximum rate.

3. SUMMARY OF RESULTS FROM INVESTIGATED ENZYME KINETICS

Glucose-6-phosphate 1-epimerase catalyses the reversible ano-
merization between the α- and β-forms of D-glucopyranose 6-phosphate.
The enzyme belongs to the upper part of the glycolytic pathway in
some organisms. The reaction was followed photometrically by coupling the β-form specific enzyme glucose 6-phosphate dehydrogenase
to the reaction (6). Eleven experiments were performed by varying
the concentrations of glucose and enzyme. Checking three different
rate laws, the mechanism given by the upper triangle in Figure 4

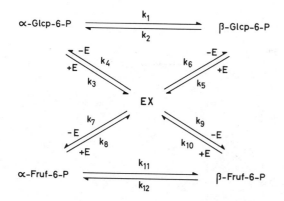

FIGURE 4: Reaction scheme for isomerisation and epimerisation catalysed by glucosephosphate isomerase. The upper triangle α-Glcp-6-P,
EX, β-Glcp-6-P represents the reaction also catalysed by glucose-
6-phosphate 1-epimerase. α-Glcp-6-P, α-D-glucopyranose 6-phosphate;
β-Glcp-6-P, β-D-glucopyranose 6-phosphate; α-Fruf-6-P, α-D-fructo-
furanose 6-phosphate; β-Fruf-6-P, β-D-fructofuranose 6-phosphate.

gave the best fit, and the rate constants k_1 to k_6 could be determined. The more widespread enzyme glucosephosphate isomerase catalyses the same reaction as the epimerase and in addition the isomerization between the 6-phosphates of glucose and fructose. The reaction scheme in Figure 4 was evaluated from 8 progress curves, 9 initial rate experiments and 14 other experiments determining steady-state concentrations when one or two of the reactants were removed by specific enzymic sink reactions. The progress curves were measured by taking samples at regular time intervals. Examples of these progress curves are given in Figure 5 together with the fitted curves. The rate constants k_3 to k_{10} could be determined by nonlinear optimization. The other rate constants were known from independent experiments (7).

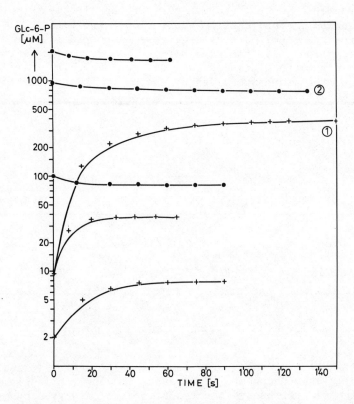

FIGURE 5: Set of fitted progress curves measured by the sample technique for the glucosephosphate isomerase. The time scale has to be multiplied by a factor of ten for curves 1 and 2. Glucose 6-phosphate in high concentration at time zero (•); fructose 6-phosphate in high concentration at time zero (+).

The kinetics of the membrane bound proton pumping enzyme bacteriorhodopsin from *Halobacterium halobium* was recorded photometrically. Progress curve analysis revealed a previously unnoticed cooperative interaction among bacteriorhodopsin molecules in the membrane (8).

Design and analysis of progress curves for the three enzymes already mentioned in this section were not undertaken with the elaborate mathematical tools given in (1). However, we applied these tools when designing and analyzing progress curve experiments for pyruvate kinases from *E.coli* and from brewer's yeast (9,15).

For both enzymes, progress curve experiments were carefully designed by the D-optimality criterion (1,13,14) and measured by the pH-stat technique. Design and measurement alternated in an iterative procedure. Together with a set of initial rate experiments, these data allowed to derive rate laws for both enzymes. The underlying general model is based on the following assumptions: (i) The enzymes consist of four protomers, which may exist in four conformational states. (ii) The transitions from one state to another are concerted. (iii) All the enzyme species are in quasi-equilibrium. Checking this general model with the experimental data, it turned out that the minimum model for pyruvate kinase from *E.coli* indicated three distinct conformational states, whereas pyruvate kinase from brewer's yeast indicated two. For both enzymes only one state is able to form the catalytically active enzyme substrate complex.

An additional interesting regulatory property of both enzymes is shown in Figure 6. This diagram is related to the notion of the Michaelis constant, which is defined as the substrate concentration $S=K_m$ for which the reaction rate v is equal to half the maximum rate V:

$$v(S) = 0.5 \, V \qquad . \qquad\qquad\qquad [1]$$

If the activity of an enzyme depends on the concentrations of n ligands C_1 to C_n

$$v = v(C_1, C_2, \ldots, C_n) \qquad , \qquad\qquad\qquad [2]$$

then one can define a half-maximum concentration $K_{0.5}$ for one of the ligands analogously to equation [1] when the concentrations of the other ligands are kept constant. Michaelis constants for multi-substrate enzymes are usually measured and reported keeping all substrates except one close to saturation. Equation [1] can be generalized as well by allowing all concentrations to vary freely with the constraint

$$v(C_1, C_2, \ldots, C_n) = 0.5 \, V \qquad . \qquad\qquad\qquad [3]$$

PK FROM S. CARLSBERGENSIS   PK FROM E. COLI

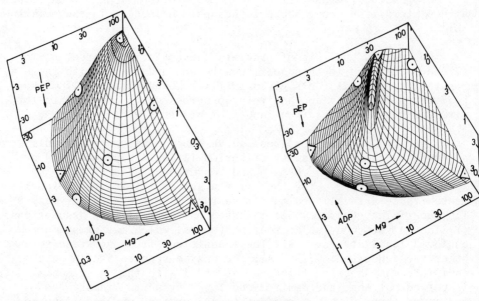

FIGURE 6: Surfaces of half-maximum rate. The coordinates represent total millimolar concentrations of the ligands PEP, ADP and Mg in logarithmic scale. For explanation see text.

___

This equation represents a hypersurface with constant half-maximum reaction rate in the n-dimensional space of the concentrations.

Figure 6 shows such surfaces of half-maximum rate with n=3 for the pyruvate kinases from *E.coli* and from brewer's yeast. The coordinate axes are the total concentrations of the ligands phosphoenolpyruvate (PEP), adenosine diphosphate (ADP) and magnesium (Mg) in logarithmic scale. The rate v(PEP,ADP,Mg) is 0.5V on the surfaces and increases along vectors, gradients of v, pointing into the figure perpendicularly to the surface. If a component of a gradient is parallel to a coordinate axis, the corresponding ligand is an activator, *i.e.*, addition of this ligand to the reaction mixture increases the reaction rate of the enzyme. If a component of a gradient is antiparallel to an axis, the corresponding ligand is an inhibitor, *i.e.*, addition of this ligand to the reaction mixture decreases the reaction rate of the enzyme. The shape of the surfaces in Figure 6 shows for both enzymes that each of the three ligands PEP, ADP and Mg may be activator in one part of the surface and inhibitor in another part.

For a detailed discussion of the regulatory properties of the enzymes in the three-dimensional space of their ligands PEP, ADP and Mg, we consider the rate law v(PEP,ADP,Mg) as a 'potential' with the corresponding gradient field and field lines. In this sense, the surface of half-maximum rate is a special surface of constant 'potential' v=0.5V.   The field lines cross the surfaces of constant rate perpendicularly, and the gradient in any point of a surface indicates the direction of maximum variation in v when going from one surface to a surface in the neighbourhood. The norm of the gradient is a measure for the steepness of the 'potential' v in the direction of the gradient. This norm has local maxima and minima on a surface of constant rate. These extrema define field lines, which we call lines of strongest regulation and lines of weakest regulation, respectively.

We found that the lines of strongest regulation appear where two of the ligands are at non-regulatory high concentration levels. These are the conditions at which the kinetics in dependence of one ligand is usually measured. The intersections of the lines of strongest regulation with the half-maximum rate surface then correspond to the K_m values of PEP, ADP and Mg. These intersections are indicated by triangles in Figure 6.

The lines of weakest regulation appear where more than one ligand is at a regulatory concentration level. Their intersections with the surface of half-maximum rate are given by circles   in Figure 6.

4. CONCLUSIONS

We may state from our experience that the measurement of prog- ress curves is a powerful supplement of classical initial rate experiments in enzyme kinetics. On the one hand, this technique saves much experimentation time, but on the other hand it requires more elaborate control of the experimental parameters than classi- cal kinetic experiments. The time saved in the laboratory must be paid in computer time for the design and analysis of the experiments. The progress curve technique is very powerful — especially when large amounts of data are inevitable — to solve the following problems in enzyme kinetics:

1.   Standard tests for regulatory properties of enzymes. Properly analyzed progress curves can quickly reveal not only changes in maximum rate, as the usual tests, but also changes in the regulatory properties of an enzyme.

2.   Studies of the influence of *milieu* factors like pH, ionic strength or temperature, if the mechanism is known.

3.   Refinement of parameters after the rate law is elaborated.

4.   Kinetic comparison of enzymes from different organisms.

5.   Detection of time-dependent effects in enzyme reactions.

6.   Investigation of enzymes from metabolic pathways with the aim
to simulate *in vivo* processes by *in vitro* synthesis of the pathways.

REFERENCES

1.   Markus, M. and Plesser, Th. (1980) In "Kinetic Data Analysis:
       Design and Analysis of Enzyme and Pharmacokinetic Experi-
       ments" (L. Endrenyi, ed.) pp. 317-339 , Plenum, New York.
2.   Orsi, B.A. and Tipton, K.F. (1979) In "Methods in Enzymology"
       (S.P. Colowick and N.O. Kaplan, eds.) Vol. 63, pp. 159-183,
       Academic Press, New York.
3.   Garfinkel, L., Kohn, M.C. and Garfinkel, D. (1977) CRC Crit.
       Rev. Bioeng. 2, 329-361.
4.   Markus, M., Hess, B., Ottaway, J.H. and Cornish-Bowden, A.J.
       (1976) FEBS Lett. 63, 225-230.
5.   Cornish-Bowden, A.J. (1976) "Principles of Enzyme Kinetics",
       pp. 142-152, Butterworths, London.
6.   Chance, E.M., Hess, B., Plesser, Th. and Wurster, B. (1975)
       Eur. J. Biochem. 50, 419-424.
7.   Plesser, Th., Wurster, B. and Hess, B. (1979) Eur. J. Biochem.
       98, 93-98.
8.   Korenstein, R., Hess, B. and Markus, M. (1979) FEBS Lett. 102,
       155-161.
9.   Boiteux, A., Doster, W., Hess, B., Markus, M., Plesser, Th.
       and Wieker, H.-J. (1979) In "Protein: Structure, Function
       and Industrial Applications" (E. Hoffmann, W. Pfeil, and
       H. Aurich, eds.) Vol. 52, FEBS 12th Meeting, Dresden 1978
       Pergamon, Oxford.
10. Plesser, Th. (1977) In "VII. Internationale Konferenz ueber
       nichtlineare Schwingungen" (G. Schmidt, ed.) Vol. II,2,
       pp. 273-280, Akademie-Verlag, Berlin.
11. Duggleby, R.G. and Morrison, J.F. (1978) Biochim. Biophys.
       Acta, 526, 398-409.
12. Hopper, M.J. (1979) "Harwell Subroutine Library. A Catalogue
       of Subroutines", Computer Science and Systems Division AERE,
       Harwell, Oxfordshire, England.
13. Markus, M. and Plesser, Th. (1976) Biochem. Soc. Trans. 4,
       361-364.
14. Fedorov, V.V. (1972) "Theory of Optimal Experiments", Academic
       Press, New York.
15. Markus, M., Plesser, Th., Boiteux, A., Hess, B. and Malcovati,
       M. (1980) Biochem. J. 189, 421-433.

DATA ANALYSIS FROM MICHAELIS-MENTEN KINETICS:

INS AND OUTS

T. Keleti

Institute of Enzymology
Biological Research Center
Hungarian Academy of Sciences
Budapest, Hungary

ABSTRACT

One can observe Michaelian kinetics in the case when the enzyme does not follow Michaelis-Menten mechanism and one can determine deviations from Michaelis-Menten kinetics with an enzyme that follows Michaelian mechanism. The discrepancies between the measured kinetics and real mechanism may be due to the existence of iso-enzymes, the non-identical behaviour of subunits in oligomeric enzymes, the association-dissociation of oligomers, the instability of the enzyme under assay conditions, the effect of ligands, interaction with proteins, other macromolecules or membrances, etc.

The kinetic parameters of a simple enzymatic reaction are usually determined by the substrate saturation curve, by measuring the initial velocity as a function of initial substrate concentration. A new diagnostic method is presented which determines, from a single progress curve, the velocity values and the corresponding substrate concentrations, using an appropriate algorithm. From these velocity-substrate concentration-value pairs, the corresponding kinetic parameters can be calculated.

If the kinetic parameters obtained from substrate saturation curves equal those determined from a single progress curve as presented above, the enzymatic reaction can be described with a simple Michaelis-Menten type mechanism. If the opposite is true, activation or inhibition by excess substrate or by product, as well as positive or negative cooperative interaction between subunits or ligands occur.

353

1.   INTRODUCTION

One of the most important discoveries of molecular biology during the past twenty years was the elucidation of the molecular mechanism of metabolic regulation through the modulation of enzymatic activity. The regulation of enzymatic activity may be classified as follows:
(i) Direct modulation, *i.e.*, inhibition, liberator effect or activation by (A) excess of substrate, (B) product, (C) other metabolites, (D) interaction with proteins, (E) with other macromolecules, (F) with membranes;
(ii) Indirect modulation, *i.e.*, inhibition or activation by kinetically governed deviation from Michaelis-type saturation caused by (A) enzyme isomerization, (B) motility of protein structure, (C) instability of enzyme molecule, (D) association-dissociation of oligomeric enzymes.

The deviation from Michaelis-type saturation often results in a sigmoidal saturation curve. An S-shaped dependence of the reaction velocity on the substrate concentration is a powerful tool for the regulation of enzymic activity. At low substrate concentrations, the enzymic activity is much lower than in the case of the regular Michaelis-Menten type hyperbolic saturation curve, *i.e.*, the enzyme "spares" the deficient metabolite. However, at higher substrate concentrations the rate of enzyme action suddenly increases. In the case of an enzyme exhibiting a sigmoidal saturation curve, over a particular range in the velocity isotherm, the reaction rate is much more sensitive to an increase in substrate concentration than for an enzyme of the same $[S]_{0.5}$ value (where $v=V_{max}/2$) and V_{max} value but exhibiting a hyperbolic isotherm. It should be noted, however, that an enzyme capable of regulation can exhibit a hyperbolic saturation curve, and that a sigmoidal saturation curve is not necessarily related to any model of regulation or even to the capability of regulation.

In the following, cases will be reviewed when one can observe Michaelian kinetics with an enzyme not following the Michaelis-Menten mechanism, and when one can determine deviations from Michaelis-Menten kinetics with an enzyme following Michaelian mechanism.

2.   SOME USEFUL DEFINITIONS

Since confusion may arise from confounding mechanism, equation and kinetics, it seems reasonable to define these terms before going into their deeper discussion.

2.1. Michaelis-Menten Mechanism

It is the simplest enzyme mechanism of a one-substate-one-product enzyme, in which only one kinetically relevant intermediary complex is involved and which is (in the time range of activity measurements) practically irreversible (1):

$$E+S \underset{k_{-1}}{\overset{k_1}{\rightleftharpoons}} ES \xrightarrow{\ k_2\ } E+P \qquad\qquad [1]$$

where k_{-1} and k_2 are first-order rate constants and k_1 is the second-order rate constant.

2.2. Michaelis-Menten Equation

This equation describes the initial rate of a reaction that follows mechanism [1], with the restriction that $k_2 \ll k_{-1}$ (rapid equilibrium assumption):

$$v_0 = V_{max}[S]/(K_S+[S]) \qquad\qquad [2]$$

where v_0 is the initial velocity, $V_{max}=k_2[E]_T$, $[E]_T$ the total enzyme concentration, $[S]$ the substrate concentration in the reaction mixture assuming $[S] \gg [E]_T$, $i.e.$ $[S]_t \approx [S_0]$; $[S]_t$ is $[S]$ at time t, $[S_0]$ is the initial substrate concentration, and $K_S=k_{-1}/k_1=[E][S]/[ES]$, $i.e.$, the thermodynamic dissociation constant of ES.

2.3. Michaelis-Menten Kinetics

An enzyme is said to follow Michaelis-Menten kinetics if the substrate saturation curve (v_0 $vs.$ $[S]$) is a rectangular hyperbola and its transformed plots (2-8) are linear in the whole substrate concentration range practically available. Theoretically, an enzyme that follows Michaelis-Menten mechanism has Michaelis-Menten kinetics which is described by the Michaelis-Menten equation.

2.4. Briggs-Haldane and van Slyke Equations

The Michaelis-Menten equation may formally be valid even if the rapid equilibrium assumption does not hold. If $k_2 \not\ll k_{-1}$ we have the Briggs-Haldane equation, with the steady-state assumption (9), and if assumption $k_2 \gg k_{-1}$ holds, the van Slyke equation (10), which are formally analogous with Eq. [2], but with a different constant in the denominator; instead of K_S we have $K_M=(k_{-1}+k_2)/k_1$ in the former and $K_k=k_2/k_1$ in the latter case.

Similarly, Eq. [2] is formally valid if there is a kinetically significant enzyme-product complex, EP, but the meanings of V_{max} and K_M are different from those given above. Though the requirements of the original Michaelis-Menten mechanism are not fulfilled, the kinetics is Michaelian.

Consequently, one can measure Michaelis-Menten kinetics with enzymes following Briggs-Haldane (steady state) or van Slyke assumptions or some non-Michaelian (*e.g.* EP-containing) mechanisms.

2.5. Artefacts in Determining Kinetic Mechanisms

The exact steady state (*i.e.*, d[ES]/dt=0) only prevails for an infinitesimal time period (11) and therefore is not amenable to current experimental techniques, especially when substrate saturation curves are determined, where substrate concentrations well below saturation are used (12). In routine work, measurements are usually carried out later, in the quasi-steady state phase. By using the quasi-steady state assumption an error is introduced in the estimation of initial velocity (13-21).

If the velocities are calculated from the quasi-steady state, the points can be fitted to a first-degree hyperbola (Michaelis-Menten kinetics) only if [S]>[E] and [E]<K_M or K_S. Deviations from the rectangular hyperbola occur if [S] is varied from values starting from [S]≤[E]. Even if [S]>[E], the rectangular hyperbola is distorted giving rise to a sigmoidal saturation curve in the case of van Slyke equation, *i.e.*, if $k_2 \gg k_{-1}$ and therefore K is a kinetic constant, $K_k = k_2/k_1$ (22).

2.6. The Basic Problem

If there are no artefacts in the evaluation of experiments, one can measure Michaelis-Menten kinetics if the enzyme follows Michaelis-Menten mechanism. However, one can observe Michaelian kinetics in the case when the enzyme does not follow Michaelis-Menten mechanism and *vice versa*, and one can determine even large deviations from Michaelis-Menten kinetics with an enzyme that follows Michaelian mechanism.*

The discrepancies between the measured kinetics and real mechanism may be due to the existence of isoenzymes, non-identical behaviour of subunits in oligomeric enzymes, the association-dissociation of oligomers, the denaturation of the protein, the effect of ligands, interaction with protein or other macromolecules,

---

*As far as the general principles of experimental determination or theoretical description of Michaelis-Menten or other type kinetics and their evaluation are concerned, we refer to a number of useful textbooks and reviews (11,12,23-50).

etc. In the following we will discuss some of the above possibilities, first those where ligand (second substrate, modifiers, macromolecules) bindings are involved, then those where some inherent
property of the enzyme (existence of isoenzymes, pre-existent asymmetry, dissociation of oligomeric enzymes) causes the deviation of
kinetics from that characteristic of the original mechanism.

3.   LIGAND-INDUCED KINETIC DEVIATIONS FROM MICHAELIS-MENTEN MECHANISM

3.1. Enzymes Acting on More Substrates

It is generally accepted that enzymes working with more than
one substrate can be treated as one-substrate enzymes if they are
saturated with the other substrate(s). If the saturation curve is
a rectangular hyperbola for all substrates, at saturating concentrations of the other substrate(s), then it is generally assumed
that the enzyme follows Michaelis-Menten mechanism. However, this
is not true. First of all, the Michaelis-Menten mechanism refers
to a one-substrate enzyme. In a two-substrate reaction, usually
two binary, and one ternary, complexes are involved. Consequently,
there are more Michaelis and dissociation constants in the rate
equation, and the (apparent) maximum velocity is also a complex
function of rate constants (and substrate concentration T*). Furthermore, if the saturation curve is determined in the presence of one
or two other substrates at saturating concentrations, this implies
that the experiments are performed with a modified enzyme, since
generally the binding of ligands (*e.g.* substrates) causes conformational changes in the enzyme protein. It has been envisaged that
the binding of a particular ligand, such as the substrate, promotes
a structural change which gives rise to the three-dimensional
structure essential for enzymatic activity, or stabilizes this
unique structure (51-53). There is ample experimental evidence
(54) proving that such changes in protein conformation really occur
during enzymic reactions.

If the enzyme works in the presence of simulataneously added
substrates, the steric changes caused by one or the other substrate
may be superimposed on the initial part of the progress curve. An
enzyme working with two or three substrates, if it follows a random
mechanism and the steady-state assumption holds, shows nonlinear
double reciprocal plots and this behaviour is inherent in the
mechanism (44,45). If the enzyme follows sequential mechanism and/
or the rapid equilibrium assumption holds, the double reciprocal
plots are linear (31,33,36,43-50,55-57), but the mechanism is not
Michaelian owing to the formation of more than one enzyme-substrate
complex and to the steric changes caused by the substrate.

---

*If apparent maximum velocity is concerned.

In conclusion: a two- or three-substrate enzyme may have
apparent Michaelis-Menten kinetics, but never follows Michaelis-
Menten mechanism.

3.2. One-Substrate Enzymes Acting in the Presence of Modifiers

The binding of a ligand may induce a structural change that
leads to three-dimensional structures which are either more or less
favourable for enzymatic activity than the structure of the non-
liganded protein. This activation or inhibition of the enzyme may
operate also through a simple interaction between the ligand and
the substrate (steric hindrance, constraint) or through the changes
in the quaternary structure (ligand-induced dissociation or association of an oligomeric enzyme, allosteric interaction, induced
misfit) (23,24,31,33,36-38,42,52,53,58,59).

In all these cases we may observe non-Michaelian kinetics even
if originally the enzyme functions according to the Michaelis-
Menten mechanism. This may be also the case if the inhibited (or
activated) enzyme follows steady-state assumption (23-25,60-66).
On the other hand, an enzyme with non-Michaelian mechanism may
simulate Michaelis-Menten kinetics in the presence of ligands
(50,67,68).

If more than one ligand acts simultaneously on the same enzyme
working with one substrate, in the simplest case the following
possibilities arise: (i) one inhibitor/activator and one liberator*;
(ii) two inhibitors/activators interact on the same enzyme. The
resulting effect of such interactions may be multifarious. The
interaction of a substrate, an inhibitor and a liberator on an
enzyme may result either in no effect on the initial velocity or
in inhibition or even activation, depending on the type of inhibition and liberation, on the mutual relation between the dissociation
constants of the three ligands and on their relative concentrations
(69-71).

The simultaneous presence of two inhibitors results only in
special cases in a simple additive effect. In most instances, the
two inhibitors either act antagonistically or synergistically (72).
Moreoever, in special cases, depending on the substrate concentration, the interaction may change from antagonism, through the simple
summation of inhibitory effects, to synergy. The interaction of
certain inhibitors may result in a greater initial velocity in the
presence of both inhibitors than in the presence of only one of
them or even in their absence (70).

---

*The liberator is a substance which may abolish the action of an
 inhibitor or activator (*i.e.*, it liberates the enzyme from inhibition or activation) without having alone any effect on enzymatic
 activity (69).

The simultaneous presence of two inhibitors or of one inhibitor and one liberator may also result in oscillating enzymatic reactions (73).

Thus, in the presence of modifiers, enzymes which follow Michaelis-Menten mechanism may exhibit non-Michaelian kinetics and those which do not follow Michaelis-Menten mechanism may present Michaelian kinetics.

3.3. Enzyme-Enzyme and Enzyme-Macromolecule Interactions

The rate of linked reaction steps catalyzed by multienzyme complexes may be considerably influenced relative to the rate of individual reactions catalyzed by single enzymes. In multienzyme complexes the conformation of the individual enzymes will differ, owing to the interactions, from that of the free enzymes, *i.e.*, their stability and/or catalytic properties change as a result of complex formation. This change can be either positive or negative, in the sense that it may facilitate or hinder the catalytic reactions (74-84). The deviations from the kinetics of individual reactions can be used to test the formation of enzyme complexes (85).

The effect of macromolecular environment on the catalytic and regulatory properties of enzymes must not be neglected either. The interaction with macromolecules may alter the steric structure of the enzyme (86,87). Similarly, the properties of an enzyme may differ depending on whether the enzyme is in a soluble or membrane-bound form (88,89).

Therefore, again, in the cases of enzyme-enzyme, enzyme-macromolecule or enzyme-membrane interactions, an enzyme originally working according to the Michaelis-Menten mechanism may display non-Michaelian kinetics and *vice versa*.

4. INHERENT KINETIC DEVIATIONS FROM MICHAELIS-MENTEN MECHANISM

4.1. Pre-existing Enzyme Conformers (Enzyme Isomerization)

From the thermodynamic viewpoint it is quite feasible for enzyme molecules in solution to exist as an equilibrium system consisting of a number of slightly different conformations (isomers) having approximately similar probabilities, thus enabling the structure of the enzyme to fluctuate between a number of possible conformations (51,90-92). This model assumes a high motility of the protein fabric (54,93-97), which is the result of the summation of a number of local fluctuations as predicted theoretically (98-108) and has indeed been demonstrated experimentally (109-120).

If two isomers of a monomeric enzyme in equilibrium catalyze
the reaction according to the Michaelis-Menten mechanism (assuming
rapid equilibrium) and the isomerization step is also in rapid
equilibrium relative to the rate-limiting step of catalysis, the
system shows simple Michaelis-Menten kinetics (121). However, if
the enzyme follows the steady-state assumption (64) or the isomerization step is rate limiting [(64), hysteretic enzymes (122,
123)], or more than two isoenzymes are functioning [mnemonic
enzymes (124)] in certain conditions non-Michaelian kinetics (sig-
moid saturation curves) may be obtained [*cf*. also (50,60,71,125)].

4.2. Asymmetry of Subunits in Oligomeric Enzymes

The effect of isomerization in enzyme kinetics may be much
more pronounced in the case of oligomeric enzymes. Theoretically,
in a tetrameric enzyme, subunits may be arranged in linear, tetra-
hedral, plane, cyclic, stirrup or triangular forms and if the enzyme
is composed of two different subunits it may have 5 to 12 isoenzymes.
Depending on permitted combinations of the four binding sites 1 to
9 isoenzymes may be found in homotetrahedrons containing asymmetric
subunits. Provided that all bonds between the subunits are possible,
we have 5 isoenzymes if both subunits are symmetric, 28 isoenzymes
if one of the subunits is asymmetric and 117 isoenzymes if both
subunits are asymmetric (126). One can imagine what complex kinetics may ensue if these isoenzymes are catalytically active and
are in equilibrium with each other, independent of the kinetic
mechanism of the individual isoenzymes.

Not only the individual subunits can be asymmetric, but also
symmetric subunits may associate to form asymmetric oligomeric
proteins (127-134). This asymmetry may cause big deviations in
kinetics leading to the half-of-the-sites reactivity (135-137) or
flip-flop mechanism (138,139).

4.3. Enzyme Inactivation

If the enzyme is unstable under assay conditions, the Michaelis-
Menten mechanism is combined with the concomitant inactivation of
the enzyme. Assuming that the substrate stabilizes the enzyme, one
gets scheme [3]:

$$
\begin{array}{ccc}
\text{E+S} \underset{k_{-1}}{\overset{k_1}{\rightleftharpoons}} \text{ES} & \overset{k_2}{\longrightarrow} & \text{E+P} \\
\downarrow k_3 & & \downarrow k_3 \\
\text{inactive E} & & \text{inactive E}
\end{array}
\qquad [3]
$$

A sigmoidicity in the saturation curve can then be observed at any k_2 and k_{-1} values (*i.e.*, in both rapid equilibrium and steady state) if k_3 is sufficiently high. In the case of van Slyke equation, the sigmoidicity is always detectable, independent of the value of k_3 (22).

Of course, the situation becomes much more complex with oligomeric enzymes. An oligomeric enzyme may show multifarious kinetics during irreversible inactivation (instead of the simple one-step process characterized by a single rate constant, k_3, as above) and consequently the perturbation of Michaelis-Menten kinetics may be more involved. The three basic mechanisms of irreversible inactivation of an oligomeric enzyme are the all-in-one, the one-intermediate and the one-by-one mechanisms (140). One obtains quite complex kinetics of irreversible inactivation of a homotetrameric enzyme if during an all-in-one process the irreversible or reversible isomerization or dissociation and/or association of the protein molecule takes place (141). Consequently, the deviations from the Michaelis-Menten kinetics becomes much more pronounced.

4.4. Association-Dissociation of Oligomeric Enzymes

Certain oligomeric enzymes can be dissociated into subpolymers upon dilution. When dissociation occurs, the interactions between subunits are cancelled. It is reasonable to assume that the change in subunit interactions affects the specific activity of the enzyme (142).

The changing behaviour with oligomeric state of an enzyme may also be reflected in the substrate saturation curve. At low protein concentrations the enzyme may show regular hyperbolic substrate saturation, whereas at high protein concentration this curve may become sigmoidal, or *vice versa* (82,143-145). The deviations from Michaelis-Menten kinetics, even if the enzyme follows Michaelis-Menten mechanism but is a dissociable oligomer, has already been predicted theoretically (42,78,146-156). The question "does any enzyme follow the Michaelis-Menten equation?" (177), is indeed a real one.

5. PROGRESS CURVES IN THE ANALYSIS OF ENZYME MECHANISM

5.1. Progress Curves

It has recently been demonstrated theoretically and experimentally that steady-state kinetic parameters can be estimated from one progress curve, just as from a set of initial velocity measurements. There are methods available based on the solution of linear equations, but these give no or only biased estimates of the standard error of parameters (33,157-161).

The principles of the Gauss-Newton method of nonlinear regression are probably familiar to most enzyme kineticists through the work of several laboratories (162-164). In the numerical integration approach, a simulated curve is compared with the experimental data, and the rate constants are adjusted until a good fit is obtained (165,166). Nonlinear regression gave satisfactory results for irreversible one-substrate-one-product enzymes (167-169) and also for similar reversible reactions (170). The kinetic parameters can be evaluated even from a single progress curve (171).

For two-substrate enzymes, the evaluation of kinetic parameters from progress curves is also possible with the aid of either a steady-state rate equation (172) or an integrated form of it (173). Consequently, progress curve analysis may be used to determine the kinetic mechanism and the values of kinetic parameters associated with an enzyme-catalyzed reaction, in the case of one- and two-substrate enzymes (174). A computer program has also been developed which employs numerical methods to solve integrated rate equations for systems that cannot be integrated analytically (175). Progress curve algorithms based on the integrated Michaelis-Menten equation have been published to calculate activities from data obtained with an automated enzyme analyzer (176).

5.2. Diagnostic Test for Kinetic Deviations from the Michaelis-Menten Mechanism

The progress curve may be used to test whether a one-substrate-one-product enzyme follows the Michaelis-Menten mechanism.

Sometimes, from initial velocity measurements one can obtain an apparently rectangular hyperbola in a given substrate concentration range even if product inhibition or inhibition by excess substrate occurs or some ligand-induced or inherent interactions effecting inhibition or activation are operative. In such cases, the comparison of parameters calculated from initial velocity measurements and from progress curves may decide whether the apparently rectangular hyperbolic saturation curve indeed reflects Michaelis-Menten mechanism or the deviation is too small to be detected by initial velocity measurements.

From a single time curve of product formation, by using a digitizer, the velocity values, $([P_2]-[P_1])/(t_2-t_1)$, and the corresponding approximate substrate concentrations, $[S_0]-([P_2]+[P_1])/2$, can be determined at given infinitesimal time intervals, by the aid of an appropriate algorithm. From these velocity-substrate concentration value pairs, the corresponding kinetic parameters can be calculated by conventional methods.

The integrated rate equation of the Michaelis-Menten mechanism is:

$$V_{max}t = [P]+K_S \ln\{[S_0]/([S_0]-[P])\} \qquad . \qquad [4]$$

If we have a mechanism with product inhibition:

$$E+S \rightleftharpoons ES \longrightarrow E+P$$

$$E+P \rightleftharpoons EP \qquad , \qquad [5]$$

the integrated rate equation will be

$$V_{max}t = [P]+K_S \ln\{[S_0]/([S_0]-[P])\}$$
$$+ (K_S/K_P)\left[[S_0] \ln\{[S_0]/([S_0]-[P])\}-[P]\right] \qquad [6]$$

where $K_P=[E][P]/[EP]$.

In the case of inhibition by excess substrate:

$$E+S \rightleftharpoons ES \longrightarrow E+P$$

$$ES+S \rightleftharpoons ES_2 \qquad , \qquad [7]$$

the integrated rate equation is as follows:

$$V_{max}t = K_S \ln\{[S_0]/([S_0]-[P])\}+[P]\{a+b(2[S_0]-[P])/2K_S\} \quad [8]$$

where $a = 1+1/\beta+1/\gamma$, $b = 1/\alpha\beta\gamma$

and $K_S=[E][S]/[E=S]$, $\beta K_S=[E][S]/[-E-S]$, $\alpha\gamma K_S=[-E-S][S]/[S-E-S]$, $\gamma K_S=[E][S]/[S-E-]$, $\alpha\beta K_S=[S-E-][S]/[S-E-S]$, $E=S$ being the enzymatically active complex and $-E-S$, $S-E-$, $S-E-S$ the inactive complexes.

These equations describe the progress curves from t=0 to t=∞. If we measure initial velocities, where $[P]\approx0$, both the Michaelis-Menten mechanism and the mechanism with product inhibition yield the same equation [2]. This equation holds also if $[S_0]$ is not high enough to produce the S-E-S complex in a sufficient quantity to influence the initial velocity. Consequently, K_S calculated from equation [2] will be analogous to that calculated from equation [4] only in the case of Michaelis-Menten mechanism, but will be different from that calculated from equations [6] or [8], *i.e.*, if inhibition by product or excess of substrate occurs.

These statements are illustrated in Figures 1-3. Figures 1A, 2A and 3A present the progress curves for Michaelis-Menten mechanism, the mechanism with product inhibition and with inhibition by excess of substrate, respectively. In all cases, it was assumed that

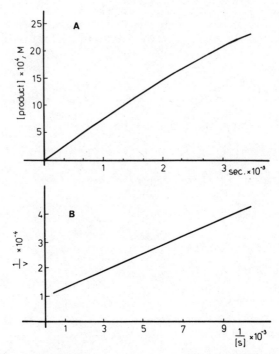

FIGURE 1: (A) Simulated progress curve of Michaelis-Menten mechanism; (B) Lineweaver-Burk plot of v-[S] pairs, $V_{max}=1\times10^{-5}M.sec^{-1}$, $K_M=4.2\times10^{-4}M$.

$[S_0]=3\times10^{-3}M$, $V_{max}=1\times10^{-5}M.sec^{-1}$ and $K_S=4\times10^{-4}M$. K_P was taken to be $1\times10^{-3}M$ in Figure 2 and a=3.5, b=1×10^{-2} in Figure 3. The progress curves were digitalized, and from the v-[S] pairs obtained the V_{max} and K_M values were determined from Lineweaver-Burk plots by the least-squares method. Figures 1B, 2B and 3B show these plots and the V_{max} and K_M values obtained. The K_M values derived from progress curves of product and excess of substrate inhibition differ considerably from those arrived at by initial velocity measurements.

In all cases, when the K_M determined from the progress curve significantly differs from that deduced from initial rate measurements (substrate saturation curve), one can suspect deviations from Michaelis-Menten mechanism (inhibition by product or excess of substrate, subunit interaction, etc.). This method cannot differentiate between rapid equilibrium, steady-state and van Slyke assumptions and cannot reveal the cause of deviations from Michaelis-Menten kinetics, but it detects qualitatively the identity or the difference of kinetic parameters determined from initial velocity and progress curves. The identity means that the stoi-

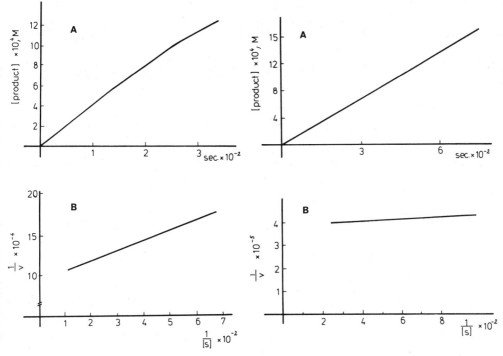

FIGURE 2: (A) Simulated progress curve of a mechanism with product inhibition; (B) Lineweaver-Burk plot of v-[S] pairs, $V_{max}=1.1\times10^{-5}M.sec^{-1}$, $K_M=1.2\times10^{-3}M$.

FIGURE 3: (A) Simulated progress curve of a mechanism with inhibition by excess substrate; (B) Lineweaver-Burk plot of v-[S] pairs, $V_{max}=2.6\times10^{-6}M.sec^{-1}$ $K_M=1.2\times10^{-5}M$.

chiometry of the assumed mechanism is correct, *i.e.*, in our case, the Michaelis-Menten mechanism is appropriate.

6. CONCLUSIONS

1. It is necessary to differentiate among Michaelis-Menten mechanism (the simplest one-substrate-one-complex-one-product mechanism) Michaelis-Menten equation (which describes the Michaelis-Menten mechanism by assuming rapid equilibrium) and Michaelis-Menten kinetics (the substrate saturation curve is a rectangular hyperbola).

2. One can measure Michaelis-Menten kinetics with enzymes following the Briggs-Haldane (steady state) or van Slyke assumptions.

3.   Deviations of the substrate saturation curves from the rectangular hyperbola occur as artefacts if [S] is varied from [S]≤[E], or in some cases of the van Slyke equation.

4.   A two- or three-substrate enzyme may have apparent Michaelis-Menten kinetics, but never follows Michaelis-Menten mechanism.

5.   In the presence of modifiers or in the cases of enzyme-enzyme, enzyme-macromolecule or enzyme-membrane interactions, enzymes which follow the Michaelis-Menten mechanism, may exhibit non-Michaelian kinetics and those which do not follow Michaelis-Menten mechanism may present Michaelian kinetics.

6.   If two isomers of an enzyme in equilibrium catalyze the reaction according to the Michaelis-Menten mechanism but the isomerization step is rate limiting, non-Michaelian kinetics may be obtained.

7.   Asymmetry of subunits or asymmetric association of symmetric subunits in an oligomeric enzyme may cause big deviations in kinetics leading to the half-of-the-sites reactivity or flip-flop mechanism.

8.   If the enzyme is unstable under assay conditions, a sigmoidicity in the saturation curve can be observed if the rate constant of inactivation is sufficiently high or in the case of the van Slyke equation.

9.   Deviations from Michaelian kinetics can be often observed even if the enzyme follows Michaelis-Menten mechanism but is a dissociable oligomer.

10. The progress curve may be used to test whether a one-substrate-one-product enzyme follows the Michaelis-Menten mechanism. The parameters calculated from initial velocity measurements and from progress curves may decide whether the apparently rectangular hyperbolic saturation curve indeed reflects Michaelis-Menten mechanism or the deviation (due to product inhibition or inhibition by excess substrate) is too small to be detected by initial velocity measurements.

REFERENCES

1. Michaelis, L. and Menten, M.L. (1913), Biochem. Z. 49, 333-369.
2. Lineweaver, H. and Burk, D. (1934) J. Amer. Chem. Soc. 56, 658-666.
3. Eadie, G.S. (1942) J. Biol. Chem. 146, 85-93.
4. Hofstee, B.J.H. (1952) Science, 116, 329-331.
5. Hanes, C.S. (1932) Biochem. J. 26, 1406-1421.
6. Eisenthal, R. and Cornish-Bowden, A. (1974) Biochem. J. 139, 715-720.
7. De Miguel Merino, F. (1974) Biochem. J. 143, 93-95.
8. Fajszi, Cs. and Endrényi, L. (1974) FEBS Lett. 44, 240-246.
9. Briggs, G.E. and Haldane, J.B.S. (1925) Biochem. J. 19, 338-339.
10. van Slyke, D.D. and Cullen, G.E. (1914) J. Biol. Chem. 19, 141-180.
11. Walter, C. (1965) "Steady State Applications in Enzyme Kinetics", Ronald Press, New York.
12. Gutfreund, H. (1972) "Enzymes: Physical Principles", Wiley-Interscience, London.
13. Walter, C. (1974) J. Theor. Biol. 44, 1-5.
14. Walter, C. and Morales, M. (1964) J. Biol. Chem. 239, 1277-1283.
15. Wong, J.T.F. (1965) J. Amer. Chem. Soc. 87, 1788-1793.
16. Walter, C. (1966) J. Theor. Biol. 15, 1-33.
17. Heineken, F.G., Tsuchiya, H.M. and Aris, R. (1967) Math. Biosci. 1, 95-113.
18. Heineken, F.G., Tsuchiya, H.M. and Aris, R. (1967) Math. Biosci. 1, 115-141.
19. Otten, H.A. and Duysens, L.N.M. (1973) J. Theor. Biol. 39, 387-396.
20. Otten, H.A. (1974) Math. Biosci. 19, 155-161.
21. Bartha, F. (1980) J. Theor. Biol. 82, 141-147
22. Fischer, É. and Keleti, T. (1975) Acta Biochim. Biophys. Acad. Sci. Hung. 10, 221-227.
23. Webb, J.L. (1963) "Enzyme and Metabolic Inhibitors", Vol. 1, Academic Press, New York.
24. Dixon, M. and Webb, E.C. (1964) "Enzymes", 2nd ed., Longmans, London.
25. Gutfreund, H. (1965) "An Introduction to the Study of Enzymes", Blackwell, Oxford.
26. Yakovlev, V.A. (1965) "Kinetika Fermentativnovo Kataliza", Izd. Nauka, Moscow.
27. Walter, C. (1966) "Enzyme Kinetics", Ronald Press, New York.
28. Laidler, K.J. (1968) "The Chemical Kinetics of Enzyme Action", Clarendon Press, Oxford.
29. Reiner, J.M. (1969) "Behaviour of Enzyme Systems", 2nd ed., Van Nostrand-Reinhold, New York.

30. Dévényi, T., Elödi, P., Keleti, T. and Szabolcsi, G. (1969)
    "Strukturelle Grundlagen der biologischen Funktion der
    Proteine", Akadémiai Kiadó, Budapest.
31. Plowman, K.M. (1972) "Enzyme Kinetics", McGraw-Hill, New York.
32. Ricard, J. (1973) "Cinétique et mécanismes d'action des
    enzymes", Doin, Paris.
33. Segel, I.H. (1975) "Enzyme Kinetics", Wiley, New York-Sydney-
    London-Toronto.
34. Wong, J.T.F. (1975) "Kinetics of Enzyme Mechanisms", Academic
    Press, London-New York-San Francisco.
35. Fromm, H.J. (1975) "Initial Rate Enzyme Kinetics", Springer,
    New York-Heidelberg-Berlin.
36. Cornish-Bowden, A. (1976) "Principles of Enzyme Kinetics",
    Butterworths, London-Boston.
37. Berezin, I.V. and Klyosov, A.A. (1976) "Prakticheskiy Kurs
    Khimicheskoy i Fermentativnoy Kinetiki", Izd. Moscow Univ.,
    Moscow.
38. Keleti, T. (1976) "Az enzimkinetika alapjai", 2nd ed.,
    Tankönyvkiadó, Budapest.
39. Ferdinand, W. (1976) "The Enzyme Molecule", Wiley, London-
    New York-Sydney.
40. Engel, P. (1977) "Enzyme Kinetics", Wiley, New York.
41. Levitzki, A. (1978) "Quantitative Aspects of Allosteric
    Mechanisms", Springer, Berlin-Heidelberg - New York.
42. Kurganov, B.I. (1978) "Allostericheskie Fermenti", Izd. Nauka,
    Moscow.
43. Cleland, W.W. (1963) Biochim. Biophys. Acta, 67, 104-137.
44. Dalziel, K. (1957) Acta Chem. Scand. 11, 1706-1723.
45. Dalziel, K. (1969) Biochem. J. 114, 547-556.
46. Fisher, J.R. and Hoagland, W.D., Jr. (1968) Adv. Biol. Med.
    Phys. 12, 163-211.
47. Peller, L. and Alberty, R.A. (1961) Progr. React. Kin. 1,
    237-260.
48. Wong, J.T.F. and Hanes, C.S. (1962) Can. J. Biochem. Physiol.
    40, 763-844.
49. Wong, J.T.F. and Hanes, C.S. (1969) Arch. Biochem. Biophys.
    135, 50-59.
50. Keleti, T., Ovádi, J. and Batke, J. (1975/76) J. Mol. Catal.
    1, 173-200.
51. Straub, F.B. and Szabolcsi, G. (1964) In "Molekularnaya
    Biologiya", pp. 182-187, Izd. Nauka, Moscow.
52. Monod, J., Wyman, J. and Changeux, J.P. (1965) J. Mol. Biol.
    12, 88-118.
53. Koshland, D.E., Jr., Némethy, G. and Filmer, D. (1966)
    Biochemistry, 5, 365-385.
54. Citri, N. (1973) Adv. Enzymol. 37, 397-648.
55. Florini, J.R. and Vestling, C.S. (1957) Biochim. Biophys.
    Acta, 25, 575-578.
56. Keleti, T. (1965) Acta Physiol. Acad. Sci. Hung. 28, 19-29.

57. Keleti, T. and Batke, J. (1965) Acta Physiol. Acad. Sci. Hung.
       28, 195-207.
58. Keleti, T. and Telegdi, M. (1966) Enzymologia, 31, 39-50.
59. Volkenstein, M.V. and Goldstein, N. (1966) Biochim. Biophys.
       Acta, 115, 478-485.
60. Botts, D.J. and Morales, M. (1953) Trans. Faraday Soc. 49,
       696-707.
61. Hearon, J.Z., Bernhard, S.A., Friess, S.L., Botts, D.J. and
       Morales, M.F. (1959) In "The Enzymes" (P.D. Boyer,
       H. Lardy and K. Myrbäck, eds.) Vol. 1, pp. 49-142, Academic
       Press, New York.
62. Walter, C. (1962) Biochemistry, 1, 652-658.
63. Frieden, C. (1964) J. Biol. Chem. 239, 3522-3531.
64. Keleti, T. (1968) Acta Biochim. Biophys. Acad. Sci. Hung. 3,
       247-258.
65. Cennamo, C. (1968) J. Theor. Biol. 21, 260-277.
66. Cennamo, C. (1969) J. Theor. Biol. 23, 53-71.
67. Staub, M. and Dénes, G. (1966) Biochim. Biophys. Acta, 128,
       82-91.
68. Faragó, A. and Dénes, G. (1967) Biochim. Biophys. Acta, 136,
       6-18.
69. Keleti, T. (1967) J. Theor. Biol. 16, 337-355.
70. Fajszi, Cs. and Keleti, T. (1976) In "Mathematical Models of
       Metabolic Regulation" (T. Keleti and S. Lakatos, eds.)
       Symp. Biol. Hung. Vol. 18, pp. 105-119, Akadémiai Kiadó,
       Budapest.
71. Keleti, T. (1975) In "Mechanism of Action and Regulation of
       Enzymes" (T. Keleti, ed.) Proc. 9th FEBS Meeting, Vol. 32,
       pp. 3-27, Akadémiai Kiadó, Budapest and North-Holland, Amsterdam.
72. Keleti, T. and Fajszi, Cs. (1971) Math. Biosci. 12, 197-215.
73. Keleti, T. (1975) In "Physical Chemistry of Oscillatory
       Phenomena", Faraday Symp. of Chem. Soc., Vol. 9, pp. 219-220,
       University Press, Aberdeen.
74. Ginsburg, A. and Stadtman, E.R. (1970) Ann. Rev. Biochem. 39,
       429-472.
75. Hess, B. and Boiteux, A. (1972) In "Protein-Protein Inter-
       actions" (R. Jaenicke and E. Helmreich, eds.) Proc. 23rd
       Mosbach Colloqu. pp. 271-297, Springer, Berlin-Heidelberg-
       New York.
76. Salerno, C., Ovádi, J., Churchich, J. and Fasella, P. (1975)
       In "Mechanism of Action and Regulation of Enzymes"
       (T. Keleti, ed.) Proc. 9th FEBS Meeting, Vol. 32, pp. 147-
       160, Akadémiai Kiadó, Budapest and North-Holland, Amsterdam.
77. Friedrich, P. (1974) Acta Biochim. Biophys. Acad. Sci. Hung.
       9, 159-173.
78. Keleti, T., Batke, J., Ovádi, J., Jancsik, V. and Bartha, F.
       (1977) Adv. Enzyme Regul. 15, 233-265.
79. Ovádi, J. and Keleti, T. (1978) Eur. J. Biochem. 85, 157-161.
80. Ovádi, J., Salerno, C., Keleti, T. and Fasella, P. (1978)
       Eur. J. Biochem. 90, 499-503.

81. Patthy, L. and Vas, M. (1978) Nature, 276, 94-95.
82. Keleti, T. (1978) In "New Trends in the Description of the
    General Mechanism and Regulation of Enzymes" (S. Damjanovich,
    P. Elödi, and B. Somogyi, eds.) Symp. Biol. Hung. Vol. 21,
    pp. 107-130, Akadémiai Kiadó, Budapest.
83. Friedrich, P., Apró-Kovács, V.A. and Solti, M. (1977) FEBS
    Lett. 84, 183-186.
84. Solti, M. and Friedrich, P. (1979) Eur. J. Biochem. 95,
    551-559.
85. Bartha, F. and Keleti, T. (1979) Oxid. Commun. 1, 75-84.
86. Jancsik, V., Keleti, T., Biczók, Gy., Nagy, M., Szabó, Z.
    and Wolfram, E. (1975/76) J. Mol. Catal. 1, 137-144.
87. Jancsik, V., Keleti, T., Nagy, M., Fenyvesi, É., Bartha, A.,
    Rudas, A., Kovács, P. and Wolfram, E. (1979) J. Mol. Catal.
    6, 41-49.
88. Katchalski, E., Silman, I. and Goldman, R. (1971) Adv. Enzymol.
    34, 445-536.
89. Solti, M. and Friedrich, P. (1976) Mol. Cell. Biochem. 10,
    145-152.
90. Linderstrøm-Lang, K.U., Schellman, J.A. (1959) In "The Enzymes"
    (P.D. Boyer, H. Lardy and K. Myrbäck, eds.) Vol. 1, 2nd ed.,
    pp. 443-510, Academic Press, New York.
91. Schellman, J.A. and Schellman, C. (1964) In "The Proteins",
    (H. Neurath, ed.) Vol. 2, 2nd ed., pp. 1-137, Academic
    Press, New York.
92. Hvidt, Aa. and Nielsen, S.O. (1966) Adv. Protein. Chem. 21,
    287-386.
93. Damjanovich, S. and Somogyi, B. (1978) In "New Trends in the
    Description of the General Mechanism and Regulation of
    Enzymes" (S. Damjanovich, P. Elödi and B. Somogyi, eds.)Symp.
    Biol. Hung. Vol. 21, pp. 159-184, Akadémiai Kiadó, Budapest.
94. Závodszky, P., Abaturov, L.B. and Varshavskiy, J.H. (1966)
    Acta Biochim. Biophys. Acad. Sci. Hung. 1, 389-402.
95. Bolotina, I.A., Markovich, D.S., Volkenstein, M.V. and
    Závodszky, P. (1967) Biochim. Biophys. Acta 132, 260-270.
96. Závodszky, P., Johansen, J.T. and Hvidt, Aa. (1975) Eur. J.
    Biochem. 56, 67-72.
97. Venyaminov, S.Yu., Rajnavölgyi, É., Medgyesi, G.A., Gergely, J.
    and Závodszky, P. (1976) Eur. J. Biochem. 67, 81-86.
98. Jencks, W.P. (1966) In "Current Aspects of Biochemical
    Energetics" (N.O. Kaplan and E. Kennedy, eds.) pp. 273-298,
    Academic Press, New York-London.
99. Damjanovich, S. and Somogyi, B. (1973) J. Theor. Biol. 41,
    567-569.
100. Careri, G. (1974) In "Quantum Statistical Mechanics in the
     Natural Sciences" (B. Kursunoglu, S.L. Mintz and S.M.
     Widmayer, eds.) pp. 15-35, Plenum Press, New York.
101. Careri, G., Fasella, P. and Gratton, E. (1975) CRC Crit. Rev.
     Biochem. 3, 141-164.
102. Welch, G.R. (1977) Progr. Biophys. Molec. Biol. 32, 103-191.

103. McCammon, J.A., Gelin, B.R. and Karplus, M. (1977) Nature, 267, 585-590.
104. Wüthrich, K. and Wagner, G. (1978) Trends Biochem. Sci. 3, 227-230.
105. Karplus, M. and McCammon, J.A. (1979) Nature, 277, 578.
106. Suezaki, Y. and Go, N. (1975) Int. J. Peptide Prot. Res. 7, 333-334.
107. McCammon, J.A., Gelin, B.R., Karplus, M. and Wolynes, P.G. (1976) Nature, 262, 325-326.
108. Cooper, A. (1976) Proc. Nat. Acad. Sci. US, 73, 2740-2741.
109. Vas, M. and Boross, L. (1970) Acta Biochim. Biophys. Acad. Sci. Hung. 5, 203-213.
110. Vas, M. and Boross, L. (1974) Eur. J. Biochem. 43, 237-244.
111. Vas, M. (1976) Acta Biochim. Biophys. Acad. Sci. Hung. 11, 105-112.
112. Telegdi, M. and Straub, F.B. (1973) Biochim. Biophys. Acta, 321, 210-219.
113. Vallee, B.L. and Riordan, J.F. (1978) In "Molecular Interactions and Activity in Proteins", Ciba Foundation Symposium 60, pp. 197-223, Exerpta Medica, Amsterdam-Oxford-New York.
114. Creighton, T.E. (1978) Progr. Biophys. Mol. Biol. 33, 231-297.
115. Brown, K.G., Erfurth, S.C., Small, E.W. and Peticolas, W.L. (1972) Proc. Nat. Acad. Sci. US, 69, 1467-1469.
116. Hull, W.E. and Sykes, B.D. (1975) J. Mol. Biol. 98, 121-153.
117. Lakowicz, J.R. and Weber, G. (1973) Biochemistry, 12, 4171-4179.
118. Eftink, M.R. and Ghiron, C.A. (1976) Biochemistry, 15, 672-680.
119. Ghose, R.C. and Englander, S.W. (1974) J. Biol. Chem. 249, 7950-7955.
120. Holowka, D.A. and Cathou, R.E. (1976) Biochemistry, 15, 3379-3390.
121. Keleti, T. (1967) Acta Biochim. Biophys. Acad. Sci. Hung. 2, 31-37.
122. Frieden, C. (1970) J. Biol. Chem. 245, 5788-5799.
123. Frieden, C. (1979) Ann. Rev. Biochem. 48, 471-489.
124. Ricard, J., Meunier, J.C. and Buc, J. (1974) Eur. J. Biochem. 49, 195-208.
125. Botts, D.J. (1958) Trans. Faraday Soc. 54, 593-604.
126. Fajszi, Cs. and Keleti, T. (1972) Biopolymers, 11, 119-126.
127. Ovádi, J., Telegdi, M., Batke, J. and Keleti, T. (1971) Eur. J. Biochem. 22, 430-438.
128. Levitzki, A. (1974) J. Mol. Biol. 90, 451-458.
129. Markovich, D.S. and Krapivinsky, G.B. (1974) Mol. Biol. (USSR) 8, 857-863.
130. Osborn, H.H. and Hollaway, M.R. (1974) Biochem. J. 143, 651-662.
131. Bühner, M., Ford, G.C., Moras, D., Olsen, K.W. and Rossmann, M.G. (1974) J. Mol. Biol. 82, 563-585.
132. Bühner, M., Ford, G.C., Moras, D., Olsen, K.W. and Rossmann, M.G. (1974) J. Mol. Biol. 90, 25-49.

133. Batke, J., Keleti, T. and Fischer, É. (1974) Eur. J. Biochem.
     46, 307-315.
134. Simon, I. (1972) Eur. J. Biochem. 30, 184-189.
135. Batke, J. (1968) FEBS Lett. 2, 81-82.
136. Malhotra, O.P. and Bernhard, S.A. (1968) J. Biol. Chem. 243,
     1243-1252.
137. Seydoux, F., Malhotra, O.P. and Bernhard, S.A. (1974) CRC
     Crit. Rev. Biochem. 2, 227-257.
138. Lazdunski, M., Peticlerc, C., Chappelet, D. and Lazdunski, C.
     (1971) Eur. J. Biochem. 20, 124-132.
139. Lazdunski, M. (1972) Curr. Top. Cell. Regul. 6, 267-310.
140. Keleti, T. (1971) J. Theor. Biol. 30, 545-551.
141. Fischer, É., Aranyi, P. and Keleti, T. (1973) Acta Biol. Med.
     Germ. 31, 153-174.
142. Batke, J. (1972) J. Theor. Biol. 34, 313-324.
143. Ovádi, J., Batke, J., Bartha, F. and Keleti, T. (1979) Arch.
     Biochem. Biophys. 193, 28-33.
144. Kagan, Z.S., Dorozhko, A.I., Kovaleva, S.V. and Yakovleva, L.I.
     (1975) Biochim. Biophys. Acta, 403, 208-220.
145. Kurganov, B.I., Wenzel, K.W., Zimmermann, G., Yakovlev, V.A.
     and Hoffmann, E. (1975) Bioorg. Khim. (USSR) 1, 632-645.
146. Kurganov, B.I. (1967) Mol. Biol. (USSR) 1, 17-27.
147. Nichol, L.W., Jackson, W.J.H. and Winzor, D.J. (1967)
     Biochemistry, 6, 2449-2456.
148. Frieden, C. (1967) J. Biol. Chem. 242, 4045-4052.
149. Kurganov, B.I. (1968) Mol. Biol. (USSR) 2, 430-446.
150. Kurganov, B.I., Kagan, Z.S., Dorozhko, A.I. and Yakovlev, V.A.
     (1974) J. Theor. Biol. 47, 1-41.
151. Kurganov, B.I. (1975) In "Mechanism of Action and Regulation
     of Enzymes" (T. Keleti, ed.) Proc. 9th FEBS Meeting,
     Vol. 32, pp. 29-42, Akadémiai Kiadó, Budapest and North-
     Holland, Amsterdam.
152. Kurganov, B.I. (1976) In "Mathematical Models of Metabolic
     Regulation" (T. Keleti and S. Lakatos, eds.) Symp. Biol.
     Hung. Vol. 18, pp. 31-45, Akadémiai Kiadó, Budapest.
153. Kurganov, B.I., Dorozhko, A.I., Kagan, Z.S. and Yakovlev, V.A.
     (1976) J. Theor. Biol. 60, 271-286.
154. Kurganov, B.I., Dorozhko, A.I., Kagan, Z.S. and Yakovlev, V.A.
     (1976) J. Theor. Biol. 60, 287-299.
155. Kurganov, B.I. (1977) J. Theor. Biol. 68, 521-543.
156. Phillips, A.T. (1974) CRC Crit. Rev. Biochem. 2, 343-378.
157. Schonheyder, F. (1952) Biochem. J. 50, 378-384.
158. Schwert, G.W. (1969) J. Biol. Chem. 244, 1278-1284.
159. Klesov, A.A. and Berezin, I.V. (1972) Biokhimiya, 37, 170-183.
160. Bizzozero, S.A., Kaiser, A.W. and Dutter, H. (1973) Eur. J.
     Biochem. 33, 292-300.
161. Atkins, G.L. and Nimmo, I.A. (1973) Biochem. J. 135, 779-784
162. Johansen, G. and Lumry, R. (1961) Compt. Rend. Trav. Lab.
     Carlsberg, 32, 185-214.

163. Wilkinson, G.N. (1961) Biochem. J. 80, 324-332.
164. Cleland, W.W. (1967) Adv. Enzymol. 29, 1-32.
165. Bates, D.J. and Frieden, C. (1973) J. Biol. Chem. 248, 7878-7884.
166. Bates, D.J. and Frieden, C. (1973) J. Biol. Chem. 248, 7885-7890.
167. Yun, S.-L. and Suelter, C.H. (1977) Biochim. Biophys. Acta, 480, 1-13.
168. Fernley, H.N. (1974) Eur. J. Biochem. 43, 377-378.
169. Nimmo, I.A. and Atkins, G.L. (1974) Biochem. J. 141, 913-914.
170. Darvey, I.G., Shrager, R. and Kohn, L.D. (1975) J. Biol. Chem. 250, 4690-4701.
171. Halwachs, W. (1978) Biotechn. Bioeng. 20, 281-285.
172. Petersen, L.C. and Degn, H. (1978) Biochim. Biophys. Acta, 526, 85-92.
173. Duggleby, R.G. and Morrison, J.F. (1977) Biochim. Biophys. Acta, 481, 297-312.
174. Duggleby, R.G. and Morrison, J.F. (1978) Biochim. Biophys. Acta, 526, 398-409.
175. Chandler, J.P., Hill, D.E. and Spivey, H.O. (1972) Comput. Biomed. Res. 5, 515-534.
176. London, J.W., Shaw, L.M. and Garfinkel, D. (1977) Anal. Chem. 49, 1716-1719.
177. Hill, C.M. Waight, R.D. and Bardsley, W.G. (1977) Mol. Cell. Biochem. 15, 173-178.

A GENERALIZED NUMERICAL DECONVOLUTION PROCEDURE

FOR COMPUTING ABSOLUTE BIOAVAILABILITY - TIME PROFILES

Victor F. Smolen

Ayerst Laboratories, Inc.
Rouses Point, N.Y. 12979
U.S.A.

ABSTRACT

The use of either body fluid measurements of drug concentration
or pharmacological response intensity data for the computation of
drug absorbed as a function of time requires a mathematical model
for the system and the ability to perform a deconvolution of the
response. Deconvolution procedures can also be used in verifying
a model's accuracy to represent any given system by comparing an
actual input into the system to that calculated by using these
methods. For simple one, two and three-compartment models, analytical expressions have been presented in the literature to determine
drug bioavailability. Such expressions are specific and restricted
to a particular configuration and order of the system. A generalized numerical deconvolution procedure has been developed to compute
the amount of drug absorbed for any order system. Its use has been
illustrated with pharmacological data, however, it applies equally
well to drug concentration measures from body fluids. The procedure
described here is simple and is easily implemented on a digital
computer. While some deconvolution schemes tend to be unstable,
the algorithm presented here is inherently stable. Furthermore,
this procedure requires a minimum of core memory and central processor time to execute.

1. INTRODUCTION

The utility of a reliable and efficient deconvolution technique
to determine the time course of systemic drug bioavailability has
been discussed (1). The task is, given a model determined from
pharmacological data and the time course of pharmacological response
on body fluid data produced by some test dosage form, determine the

systemic drug availability as a function of time. Another use for
a reliable deconvolution technique is in the verification of the
pharmacokinetic model itself (2,3).

In another chapter (4), two general methods of model identification are discussed: (i) nonlinear least-squares fitting in the
time domain, and (ii) a pulse testing method to identify the frequency response of the system. For both methods, the model of the
system was determined from the impulse response (approximated by a
bolus intrvenous injection) of the system. Such an experimental
procedure only can assure us that the system is adequately modelled
for impulse inputs to the system. Since the system under consideration is not strictly linear, that is, the pharmacological response
intensity can in general be a nonlinear function of the biophasic
drug level (5), there is no assurance that the system model remains
the same when a different input is introduced into the system.
Hence, the model developed from the impulse response of the system
must be verified with respect to other kinds of inputs into the
system. The model is considered valid only if it predicts the system behaviour to other kinds of inputs as well. In this paper,
model validity is verified by using a step input (zero order infusion). To do this, a reliable deconvolution procedure must be
available.

In the Laplace transform domain, the input, $x(t)$ and the output, $y(t)$ are related by the expression

$$\frac{Y(s)}{X(s)} = G(s) \qquad\qquad\qquad\qquad [1]$$

where $X(s)$ is the Laplace transform of the input $[x(t)]$, $Y(x)$ is
the Laplace transform of the output $[y(t)]$, and $G(s)$ is the transfer
function of the system (6-8). When the system transfer function
is known, the output response of the system to any given input
function is obtained from

$$Y(s) = G(s) * X(s) \qquad . \qquad\qquad\qquad [2]$$

When dealing with drug input and pharmacological response, the
output $[y(t)]$ becomes the relative biophasic drug level $[f(I)]$.
In the time domain, Eq. [2] is equivalent to the convolution integral

$$y(t) = \int_0^t g(t-\tau)x(\tau)d\tau = \int_0^t g(t)x(t-\tau)d\tau \qquad . \qquad [3]$$

In the case of drug absorption analysis, the input to the system is usually not known, especially if the drug is absorbed in the
gastrointestinal tract from a capsule or tablet dosage form. The
utility of the modelling approach lies in being able to calculate

the quantity of drug that was absorbed into the system as a function of time from a given dosage form by merely observing the output pharmacological response and a predetermined system model. This is the inverse problem of convolution; it is sometimes called deconvolution. Symbolically, $X(s)$ must be calculated, given the $G(s)$ and $Y(s)$ for a system, that is,

$$X(s) = \frac{1}{G(s)} \cdot Y(s) \qquad . \qquad\qquad [4]$$

Let

$$G'(s) = \frac{1}{G(s)} \qquad\qquad , \qquad\qquad [5]$$

then substituting Eq. [5] into [4] we obtain

$$X(s) = G'(x) * Y(s) \qquad . \qquad\qquad [6]$$

In the time domain, Eq. [6] is equivalent to a convolution integral similar to the one given in Eq. [3], that is,

$$x(t) = \int_0^t g'(t-\tau)y(\tau)d\tau = \int_0^t g'(\tau)y(t-\tau)d\tau \quad . \qquad [7]$$

Analytical expressions for deconvolution, as expressed by Eq. [7], corresponding to second and third-order systems have been derived and presented in the literature (9-14). However, it is not always simple or possible to determine analytical expressions for the deconvolution integral. In that case numerical methods must be used. Various deconvolution methods using approximation techniques (15,16), an analog computer (17), or numerical techniques (18) have also been presented in the literature. The problem of numerically evaluating Eq. [7] is not an easy one because deconvolution tends to be unstable. A literature review of the various deconvolution techniques generally used in engineering applications has been presented by Rodeman and Yao (19).

2. THEORETICAL PROCEDURE

2.1. Simple Numerical Technique

In order for us to be able to evaluate the convolution integral given in Eq. [7], $g'(t)$ must be known. This may not always be easy. Hence, a more general approach involves evaluating $x(t)$, when $g(t)$ and $y(t)$ are known, from Eq. [3] (20). Suppose that $g(t)$ and $y(t)$ are known at equally spaced intervals of Δt. If this is not the case, then the observed $y(t)$ should be smoothed and evaluated at these points. Let the interval, 0 to t, be divided into N equal parts such that

$$t = N \cdot \Delta t \qquad . \qquad\qquad [8]$$

To simplify the notation, let

$$y_j = y(j \cdot \Delta t) \qquad \text{for} \qquad j = 0, 1, 2, \ldots, N \quad , \qquad [9a]$$

and

$$g_j = g(j \cdot \Delta t) \qquad \text{for} \qquad j = 0, 1, 2, \ldots, N \quad , \qquad [9b]$$

and

$$x_j = x(j \cdot \Delta t) \qquad \text{for} \qquad j = 0, 1, 2, \ldots, N \quad . \qquad [9c]$$

The product in the convolution integral, Eq. [3], for $n \leq N$ can be represented as

$$g[(n-j)\Delta t] \cdot x(j \cdot \Delta t) = g_{n-j} \cdot x_j \qquad \text{for} \qquad t = n\Delta t \quad . \qquad [10]$$

Figure 1 shows g_j, y_j and x_j for the entire interval. It should be pointed out here that in practice x_j is not yet known, only the values of g_j and y_j are given. The trapezoid approximation for an integral gives Eq. [3] in terms of a sum, that is,

$$y_n = \sum_{j=1}^{n} \frac{(g_{n-j+1}x_{j-1} + g_{n-j}x_j)}{2} \Delta t, \qquad \text{for } n = 1, 2, 3, \ldots, N \quad [11]$$

Multiplying by $2/\Delta t$ and letting $y_n' = 2y/\Delta t$, the following equations are obtained:

$$y_1' = g_0 x_1 + g_1 x_0$$
$$y_2' = g_0 x_2 + 2g_1 x_1 + g_2 x_0$$
$$y_3' = g_0 x_3 + 2g_1 x_2 + 2g_2 x_1 + g_3 x_0 \qquad\qquad [12]$$
$$\vdots$$
$$y_N' = g_0 x_N + 2g_1 x_{N-1} + \cdots + 2g_{N-1} x_1 + g_N x_0 \quad .$$

In the set of equations given by Eq. [12], the y's and the g's are known, and the problem requires the evaluation of the x's. Note that there are only N equations and N+1 unknowns $x_0, x_1, x_2, \ldots, x_N$ in the above set of equations. If $g_0 \neq 0$, one of the x_j, say x_0, must be known. Usually though $g_0 = 0$ so the set reduces to N equations in the N unknowns $x_0, x_1, x_2, \ldots, x_{N-1}$. Since the system matrix is lower triangular, that is,

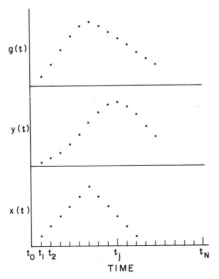

FIGURE 1:   Typical curves of the transfer function, the output
            response and the deconvolved input function as obtained
            from simple numerical deconvolution.

$$
\begin{bmatrix} y'_1 \\ y'_2 \\ y'_3 \\ \vdots \\ y'_N \end{bmatrix} = \begin{bmatrix} g_1 & 0 & & \cdots & 0 \\ 2g_1 & g_2 & & \cdots & 0 \\ 2g_1 & 2g_2 & g_3 & \cdots & 0 \\ \vdots & \vdots & \vdots & & \vdots \\ 2g_1 & 2g_2 & 2g_3 & \cdots & g_N \end{bmatrix} \begin{bmatrix} x_0 \\ x_1 \\ x_2 \\ \vdots \\ x_{N-1} \end{bmatrix} , \qquad [13]
$$

this system of equations can be iteratively solved for the x_j's.
However, it turns out that this method seldom works well because
the estimates of $x(t)$, x_j, obtained by this method tend to oscillate
as N increases. A more stable numerical scheme was developed
during the course of this research for estimating the x_j's. This
method involves starting with the deconvolution expression in the
frequency domain, Eq. [4], and formulating it in terms of vector-
matrix notation (21). The vector-matrix equations are then integrated to obtain the estimates of the x_j's. Since this method involves the integration of equations, it tends to be more stable.

2.2. Vector-Matrix Formulation for Deconvolution

For purposes of clarity of presentation, a second-order trans-

fer function will be considered for this section. The presentation
here is perfectly general and can be extended to any order transfer
function. Consider the general second-order transfer function of
the form

$$\frac{Y(s)}{X(s)} = G(s) = \frac{K(Ds^2+Es+F)}{As^2 + Bs + C} \qquad . \qquad [14]$$

Equation [14] can be rearranged to the form

$$\frac{X(s)}{Y(s)} = \frac{As^2 + Bs + C}{K(Ds^2+Es+F)} \qquad . \qquad [15]$$

Equation [15] can be further rearranged to the form

$$(Ds^2 + Es + F)X(s) = \frac{As^2 + Bs + C}{K} Y(s) \qquad . \qquad [16]$$

Utilizing the basic definition of the Laplace transform (6), Eq. [16]
can be expressed in the time domain as

$$D \cdot \frac{d^2x}{dt^2} + E \cdot \frac{dx}{dt} + F \cdot x = \frac{1}{K} \left| A \cdot \frac{d^2y}{dt^2} + B \cdot \frac{dy}{dt} + C \cdot y \right| . $$
$$[17]$$

Now, let

$$x = x_1 \qquad\qquad [18a]$$

and

$$\frac{dx}{dt} = x_2 = \dot{x}_1 \qquad . \qquad [18b]$$

Then we can rewrite Eq. [17] as a system of two equations (21),
thus

$$\frac{dx}{dt} = \dot{x}_1 = x_2 \qquad\qquad [19a]$$

$$\frac{d^2x}{dt^2} = \dot{x}_2 = \frac{1}{D} [-E \frac{dx}{dt} - Fx] + \frac{1}{D \cdot K} (A \frac{d^2y}{dt^2} + B \frac{dy}{dt} + Cy). \qquad [19b]$$

Or, in matrix notation, Eq. [19] can be written as

$$\begin{pmatrix} \dot{x}_1 \\ \dot{x}_2 \end{pmatrix} = \begin{pmatrix} 0 & 1 \\ -\frac{F}{D} & -\frac{E}{D} \end{pmatrix} \begin{pmatrix} x_1 \\ x_2 \end{pmatrix} + \begin{pmatrix} 0 \\ \frac{1}{D \cdot K}(A \frac{d^2y}{dt^2} + B \frac{dy}{dt} + Cy) \end{pmatrix} \qquad . \qquad [20]$$

Equation [20] represents the vector-matrix formulation of the deconvolution problem. It should be pointed out here again that in the
deconvolution procedure we have at our disposal the output function,
y(t), and the system model, g(t), that is, the coefficients A, B,
C, D, E, F, and K. We want to compute x(t). Equation [20] can be
directly integrated to obtain $x_1(t)$, and hence x(t). One problem
with this procedure is that Eq. [20] requires numerical derivatives
of the output function, y(t), to be taken at every step of the integration procedure. Unfortunately, it turns out that this numerical differentiation is at the root of the instability of the deconvolution procedure.

2.3. Solving a Related Problem

The instability problem can be avoided by not directly solving
Eq. [20], but instead solving a related problem. This procedure
has the advantage that no numerical derivatives of y(t) are required
to be taken explicitly. The implicit differentiation in this method
is much more stable numerically. First, Eq. [15] is rearranged to
form

$$X(s) = \frac{As^2 + Bs + C}{K(Ds^2 + Es + F)} \cdot Y(s) \qquad . \qquad [21]$$

Also define

$$H(s) = \frac{Y(s)}{K(Ds^2 + Es + F)} \qquad ; \qquad [22a]$$

then,

$$X(s) = (As^2 + Bs + C)H(s) \qquad . \qquad [22b]$$

Now, instead of directly evaluating X(s) by Eq. [21], H(s) will be
obtained from Eq. [22a] and X(s) will be evaluated from Eq. [22b].
It will be shown that by following this procedure we avoid the
problem of directly taking derivatives of y(t).

Let us rearrange Eq. [22a] to the form

$$(Ds^2 + Es + F)H(s) = Y(s)/K \qquad . \qquad [23]$$

Again, inverting Eq. [23] into the time domain yields

$$(D\frac{d^2h}{dt^2} + E\frac{dh}{dt} + Fh) = \frac{1}{K} \cdot y(t) \qquad [24]$$

where h(t) is the time domain representation of H(s). As before,
define

$$h = h_1 \qquad\qquad [25a]$$

and

$$\frac{dh}{dt} = h_2 = h_1 \qquad . \qquad [25b]$$

Then we can rewrite Eq. [24] as a system of two equations, thus we have

$$\frac{dh}{dt} = \dot{h}_1 = h_2 \qquad\qquad [26a]$$

$$\frac{d^2h}{dt^2} = \dot{h}_2 = \frac{1}{D} [-E \frac{dh}{dt} - Fh] + \frac{1}{D \cdot K} \cdot y(t) \qquad . \qquad [26b]$$

In matrix notation, Eq. [26] can be written as

$$\begin{bmatrix} \dot{h}_1 \\ \dot{h}_2 \end{bmatrix} = \begin{bmatrix} 0 & 1 \\ -\dfrac{F}{D} & -\dfrac{E}{D} \end{bmatrix} \begin{bmatrix} h_1 \\ h_2 \end{bmatrix} + \begin{bmatrix} 0 \\ \dfrac{1}{D \cdot K} \cdot y(t) \end{bmatrix} \qquad . \qquad [27]$$

Comparing Eqs. [20] and [27] we observe that the solution of Eq. [27] does not require taking the derivatives of $y(t)$. Since transfer functions are given in deviation variable form, $x(t)$ and $y(t)$ are zero at time, t=0. Also, since $H(s)$ was defined in Eq. [22a] in terms of $Y(s)$ and part of the original transfer function, $h(t)=0$ at t=0. Using this as an initial condition, Eq. [27] can be integrated forwards in time from t=0 to some final time, t=T, to yield $h(t)$. While this process of integration is proceeding, the values h_1, h_2, $\dot{h}_1$, and $\dot{h}_2$, as defined by Eq. [27], are available at every time t.

The value of $x(t)$ can be computed during the process of integration of Eq. [27]. Inverting Eq. [22b] to the time domain produces the relationship

$$x(t) = A \frac{d^2h}{dt^2} + B \frac{dh}{dt} + C h(t) \qquad . \qquad [28]$$

Utilizing relationships [26b], [26a] and [25a] we can express Eq. [28] as

$$x(t) = A \cdot \dot{h}_2 + B \cdot \dot{h}_1 + C \cdot h_1 \qquad . \qquad [29]$$

As pointed out earlier, $\dot{h}_2$, $\dot{h}_1$, and h_1 are all available at every time increment of integration, t, during the integration of Eq. [27].

Thus, we have been able to calculate $x(t)$ by solving a related problem involving $h(t)$, which requires no explicit differentiation of the output function, $y(t)$. It should be noted that there is an implicit differentiation of $y(t)$ that takes place in this method; however, this process is much more stable than explicit differentiation.

As mentioned earlier, though developed for a second-order transfer function, this method is entirely general and can be applied to any order system. One very important point touched upon earlier will now be discussed.

2.4. Adding Zeros to the Transfer Function

It will be recalled that at the beginning of the discussion on vector-matrix notation for deconvolution, a transfer function was defined (Eq. [14]) where the numerator was of the same order as the denominator. For real systems, the order of the numerator of a transfer function must be less than, or at most equal to, the order of the denominator. This however poses a problem. When Eq. [14] is rearranged to obtain Eq. [15], the numerator of the original transfer function is now in the denominator. If the numerator was originally of lower order than the denominator in Eq. [14], a condition will exist where the denominator of Eq. [15] is of lower order than the numerator.

This problem is avoided by adding artificial zeros to any system transfer function in such a way as not to change significantly its behaviour, so that the order of the numerator will become equal to that of the denominator before applying this method. In practice, one less zero than the difference between the order of the denominator and the numerator must be added. This is so because even though the transfer function usually relates the output response to the instantaneous rate of input, quite often the cumulative amount of drug absorbed, $A(t)$, is desired. Symbolically,

$$A_t = \int_0^t x(t)dt \qquad . \qquad [30]$$

In the Laplace transform domain,

$$A_t(s) = X(s)/s$$

or,

$$X(s) = s \cdot A_t(s) \qquad . \qquad [31]$$

Then using Eq. [30] we can rewrite Eq. [14] as

$$\frac{Y(s)}{A_t(s)} = s \cdot G(s) \qquad . \qquad\qquad [32]$$

This operation has the effect of adding a zero at the origin in the complex plane.

The other zeros must be added in such a way as not to affect significantly the transient behaviour of the original system. This is accomplished by first identifying the highest frequency present in either the numerator or denominator of the transfer function. This corresponds to the fastest (*i.e.*, smallest) time constant of the system. The zeros added are then located at a frequency an order of magnitude higher than the highest frequency in the system. This has the effect of adding a transient to the system which has a time constant an order of magnitude faster than the fastest time constant in the original system, which, in turn, means that the term added has a time constant an order of magnitude smaller than the smallest significant time constant.

The device of adding zeros can be illustrated by considering, as an example, a second-order system in factored form given by

$$G(s) = \frac{K'(\tau_1 s+1)}{(\tau_2 s+1)(\tau_3 s+1)} = \frac{Y(s)}{X(s)} \qquad\qquad [33]$$

where τ_1, τ_2 and τ_3 are the time constants and K is the gain for this system. Inverting $G(s)$ for the purposes of the deconvolution we obtain

$$\frac{X(s)}{Y(s)} = \frac{1}{G(s)} = \frac{(\tau_2 s+1)(\tau_3 s+1)}{K'(\tau_1 s+1)} \qquad . \qquad\qquad [34]$$

Say, for this example, that τ_3 is the smallest time constant, then we add a term in the denominator of Eq. [34] of the form $(\tau s+1)$ where $\tau \ll \tau_3$, to produce

$$\frac{X(s)}{Y(s)} = \frac{1}{G(s)} = \frac{(\tau_2 s+1)(\tau_3 s+1)}{K'(\tau_1 s+1)(\tau s+1)} \qquad . \qquad\qquad [35]$$

The fact that τ is an order of magnitude smaller than τ_3 causes the dynamic behaviour produced by Eq. [35] to be equivalent to that of Eq. [34] for the signal of the highest frequency of interest, *i.e.*, for signals near $w=1/\tau_3$. The concept of adding zeros to the transfer function without significantly affecting the dynamic calculations can be visualized if one considers the Bode diagram for the system (4,6,7). This technique has been found to be satisfactory and does not affect the system transfer function appreciably, yet it provides the additional zeros that are required for the numerical solution.

2.5. Computational Algorithm

The method used here to integrate Eq. [27] is the modified Euler technique, which is the same as a second-order Runge-Kutta method. A higher-order integration scheme would yield no significant increase of accuracy for this problem, and would require more computer time. An outline of this method will be presented here. Consider a single first-order differential equation of the form

$$\frac{dy}{dt} = f[y(t),t] \qquad \qquad \qquad [36]$$

The usual Euler method uses the algorithm which requires only one derivative evaluation at the initial point of the interval

$$y(t+\Delta t) = y(t) + \Delta t \cdot f[y(t),t] \qquad . \qquad [37]$$

The modified Euler method requires the evaluation of the derviative twice, once at the initial point of the interval and once at the end-point of the interval. The final derivative used in the algorithm is an average of these two evaluations, thus

$$y_1(t+\Delta t) = y(t) + \Delta t \cdot f(y,t) \qquad\qquad [38a]$$

$$y_2(t+\Delta t) = y(t) + \Delta t \cdot f[y(t+\Delta t,t+\Delta t] \qquad . \qquad [38b]$$

Hence

$$y(t+\Delta t) = y(t) + \frac{\Delta t}{2} \cdot \{f(y,t)+f[y(t+\Delta t),t+\Delta t]\} \qquad . \qquad [39]$$

3.   EXPERIMENTAL APPLICATION AND DISCUSSION

The numerical deconvolution technique described here was applied to experimental data gathered during a study of the effect of chlorpromazine on various pharmacological responses (2,3). The application of the technique will be illustrated here with the response of pupil diameter in rabbits. The experimental procedure is discussed briefly in another paper in this volume (4) and in more detail elsewhere (1-3). As pointed out by Kuehn *et al.* (1), a reliable deconvolution technique can be used for both the verification of the pharmacokinetic models and for the calculation of the time course of systemic drug bioavailability for test dosage forms. Another paper (4) describes in some detail the theoretical aspects of the time domain and frequency domain techniques which are used to obtain the pharmacokinetic models presented by Kuehn *et al.* (1) for the drug chlorpromazine. The models obtained were determined from known bolus intravenous (impulse) inputs of the drug. However, since the systems under study are not strictly linear, but only approximately linear over a limited range, model validity must be verified by using them with a deconvolution scheme to back-calculate a known, *but significantly different* input of drug. To this end, a direct comparison was made of experimentally known drug inputs obtained by programmed intravenous infusion with results computed from monitoring the pharmacological response intensities during the intravenous administration of the drug (Figure 2). These observed time variations of drug response intensities are converted into their corresponding relative biophasic drug level *vs.* time profile by using the dose-effect curve (1,4). The cumulative amount of drug infused (or absorbed), A_t, is then

obtained by deconvoluting the biophasic drug levels, $f(I)$, with the inverse of the transfer function, $G(s)$. A second-order model was determined to be satisfactory for the miotic response of rabbits to chlorpromazine (4). Its form is

$$G(s) = \frac{A_1'}{s+m_1} + \frac{A_2'}{s+m_2} = \frac{K_1}{(\tau_1 s+1)(\tau_2 s+1)} = \frac{Y(s)}{X(s)} \qquad [40]$$

where

$$K_1 = A_1'/m_1 + A_2'/m_2$$

and

$$\tau_1 = 1/m_1 \quad , \quad \tau_2 = 1/m_2 \quad .$$

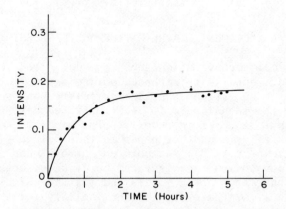

FIGURE 2: Intensity of pupillary diameter decrease in a single rabbit (No. 17) in response to a slow intravenous infusion of chlorpromazine. The infusion rate is 0.50 mg/kg/hr.

The model parameters, as given in another paper in this volume (4) by pulse testing, are $A_1 = 2.35755$, $A_0' = -2.35755$, $m_1 = 0.008424$ and $m_2 = 0.02576$. We see from Eq. [40] that even if we calculate the cumulative input, $A(t)$, one zero must still be added. That is, inverting Eq. [40], the expression for $A(t)$ is

$$\frac{A(s)}{Y(s)} = \frac{(\tau_1 s+1)(\tau_2 s+1)}{K_1 s} = \frac{1}{sG(s)} \qquad . \qquad [41]$$

A zero of the form, $\tau s+1$, is added to the denominator of Eq. [41], with the constraint that $\tau \ll 1/m_2 = 38.81$. After adding this zero and rearranging, we obtain

$$\frac{A(s)}{Y(s)} = \frac{s^2 + (m_1+m_2)s + m_1 m_2}{(K_1' m_2 + K_2' m_1)(\tau s^2 + s)} \qquad . \qquad [42a]$$

Relating the coefficients in Eq. [42] to those in Eq. [15] we obtain $A=1$, $B=m_1+m_2$, $C=m_1 m_2$, $D=\tau$, $E=1$, $F=0$, $K=K_1' m_2 + K_2' m_1$.

The deconvolution is accomplished numerically by using Eqs. [27] and [29]. Results are presented in Figure 3 for both the pulse testing and multifit methods. Results for many experimental runs are given elsewhere (1-3). The ability of the model to compute adequately the zero-order infusion rate producing the output re-

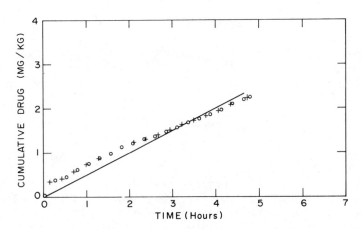

FIGURE 3:   Experimentally determined amounts of chlorpromazine slowly infused to a single rabbit (No. 17) as a function of time. The solid line represents the actual amount of chlorpromazine infused during the experiment, 0.50 mg/kg/hr. Calculated values are shown by pulse test (o) and Multifit (+).

sponse, as shown in Figure 3, attests to its validity. This indicates the ability of the model to determine the rate of bioavailability for other inputs of drug which are also significantly different from bolus i.v. inputs, for example that produced by some solid oral dosage form. The results in Figure 3 also indicate the essential equivalence of the models as determined by the time domain technique "multifit" and the frequency domain technique "pulse testing". The equivalence indicated by model verification with slow i.v. infusion would be expected since these models appeared to produce an equally good simulation of the rapid i.v. data used for model determination (4).

4. CONCLUSIONS

A general numerical deconvolution procedure has been presented. It can be used to evaluate drug delivery systems. The time course of the amount of drug absorbed and the verification of pharmacokinetic models are two significant uses for deconvolution. This method, in general, can be applied to either pharmacological response data or drug concentration measures from body fluids. As an example, this method was used to verify a pharmacokinetic model derived from the miotic response produced by chlorpromazine in rabbits. The method is well suited to use with a digital computer. The topics of vector-matrix formulation, implicit differentiation, adding zeros for numerical stability and the computational algorithm have been discussed in detail. While some deconvolution schemes tend to be unstable, this method has the advantages of simplicity, efficient computer use and numerical stability.

ACKNOWLEDGEMENTS

The substantial contributions to this work by Dr. Ashok Jhawar, Dr. Paul Khuehn and Dr. William Weigand, all formerly at Purdue University, West Lafayette, Indiana, are very gratefully acknowledged.

REFERENCES

1. Kuehn, P.B., Jhawar, A.K., Weigand, W.A. and Smolen, V.F. (1976) J. Pharm. Sci. 65, 1593.
2. Kuehn, P.B. (1974) "Pharmacological Response Kinetics of Chlorpromazine in Rabbits", Ph.D. Thesis, Purdue University.
3. Jhawar, A.K. (1974) "Mathematical Modelling of Physiological Data", Ph.D. Thesis, Purdue University.
4. Smolen, V.F. (1980) In "Kinetic Data Analysis: Design and Analysis of Enzyme and Pharmacokinetic Experiments" (L. Endrenyi, ed.) pp. 209-233, Plenum, New York.
5. Smolen, V.F., Barile, R.G. and Theofanous, T.G. (1972) J. Pharm. Sci. 61, 467.

6. Coughanowr, D.R. and Koppel, L.B. (1965) "Process Systems
    Analysis and Control", McGraw-Hill, New York.
7. Harriott, P. (1964) "Process Control", McGraw-Hill, New York.
8. Luyben, W.L. (1973) "Process Modelling, Simulation and Control
    for Chemical Engineers", McGraw-Hill, New York.
9. Schoenwald, R.D. and Smolen, V.F. (1971) J. Pharm. Sci. 60,
    1039.
10. Smolen, V.F. (1971) J. Pharm. Sci. 60, 354.
11. Smolen, V.F. (1971) J. Pharm. Sci. 60, 878.
12. Smolen, V.F. (1972) Can. J. Pharm. Sci. 7, 1.
13. Smolen, V.F. and Schoenwald, R.D. (1971) J. Pharm. Sci. 60,
    96.
14. Wagner, J.G. (1975) "Fundamentals of Clinical Pharmacokinetics",
    pp. 57-125, Drug Intelligence Publ., Hamilton, Ill.
15. Loo, J.C.K. and Riegelman, S. (1968) J. Pharm. Sci. 57, 918.
16. Rowland, M., Riegelman, S. and Epstein, W.L. (1968) J. Pharm.
    Sci. 57, 984.
17. Silverman, M. and Bunger, A.S.V. (1961) J. Appl. Physiol. 16,
    911.
18. Rescigno, A. and Segre, G. (1966) "Drug and Tracer Kinetics",
    pp. 57-137, Blaisdell, New York.
19. Rodeman, R. and Yao, J.T.P. (1973) "Structural Identification
    -- Literature Review", School of Civil Engineering, Purdue
    University.
20. Jacquez, J.A. (1972) "Compartmental Analysis in Biology and
    Medicine", American Elsevier, New York.
21. Koppel, L.B. (1968) "Introduction to Control Theory", Prentice-
    Hall, Englewood Cliffs, N.J.

INFLUENCE OF VARIABILITY IN PLASMA CONCENTRATION ON DOSE PREDICTION
ASSUMING MICHAELIS-MENTEN KINETICS

A model study for phenytoin with dose prediction according to
Richens and according to Ludden *et al.* and Martin *et al.*

E.A. van der Velde,* O. Driessen,** and J. Hermans*

  *Department of Medical Statistics
 **Department of Pharmacology
  University of Leiden
  Wassenaarseweg 80 and 72
  Leiden, The Netherlands

ABSTRACT

Several dose prediction methods have been published for phenytoin in the treatment of epilepsy patients. All methods assume
Michaelis-Menten kinetics. In a model study, the performance of
these methods is investigated. The performance is measured by comparing a target concentration with the concentration resulting in a
model patient when applying a predicted dose. Methods requiring
two trial doses in order to come to a predicted dose turned out to
be usually safe only if the measurement error was very low. A
method, requiring one trial dose and assuming a fixed value for K_m
turned out to be less influenced by the measurement error and was
especially satisfactory if the patient's K_m did not deviate too much
from the K_m-value assumed in the method.

1.   INTRODUCTION

Patients with epilepsy are often treated with phenytoin. Some
of these patients need rather elevated plasma concentrations, about
15 µg/ml, to suppress epilepsy symptons. Concentrations about
20 µg/ml are generally avoided as symptoms of toxicity will occur
frequently at such plasma concentrations. It was observed that by
augmenting the dose, the plasma concentration may increase more than
proportionally (1,2). Especially at high dose ranges almost all
patients showed this phenomenon clearly. Hence it is quite critical
to predict for an arbitrary patient a dose that will yield a plasma
concentration of about 15 µg/ml, without overshooting a value of
20 µg/ml.

In the literature several prediction methods (3-7) are advocated for this purpose. The more than proportional increase of phenytoin plasma concentration is attributed to the fact that, in the clinical dose range, the phenytoin metabolizing enzyme system is saturated. In view of this supposed saturation, all prediction methods apply Michaelis-Menten kinetics.

Patients are administered phenytoin mostly several times daily and because of the long biological half-life of the drug, a steady state is supposed to exist. Hence, the amount of phenytoin metabolized in a dosage interval equals the quantity administered in that interval. In the steady state, the Michaelis-Menten relation between dose (D) and plasma concentration (C) is given by:

$$D = V_{max} \, C/(K_m + C) \hspace{3cm} [1.1]$$

or, equivalently,

$$C = K_m D/(V_{max} - D) \hspace{1cm} . \hspace{2cm} [1.2]$$

All dose prediction methods are based on the following reasoning. Equation [1.1] can be used to calculate the dose (D) required to reach the specified target concentration (C) of 15 µg/ml, as soon as information is available about V_{max} and K_m. To obtain this information about V_{max} and K_m, one or more low trial doses are applied and the resulting plasma concentrations are measured. If (at least) two trial doses are applied, values for V_{max} and K_m can be solved algebraically from Eq.'s [1.1] or [1.2]. The prediction methods mentioned in references (4-6) are based on this approach. References 3 and 7 discuss a prediction method which only requires one (low) trial dose. However, an extra assumption is made in this method, namely a fixed value of K_m: either 4 µg/ml (3) or 6 µg/ml (7). For the assumed value of K_m and the trial dose with its corresponding measured plasma concentration, Eq.'s [1.1] or [1.2] can again be solved algebraically for V_{max}.

From a previous study in 26 patients (8) it was concluded that prediction in general was hazardous, while the method assuming a fixed K_m value gave for these 26 patients relatively better results than the methods based on (at least) two trial doses. The impression was that the performance of prediction was at least strongly related to variability in plasma concentrations. Therefore, a model study was started to investigate the influence of the several parameters of the different prediction methods, and especially the influence of the variability in plasma concentration.

In the present model study three important factors will be considered.

1.   The first one is the "model patient", represented by assumed K_m and V_{max} values. Only the results of a patient with a V_{max} value of 450 mg/24 h are presented. However, as shown in the Appendix, this imposes no restriction on the results of the study. The method using one trial dose assumes that all patients have the same fixed K_m value. However, as this is a simplifying assumption, we studied also the performance of this method for patients with a K_m value different from the assumed one. The method that uses two trial doses needs no assumptions about the patient's K_m value. It is shown in the Appendix that basically the performance of the prediction method does not depend on the patient's K_m value.

2.   A second factor is the trial dose with its resulting plasma concentration. It will be difficult to predict the dose yielding a target concentration of 15 µg/ml if the trial dose led to a concentration of 4 µg/ml, whereas if the dose giving a concentration of 12 µg/ml is already known from the trial period, prediction will be less hazardous. Also, applying methods requiring two trial doses, the distance of the resulting two plasma concentrations appears to be important for accurate prediction.

     Trial doses were chosen in such a way that the corresponding plasma concentrations were in the range of 4 to 12 µg/ml. With Eq. [1.1] or [1.2], these trial doses can readily be found for a given "model patient". This range contains most of the plasma concentrations which a physician will use for predictions.

3.   The third factor is the variability as assumed around the plasma concentration resulting from the trial dose. Given the patient's values for K_m and V_{max}, a trial dose leads, according to Eq. [1.1], to a unique plasma concentration. However, due to analytical and biological variability, the observed concentration will differ from this unique concentration calculated from the Michaelis-Menten equation. The magnitude of this variability will, of course, influence prediction. In the present model study this variability was introduced by assuming a lognormal distribution of observed concentrations around the Michaelis-Menten derived one. This variability will loosely be called measurement error.

     To evaluate in the model study the performance of the prediction methods, the following procedure was followed:

     All prediction methods lead, for a given target concentration, to a predicted dose. Having this predicted dose and knowing the patient's characteristics (assumed V_m and K_m), one can calculate from Eq. [1.2] the concentration that will actually result. The comparison of the target concentration with this resulting concentration enables one to evaluate the prediction methods. In this study, a range of resulting concentrations around the target concentration (15 µg/ml) was considered as acceptable. As stated above, a concentration of more than 20 µg/ml is usually toxic and therefore resulting concentrations above 20 µg/ml are considered as unacceptable. On the other hand, concentrations less than 12 µg/ml are considered in the study as too low. The evaluation will be based on the percentages of predictions with acceptable concentrations (between 12 and 20 µg/ml), with too low concentrations (less than 12 µg/ml) and with toxic concentrations (above 20 µg/ml). These percentages can be examined in the model study as a function of (i) the assumed V_{max} and K_m, *i.e.*, the patient's characteristics, (ii) the chosen trial dose, and (iii) the variability of the concentration measurements.

2. METHODOLOGY OF MODEL STUDY

2.1. Model Study Based on Richens' Method

First, Richens' method for the prediction of doses yielding desired plasma concentrations will be recapitulated. Next, the structure of the model study will be given.

2.1.1. Richens' Method. The three assumptions underlying the method are: The Michaelis-Menten kinetics is valid, the patient is in a steady state, and the K_m value is the same for all patients.

Richens presented his methods in the form of nomograms, one for a fixed K_m=4 µg/ml (3) and one for a fixed K_m=6 µg/ml (7). The latter one is presented in the literature by Rambeck *et al.* (7) but will be called by us Richens' second method. The reasoning behind the nomograms will be presented in an algebraical way, because these algebraic equations had to be used in the model study.

The method requires one plasma concentration C to be measured at a chosen low trial dose, D. Substitution of these D and C values into Eq. [1.1], together with the fixed K_m=4 µg/ml of the first nomogram, gives an estimate of V_{max}:

$$\hat{V}_{max} = \frac{4 + C}{C}\ D \qquad . \qquad\qquad\qquad [2.1]$$

With this V_{max}, the target concentration (C_T) and (again) the fixed K_m=4 µg/ml, one finds the required prediction dose (D_T) from

$$D_T = \hat{V}_{max}\ C_T/(4 + C_T) \qquad . \qquad\qquad\qquad [2.2]$$

For the second nomogram, the fixed K_m=4 µg/ml is replaced by the fixed K_m=6 µg/ml in the equations [2.1] and [2.2].

2.1.2. Structure of the Model Study.

Step 1. Choose values for V_{max} and K_m.

These two values of V_{max} and K_m characterize a patient. In practice, these patient characteristics are unknown; in this study they will be called the patient's V_{max} and K_m values.

Step 2. Choose a *trial* dose (D^{tr}) such that the corresponding theoretical concentration (C^{tr}), calculated according to

$$C^{tr} = K_m D^{tr}/(V_{max} - D^{tr}) \quad , \qquad [2.3]$$

is within the range of 4 µg/ml to 12 µg/ml. This C^{tr} represents the concentration which will be obtained if no variability is involved.

Step 3. Simulate *observed* concentrations C^{obs} by introducing variability around the theoretical value C^{tr}. We assumed the logarithm of the observed concentration to be normally distributed with the value of log C^{tr} as its mean value and some constant, named SIGMA, as its standard deviation. The distribution of the concentrations C^{obs} is then lognormal with a standard deviation approximately

$$\text{SIGMA} \times C^{tr} \qquad [2.4]$$

In this way, the model mimics measurement variability by implying a skewed frequency distribution for the C^{obs}, centered around the theoretical C^{tr} value. The variability is represented by the standard deviation which is proportional to this C^{tr}. The magnitude of this variability is determined by the constant SIGMA.

In practice, concentrations are rounded off to one digit after the decimal point. To follow practice closely, we made from the lognormal frequency distribution over the range

$$\log C^{tr} \pm (3 \times \text{SIGMA})$$

a histogram with a class width of 0.1. Each class of the histogram corresponds to a possible C^{obs}, while the relative frequency gives the probability that this C^{obs} will occur under the model conditions assumed (Figure 1).

Step 4. Prediction of the *target* dose (D_T).

For the concentration of each class, from the C^{obs} histogram, together with the associated trial dose (D^{tr}), the estimated value of $\hat{V}_{max}$ can be obtained from Eq. [2.1], while next the target dose

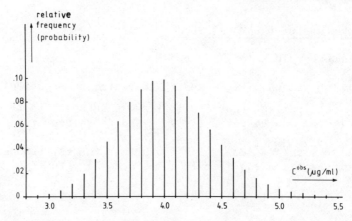

FIGURE 1:   Relative frequency distribution for C^{obs} corresponding
to C^{tr}=4 µg/ml and SIGMA=0.10.

(D_T) can be derived from Eq. [2.2]. This leads to a histogram of
predicted D_T values, with as many classes as the C^{obs} histogram and
exactly the same relative frequencies (Fig. 2A). However, the
smallest pharmaceutical formulation available is the 25 mg tablet.
To make the predicted dose values realistic ones, we rearranged its
histogram into one with a class width of 25 mg (Fig. 2B).

Step 5.   Calculation of the *resulting* concentration (C^{res}).

For the dose of each class of the D_T histogram (Fig. 2B), together with the patient's V_{max} and K_m values, the resulting concentrations can be calculated:

$$C^{res} = K_m D_T / (V_{max} - D_T) \quad .$$ [2.5]

The histogram in Figure 3 summarizes these resulting plasma concentrations.   The relative frequencies equal those of Fig. 2B.

Step 6.   Calculation of the percentages of resulting concentrations
which are too low, acceptable or toxic.

It was stated *a priori* that the desired concentration was
15 µg/ml, while a concentration between 12 and 20 was taken as still
acceptable. From the histogram of C^{res} values (Fig. 3), one can now
derive the percentages of cases where the concentrations turn out
to be too low (C^{res}<12, P_{LOW}), acceptable ($12 \leq C^{res} \leq 20$, P_{ACC}), or
toxic (C^{res}>20, P_{TOXIC}).   These P values will be measures of the
performance of the prediction. We shall finally report P_{ACC} and
the percentage of toxic predictions among the not acceptable predictions:

$$P_{T/NA} = \frac{P_{TOXIC}}{P_{LOW} + P_{TOXIC}} \cdot 100 \qquad . \qquad [2.6]$$

Summarizing: From an input of a patient's V_{max} and K_m values, a trial dose (D^{tr}) and a measurement error (SIGMA), we finally obtain values for the performance measures P_{ACC} and $P_{T/NA}$. The consequences on the prediction performance of the variation of the input parameters can then be studied.

2.2. Model Study Based on the Method with Two Trial Doses and Corresponding Concentrations

The approach of the model study based on Richens' method will be followed as far as possible. The dose prediction method using

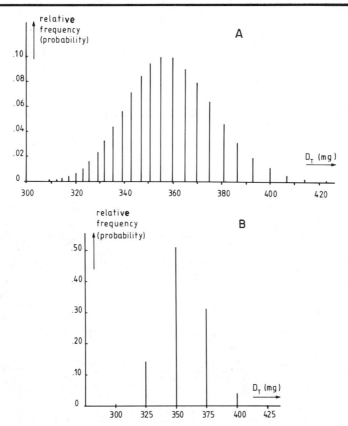

FIGURE 2: Relative frequency distribution of the predicted dose, D_T. (A) Distribution derived from the distribution of C^{obs} given in Figure 1. (B) Distribution obtained from Figure 2A after rounding the doses to multiples of 25 mg.

measured concentrations at two trial doses will be explained. Restricting to *two* trial doses means that now the methods of Ludden
et al. (4) and Martin *et al.* (5) are identical. The explanation
will again be given in an algebraical way. After this, explanation
the structure of the model study will be outlined.

2.2.1. Description of the Prediction Method. It is assumed that
the dose D and plasma concentration C in the steady state are related by equation [1.1], where K_m and V_{max} are parameters characterizing the patient. Contrary to Richens' method, nothing is assumed
about K_m. Two plasma concentrations C_1 and C_2 become available at
two different doses D_1 and D_2 during two different trial periods.

   Estimates $\hat{V}_{max}$ and $\hat{K}_m$ of V_{max} and K_m, respectively, are obtained by solving simultaneously the two equations:

$$C_1 = \hat{K}_m D_1 / (\hat{V}_{max} - D_1)$$

$$[2.7]$$

and $$C_2 = \hat{K}_m D_2 / (\hat{V}_{max} - D_2) \quad .$$

With some simple algebra one finds:

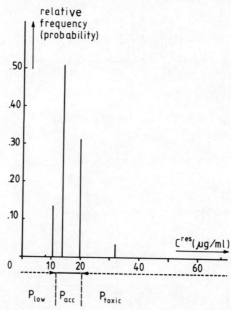

FIGURE 3: The relative frequency distribution of the resulting
concentrations (C^{res}) as derived from Figure 2B. The intervals are
indicated from which the performance measures P_{ACC}, P_{LOW} and P_{TOXIC}
are calculated.

$$\hat{V}_{max} = D_1 D_2 \frac{C_2 - C_1}{C_2 D_1 - C_1 D_2}$$

[2.8]

and $\qquad \hat{K}_m = C_1 C_2 \dfrac{D_2 - D_1}{C_2 D_1 - C_1 D_2}$.

The dose D_T required for a target concentration C_T can now be predicted by

$$D_T = \hat{V}_{max} C_T / (\hat{K}_m + C_T)$$

[2.9]

2.2.2. Structure of the Model Study

Step 1. Choose values for V_{max} and K_m.

The same as has been said in step 1 in Section 2.1.2 holds here.

Step 2. Choose two trial doses D_1^{tr} and D_2^{tr} $(D_2^{tr} > D_1^{tr})$ such that the corresponding concentrations C_1^{tr} and C_2^{tr}, calculated according to

$$C_1^{tr} = K_m D_1^{tr} / (V_{max} - D_1^{tr})$$

[2.10]

and $\qquad C_2^{tr} = K_m D_2^{tr} / (V_{max} - D_2^{tr})$

are within the range of 4 µg/ml to 12 µg/ml. These C^{tr} values would be obtained if no measurement errors were involved.

Step 3. Variability around C_1^{tr} as well as around C_2^{tr} is introduced in exactly the same way as it is done in step 3 of Section 2.1.2. For C_1^{tr} and C_2^{tr}, the same value of SIGMA is chosen. As a result, we now have two histograms, one for C_1^{obs} and one for C_2^{obs}.

Step 4. Prediction of the target dose (D_T).

Let C_1^{obs} be the concentration of one class from the C_1^{obs} histogram and C_2^{obs} the concentration of one class from the C_2^{obs} histogram. Let P_1 and P_2 be the probabilities (relative frequencies) associated with these two classes. From C_1^{obs} and C_2^{obs} and their corresponding trial doses, D_1^{tr} and D_2^{tr}, one can calculate $\hat{V}_{max}$, $\hat{K}_m$ and finally the target dose D_T in the way described in Section 2.2.1. To this D_T corresponds the product of probabilities $P_1 P_2$. By calculating for all possible pairs of C_1^{obs} and C_2^{obs} from the two histograms the value of D_T and the corresponding probability,

one obtains a histogram of predicted D_T values. Again the histogram
is rearranged into one with class width of 25 mg.

In practice, the physician will not execute the prediction if
the data he has at his disposal are incompatible with the Michaelis-
Menten concept. This is the case if C_2^{obs} is not high enough in
relation to C_1^{obs} and the corresponding trial doses. To say it
exactly: if $C_2^{obs}/D_2^{tr} \le C_1^{obs}/D_1^{tr}$ then $\hat{K}_m$ is not positive according
to Eq. [2.8]. In such cases, prediction is not executed in the
model study.

Steps 5 and 6. To calculate the *resulting* concentration (C^{res}) and
the performance measures (P_{ACC} and $P_{T/NA}$), the procedure is the
same as in steps 5 and 6 in Section 2.1.2.

Summarizing: From an input of a patient's V_{max} and K_m values,
two trial doses (D_1^{tr} and D_2^{tr}), and a measurement error (SIGMA), we
finally obtain values for the performance measures P_{ACC} and $P_{T/NA}$.
Variation of these input values enables us again to study the consequence on the prediction performance.

2.3. Choice of the Input Values

2.3.1. K_m and V_{max}. It is shown in the Appendix that the patient's
V_{max} value is irrelevant for Richens' method as well as for the
method with the two trial doses. Therefore, the choice of V_{max} was
always 450 mg/24 h.

The patient's K_m value is relevant for the performance of
Richens' method. Its performance was studied for the following
series of K_m values: K_m=3, 3.5, 4, 5, 6 and 8 µg/ml. This was
done for both verions of that method as mentioned in Section 2.1.1.

For the method with the two trial doses, the patient's K_m value
is basically irrelevant. However, as explained in the Appendix,
some influence of the patient's K_m on the performance measures
remains present in our model study. Therefore, the performance of
the method with two trial doses is studied for two patient's K_m
values: 4 µg/ml and 6 µg/ml.

2.3.2. Trial Doses. Only trial doses corresponding to trial concentrations within the range of 4-12 µg/ml were used in the model
study. The upper limit of 12 µg/ml was chosen because, in practice,
a physician will usually not need a dose prediction method if the
observed concentration is already above this limit. The lower limit
was chosen because the analytical errors might be excessive for

concentrations smaller than 4 µg/ml. The lognormal distribution imposed on C^{obs} would then be inappropriate.

For Richens' method, the trial doses were obtained for each of the patient's K_m values (as chosen in Section 2.3.1) by substituting C^{tr}=4, 6 and 10 µg/ml in Eq. [1.1] and rounding these values off to multiples of 25 mg/24 h. The plasma concentrations corresponding to these doses are given in Table 1. These concentrations show, of course, some variation around the intended values of 4, 6 or 10 µg/ml due to the rounding off to multiples of 25 mg/24 h of the doses.

For the second version, only the results for C^{tr}=4 µg/ml will be reported in a comparative diagram.

For the method with two trial doses, we obtained a pair, D_1^{tr} and D_2^{tr}, by substituting C_1^{tr} and C_2^{tr} into Eq. [1.1] and rounding these values off to multiples of 25 mg/24 h. Our choice of trial doses D_1^{tr} and D_2^{tr} can be read from the second column of Table 3 (or Table 4). The corresponding trial concentrations can again be read from Table 1. These trial concentrations show, of course, also some variation around the intended ones.

2.3.3. SIGMA. The values of SIGMA used in the model study are 0.01, 0.05, 0.10 and 0.20. To obtain a succinct presentation, the performance of Richens' method will be reported only for SIGMA=0.05, 0.10 and 0.20, while the performance of the methods of Ludden *et al.* (4) and Martin *et al.* (5) will be reported only for SIGMA=0.01, 0.05 and 0.10. From the discussion it will become clear that this results in no serious loss of information. There it will also be made clear that the prediction methods are far too hazardous when one is confronted with a variability greater than 0.20.

TABLE 1:   Trial concentrations corresponding to the trial doses after rounding off to multiples of 25 mg/24 h

Intended Trial Concentrations	Patient's K_m (in µg/ml)					
	3.0	3.5	4.0	5.0	6.0	8.0
4 µg/ml	3.8	4.4	4.0	4.0	3.8	4.0
6 µg/ml	6.0	5.5	6.3	6.3	6.0	6.4
10 µg/ml	10.5	9.1	10.4	10.0	9.4	10.0
12 µg/ml	10.5	12.3	14	13	12	12.6

The very low value of SIGMA=0.01 (too low to be expected in practice) was incorporated in this model study for theoretical reasons.

3.   RESULTS

The results obtained for Richens' method with K_m=4 µg/ml are presented in Table 2. For a given row, the performance measures P_{ACC} and $P_{T/NA}$ are given for the values of SIGMA as indicated in

TABLE 2: P_{ACC} and $P_{T/NA}$ for Richens' method with fixed K_m=4 µg/ml.

| | | | P_{ACC} | | | $P_{T/NA}$ | | |
| | | | S | I G | M A | | | |
C^{tr}	D^{tr}	K_m	.05	.10	.20	.05	.10	.20
4	250	3	2	15	19	100	99	85
4	250	3.5	85	70	50	100	98	80
4	225	4	98	82	52	1	21	39
4	200	5	60	55	44	0	1	16
4	175	6	36	43	41	0	0	9
4	150	8	0	2	14	0	0	1
6	300	3	21	33	29	100	98	82
6	275	3.5	96	80	58	100	99	80
6	275	4	100	91	63	-	16	37
6	250	5	97	82	59	0	2	21
6	225	6	93	76	57	0	1	16
6	200	8	12	29	38	0	0	2
10	350	3	86	68	45	100	93	72
10	325	3.5	100	93	72	-	99	82
10	325	4	100	97	76	-	9	33
10	300	5	100	99	79	-	18	38
10	275	6	100	99	82	-	55	52
10	250	8	100	97	81	-	0	15

For the choice of the patient's K_m values, trial doses and concentrations and SIGMA values see Section 2.3.
$P_{T/NA}$ is denoted by - if P_{ACC} = 100 per cent

the heading of the columns. By comparing the values of these performance measures in a row, one can see how they are influenced by the magnitude of SIGMA. The table is subdivided into three blocks. All rows within a block correspond to the same intended trial concentration.

In Figure 4, the results of the first block (C^{tr}=4 µg/ml) are graphically represented. In Figure 4A, the P_{ACC} is plotted against the patient's K_m-value. Points corresponding with one value of SIGMA are connected by a dotted line. In Figure 4B, the same is done for $P_{T/NA}$.

In Figure 4A for each SIGMA, the corresponding curve shows its maximum at the patient's K_m=4 µg/ml. This was to be expected as, in such a case, the assumed value of K_m in the method is correct in the sense that it is equal to the actual patient's K_m-value. Moreover, it is not surprising that for this patient's K_m=4 µg/ml, the value of P_{ACC} decreases if SIGMA increases: when the plasma concentration is measured less accurately, it is more likely that the C^{obs} is considerably too high or too low leading to too low or too high predicted doses, respectively.

However, if the patient's K_m equals 8 µg/ml, a higher value of SIGMA seems to be favourable. To understand this phenomenon, first consider what would have happened if SIGMA=0.00. If the patient's K_m=8 µg/ml and V_{max}=450 mg/24 h then the trial dose (D) corresponding to C^{tr}=4 µg/ml is 150 mg/24 h (see Tables 1 and 3). In the case of no variability (SIGMA=0.00), $C^{obs}=C^{tr}$ and one can find, using Richens' first method, that D_T=225 mg/24 h, leading in reality to C^{res}=8 µg/ml, which is too low. Thus if a patient's K_m=8 µg/ml, C^{tr}=4 µg/ml and SIGMA=0.00 then the resulting concentration is always too low. As for a given trial dose, a lower observed concentration leads to a higher predicted dose (D_T), and an acceptable prediction in the case of a patient with K_m=8 µg/ml is only due to lack of accuracy.

If the patient's K_m value is equal to 3 µg/ml, it can be shown in the same way that the prediction is always too high if there is no variability. An acceptable prediction will be obtained only if C^{obs} is considerably higher than C^{tr}. From these considerations it will also be clear that $P_{T/NA}$ will be high for low values of patients' K_m, and will decrease with increasing K_m value. This is clearly illustrated in Figure 4B.

The results of the last column of Table 2 (SIGMA=0.20) are plotted in Figure 5. In Figure 5A one can see that P_{ACC} is

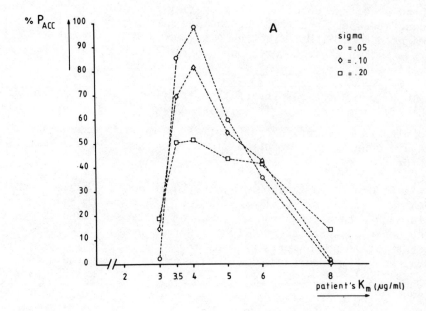

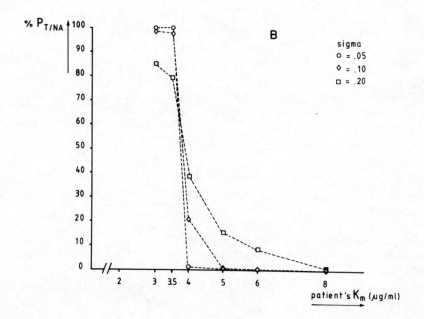

FIGURE 4: Richens' method with fixed K_m=4 µg/ml and approximate trial concentration of C^{tr}=4 µg/ml. (A) P_{ACC}, and (B) $P_{T/NA}$ for several values of the patient's K_m. Points corresponding to the same value of SIGMA are connected.

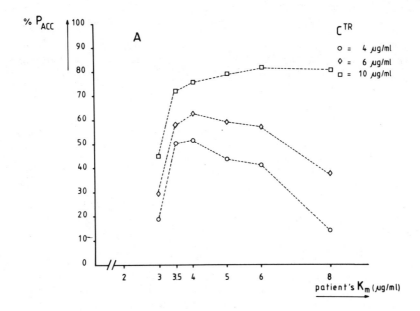

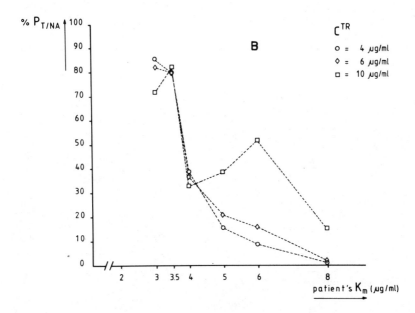

FIGURE 5: Richens' method with fixed K_m=4 µg/ml and SIGMA=0.20.
(A) P_{ACC}, and (B) $P_{T/NA}$ for several values of the patient's K_m.
Points corresponding to the same approximate trial concentration
are connected.

(generally) lower if the patient's K_m is not equal to the value
assumed in the method. Figure 5B illustrates that $P_{T/NA}$ (generally)
decreases with increasing patient's K_m-value. For $C^{tr}=10$ µg/ml an
exception occurs, which can be shown to be related to the rounding
procedure of D_T and the choice of the rather rigorous evaluation
criterion (e.g., $C^{res}=11.9$ is too low while $C^{res}=12.0$ is considered
to be as good as $C^{res}=19.9$). Such irregularities, and even more
severe ones, would have been encountered in most of the foregoing
figures if P_{ACC} and $P_{T/NA}$ had been plotted for intermediate values
of patient's K_m. This phenomenon will not be discussed further in
the present paper. However, one has to realize that quite important
discontinuities can be induced in the course of P_{ACC} and $P_{T/NA}$ as a
function of K_m by rounding off the administered dose. As a reminder
to the irregularities, the connecting lines are dotted in all these
figures.

Concerning the performance of the second version of Richens'
method, similar tendencies were found as observed in the first version. Presentation will be restricted to the results with
$C^{tr}=4$ µg/ml (Figure 6). Clearly, this version works best if the
patient's K_m is now equal to 6 µg/ml. The curves can easily be
compared with the corresponding ones in Figure 4.

TABLE 3: P_{ACC} and $P_{T/NA}$ for the method with two trial doses.

$C_1^{tr}-C_2^{tr}$	$D_1^{tr}-D_2^{tr}$	P_{ACC}			$P_{T/NA}$		
				S I G M A			
		.01	.05	.10	.01	.05	.10
4 - 6	225-275	100	68	40	-	45	44
4 - 8	225-300	100	94	65	-	46	48
4 - 10	225-325	100	100	90	-	-	54
4 - 12	225-350	100	100	99	-	-	65
6 - 8	275-300	100	68	43	-	-	38
6 - 10	275-325	100	99	81	-	-	62
6 - 12	275-350	100	100	98	-	-	88

Patient's $V_{max}=450$ mg/24 h, $K_m=4$ µg/ml. For the choice of the
trial doses see Section 2.3.

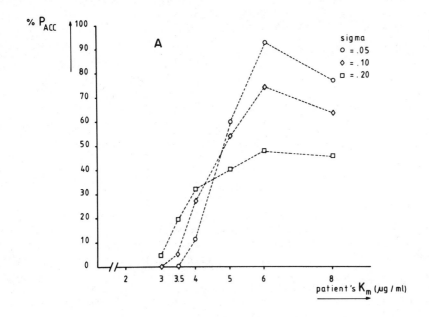

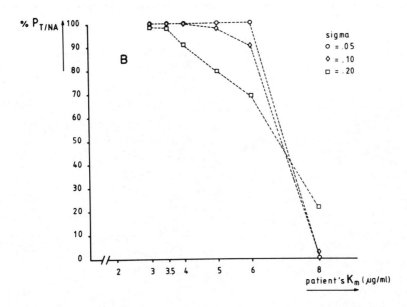

FIGURE 6: Richens' method with fixed K_m=6 µg/ml and approximate trial concentration of C^{tr}=4 µg/ml. (A) P_{ACC}, and (B) $P_{T/NA}$ for several values of the patient's K_m. Points corresponding to the same value of SIGMA are connected.

Tables 3 and 4 contain the results for the method with two trial doses. In Table 3, the results are given for a patient with V_{max}=450 mg/24 h and K_m=4 µg/ml. In Table 4, the values are V_{max}=450 mg/24 h and K_m=6 µg/ml.

It must be remembered that the prediction is stopped if $\hat{K}_m$ turns out to be negative. The probability that this occurs increases with increasing variability in the C^{obs}, *i.e.*, with increasing SIGMA. Therefore, the value of P_{ACC} need not decrease monotonically with increasing SIGMA and the value of $P_{T/NA}$ need not increase monotonically.

4. DISCUSSION

Only the main points of the study will be discussed. It must be remembered that the model study started with the assumption that a steady-state plasma concentration existed and Michaelis-Menten kinetics were valid. Hence these points are not considered now.

In balancing the results of the study, evaluation standards had to be introduced. In the first approach, a prediction method was considered to be satisfactory for one combination of input values if 8 out of 10 predictions would yield concentrations in the

TABLE 4: P_{ACC} and $P_{T/NA}$ for the method with two trial doses.

		P_{ACC}			$P_{T/NA}$		
		S I G M A					
C_1^{tr}-C_2^{tr}	D_1^{tr}-D_2^{tr}	.01	.05	.10	.01	.05	.10
4 - 6	175-225	96	63	41	100	83	59
4 - 8	175-250	100	79	60	-	99	83
4 - 10	175-275	100	91	75	-	100	98
4 - 12	175-300	100	98	85	-	100	100
6 - 8	225-250	93	59	44	100	84	51
6 - 10	225-275	100	81	69	-	100	97
6 - 12	225-300	100	96	81	4	100	100

Patient's V_{max}=450 mg/24 h, K_m=6 µg/ml. For the choice of the trial doses see Section 2.3.

acceptable range of 12-20 µg/ml. This criterion was applied to evaluate the prediction methods studied.

4.1. The Influence of the Magnitude of the Variability in Plasma Concentration on the Accuracy of Dose Prediction

With a variability of only 1% (SIGMA=0.01) around the plasma concentration C^{tr}, the methods of Ludden $et\ al.$ and Martin $et\ al.$ give excellent dose predictions, as shown in Tables 3 and 4. Theoretically this was to be expected. However, a variability of 10 per cent reduces the number of acceptable predictions appreciably. It should be noted that mispredictions will then frequently yield toxic plasma concentrations, as is shown by the high $P_{T/NA}$ values of Tables 3 and 4. The percentage of acceptable predictions was dissatisfying low (less than 80 per cent) when the variability around C^{tr} was above 10%. For this reason, tabulations with a variability greater than 10% were not performed.

Observing the results for estimating the influence of variability, it becomes also clear that a value of C_2^{tr} near the target concentration (C_T) of 15 µg/ml is favourable for the accuracy of prediction.

Looking at Table 2 and the corresponding figures it appears that in evaluating Richens' methods the situation is more complex because there is an interaction between the SIGMA and K_m values. For a small variability (SIGMA=0.05), results can be summarized as follows: For patients with a K_m value not too far from the fixed K_m value used in the method, the doses desired will be adequately predicted. Patients with a K_m value smaller than the one used in the nomogram will very likely get toxic plasma concentrations, whereas patients with greater K_m values will very likely get too low plasma concentrations. For a variability from 10% onwards, the performance of the method is considerably lower. At high levels of variability, the K_m-value also influences the prediction. As in the methods of Ludden $et\ al.$ and Martin $et\ al.$, the beneficial influence of a C^{tr} concentration near the target concentration can be seen.

4.2. The Influence of V_{max} and K_m Values of Patients on the Accuracy of Dose Prediction

As shown in the Appendix, the magnitude of the V_{max} value of a patient is irrelevant for the result of the dose prediction in the method of Richens. However, the K_m value of a patient, in relation to the fixed K_m value used in the method (nomogram), appears to be very important for the resulting prediction, as outlined before.

In the methods of Ludden *et al.* and Martin *et al.*, neither K_m nor V_{max} values essentially influence the results of prediction in a systematical way. In this model study, the results for patients' $K_m=4$ µg/ml (Table 3) and $K_m=6$ µg/ml (Table 4) are not identical, since doses are not administered in any quantity but in 25 mg steps. From Table 1 it can be inferred that sometimes quite a difference exists between intended trial concentrations and actual trial concentrations, and that these differences are not the same for patients with K_m-values of 4 and 6 µg/ml. A second cause for the slight resting influence of K_m on prediction is given in the Appendix.

4.3. The Influence of the Trial Dose on the Accuracy of Prediction

From Tables 2, 3, and 4 it can be seen that a trial dose with an associated plasma concentration near the target concentration of 15 µg/ml is favourable for a more accurate dose prediction. From Tables 3 and 4 it can be seen that the number of acceptable predictions will increase if the two C^{tr} plasma concentrations are more widely spaced.

4.4. Comparison of the Two Methods of Richens

Richens constructed two nomograms (3,7). In the first one, a fixed K_m value of 4, in the second a K_m value of 6 µg/ml is used. With the second nomogram, a larger dose is predicted to reach the target concentration of 15 µg/ml than with the first nomogram. This means that the probability that the predicted dose yields a toxic concentration must be greater with the second nomogram than with the first one. By using Richens' method with a fixed K_m value of 6 µg/ml, more patients will get too high plasma concentrations than by using Richens' first nomogram. Indeed, this was observed in practice (8). On the other hand, if the patient's K_m-value is 6 µg/ml, the second version is superior to the first one as can be seen by comparing Figure 4 with Figure 6. In practice, one should have information about the distribution of the K_m values in the population. The revision of the first nomogram was argued because the mean K_m-value in 127 residential patients was 6 µg/ml and it was found "superior to the original" nomogram in a further 78 patients. In reference (8), K_m-values ranging from 3 to 12 µg/ml (with a mean value of 6 µg/ml) are reported.

4.5. Comparison of the Method of Richens with the Method of Ludden *et al.* and Martin *et al.*

Examination of the results showed that for a very small variability (SIGMA=0.01), the methods of Ludden *et al.* and Martin *et al.* always give excellent predictions, while for Richens' method this

is usually true if the patient's K_m is about equal to the K_m assumed
in the method. For greater values of SIGMA, Richens' method becomes
more and more superior to the other methods, especially when the
percentage of toxic predictions among the non-acceptable predictions
is considered.

4.6. Prediction Methods and Practical Considerations

From the foregoing it is clear that the choice between Richens'
method and the two other ones highly depends on the variability in
observed plasma concentrations occurring in practice. If it is more
than 10 per cent, Richens' method should usually preferred, especially if one has evidence that the patient's K_m value is not less than
the K_m value assumed in the nomogram. From the literature, a variability in phenytoin steady-state plasma concentrations measurements
of 10 to 20 per cent or even more is known. This variability is
due to analytical error (9,10) as well as physiological fluctuation
(11). In a previous publication (8), a better performance of Richens'
method was demonstrated, in agreement with the foregoing considerations.

5. CONCLUSIONS

Dose predictions based on the Michaelis-Menten relation and on
the *extrapolation* from only two observations of doses with associated
steady-state phenytoin plasma concentrations turned out to be dangerous. The variability of the observed plasma concentration is a
crucial element in the accuracy of prediction. As shown, a well
chosen fixed K_m value can be of value in situations where predictions
must be made and where plasma concentrations are subject to great
variability.

ACKNOWLEDGEMENT

We are very grateful to Mrs. E.W.M. Arents-Bos for the accurate
typing of the manuscript.

REFERENCES

1. Bochner, F., Hooper, W.D., Tyrer, J.H. and Eadie, M.J. (1972)
       J. Neurol. Neurosurg. Psych. 35, 873-876.
2. Remmer, H., Hirschmann, J. and Greiner, I. (1969) Deut. Med.
       Wochenschr. 94, 1265-1272.
3. Richens, A. (1975) Epilepsia, 16, 627-646.
4. Ludden, T.M., Hawkins, D.W., Allens, J.P. and Hoffman, S.F.
       (1976) Lancet, I, 307-308.
5. Martin, E., Tozer, Th.N., Sheiner, L.B. and Riegelman, S.
       (1977) J. Pharmacokin. Biopharm. 5, 579-596.
6. Mullen, P.W. (1978) Clin. Pharmacol. Ther. 23, 228-232.

7. Rambeck, B., Boenigk, H.E., Dunlop, A., Mullen, P.W., Wadsworth, J. and Richens, A. (1979) Therap. Drug Monit. <u>1</u>, 325-333.
8. Driessen, O., Van der Velde, E.A. and Höppener, R.J. (1980) Eur. Neurol. <u>19</u>, 103-114.
9. Dijkhuis, I.C. (1979) Thesis, Leiden.
10. McCormick, W., Ingelfinger, J.A., Isakson, G. and Goldman, P. (1978) New. Engl. J. Med. <u>299</u>, 1118-1121.
11. Driessen, O., Höppener, R. and Van der Velde, E.A. (1980) Eur. Neurol., in press.

APPENDIX

It will be shown in this appendix that, for the performance of Richens' method, the value of the patient's V_{max} is essentially irrelevant and that, for the method with two trial doses, the values of both K_m and V_{max} are irrelevant for the present model study. It will be shown first that in the case of *Richens' method*, without loss of generality, one value of V_{max} can be used. In an algebraical way the reasoning corresponding to Richens' nomogram based on K_m=4 μg/ml will be given; considerations for the nomogram based on K_m=6 μg/ml are, of course, completely equivalent.

Let D^{tr} be the trial dose administered to a patient characterized by V_{max} and K_m. Equation [2.3], from which the corresponding C^{tr} is calculated, can be rewritten as

$$D^{tr} = V_{max} C^{tr}/(K_m + C^{tr}) \quad . \qquad [A.1]$$

From the C^{tr} value, C^{obs} values are obtained according to step 3 of Section 2.1.2 by assuming some measurement variability around C^{tr}. In this step only SIGMA can be varied. So, C^{obs} depends only on C^{tr} and the variability SIGMA.

The dose prediction process takes place in two steps. First, the patient's V_{max} is estimated according to Eq. [2.1] by

$$\hat{V}_{max} = \frac{4 + C^{obs}}{C^{obs}} D^{tr} \quad . \qquad [A.2]$$

Then, with this result, the prediction D_T is obtained by Eq. [2.2]:

$$D_T = \hat{V}_{max} C_T/(4 + C_T) \quad . \qquad [A.3]$$

For this prediction it follows, by Eq. [2.5]:

$$C^{res} = \frac{K_m D_T}{V_{max} - D_T} = K_m \frac{D_T/V_{max}}{1 - D_T/V_{max}} \quad . \qquad [A.4]$$

In this last equation, V_{max} and K_m are the "true" parameter values and not the estimated ones, as we have to calculate the concentration which will be really reached after the administration of D_T.

Now suppose that V_{max} is multiplied by a factor f while K_m, C^{tr}, SIGMA and C_T are kept fixed. Then we see by Eq. [A.1] that D^{tr} is multiplied by the same factor f. The value for C^{obs} remains fixed, because it depends only on C^{tr} and SIGMA. So, it follows from Eq. [A.2] that the multiplication of D^{tr} by f induces a multiplication of $\hat{V}_{max}$ by f, which, by Eq. [A.3], leads to a D_T value which is f-fold higher. However, in Eq. [A.4] one can see that only D_T/V_{max} is of importance. Therefore, the factor f cancels out in C^{res}. So C^{res} and therefore also P_{ACC} and $P_{T/NA}$ are not influenced by a change of V_{max} as long as K_m, C^{tr}, C_T and SIGMA are unchanged.

We note that by rounding to multiples of 25 mg/day of the doses, C^{tr} can not always remain unchanged, since for a specified K_m and V_{max} value there need not be a dose (in multiples of 25 mg/day) corresponding exactly to C^{tr}. Therefore, our statement about the irrelevancy of V_{max} holds only approximately.

For the method with *two trial doses*, we also start with recalling the equations used in the prediction.

D_1^{tr} and D_2^{tr} are the two trial doses corresponding to C_1^{tr} and C_2^{tr}. As in Eq. [A.1], we now have

$$D_1^{tr} = V_{max} \, C_1^{tr}/(K_m + C_1^{tr})$$

[A.5]

and

$$D_2^{tr} = V_{max} \, C_2^{tr}/(K_m + C_2^{tr}) \quad .$$

The estimates $\hat{V}_{max}$ and $\hat{K}_m$ are, according to Eq. [2.8]:

$$\hat{V}_{max} = D_1^{tr} \, D_2^{tr} \, \frac{C_2^{obs} - C_1^{obs}}{C_2^{obs} D_1^{tr} - C_1^{obs} D_2^{tr}}$$

[A.6]

$$\hat{K}_m = C_1^{obs} \, C_2^{obs} \, \frac{D_1^{tr} - D_2^{tr}}{C_2^{obs} D_1^{tr} - C_1^{obs} D_2^{tr}} \quad .$$

These values are substituted into Eq. [2.9] which is:

$$D_T = \hat{V}_{max} \, C_T / (\hat{K}_m + C_T) \quad .$$ [A.7]

As in Richens' method, C^{res} is obtained by

$$C^{res} = K_m \frac{D_T/V_{max}}{1 - D_T/V_{max}} \quad .$$ [A.8]

Subsequent substitution of Eq. [A.5] into Eq.'s [A.6], [A.7] and [A.8] leads, after some (tedious) algebra, to

$$C^{res} = \frac{(C_2^{obs}-C_1^{obs})C_1^{tr}C_2^{tr} \cdot C_T}{C_1^{obs}C_2^{obs}(C_2^{tr}-C_1^{tr}) - (C_1^{obs}C_2^{tr}-C_2^{obs}C_1^{tr})C_T} \quad .$$ [A.9]

In this equation, the V_{max} and K_m cancel out. So, having fixed values of C_1^{tr}, C_2^{tr}, C_T and SIGMA (and consequently of C_1^{obs} and C_2^{obs}), the C^{res} histogram is not influenced by variation in V_{max} and K_m. In other words, without loss of generality, one value of V_{max} and K_m can be chosen.

Due to two effects this statement holds only approximately. In the first place, rounding to multiples of 25 mg/24 h of the doses has similar effect as described for Richens' method. The second effect is evoked by stopping the prediction if $\hat{K}_m < 0$. The probability that this occurs depends on the curvature of the relation between dose and concentration for the patient, or in other words on the patient's K_m value.

CONTRIBUTOR INDEX